Cancer Treatment and Research

Volume 183

Series Editor

Steven T. Rosen, Duarte, CA, USA

This book series provides detailed updates on the state of the art in the treatment of different forms of cancer and also covers a wide spectrum of topics of current research interest. Clinicians will benefit from expert analysis of both standard treatment options and the latest therapeutic innovations and from provision of clear guidance on the management of clinical challenges in daily practice. The research-oriented volumes focus on aspects ranging from advances in basic science through to new treatment tools and evaluation of treatment safety and efficacy. Each volume is edited and authored by leading authorities in the topic under consideration. In providing cutting-edge information on cancer treatment and research, the series will appeal to a wide and interdisciplinary readership. The series is listed in PubMed/Index Medicus.

More information about this series at https://link.springer.com/bookseries/5808

Priya Hays
Editor

Cancer Immunotherapies

Solid Tumors and Hematologic Malignancies

 Springer

Editor
Priya Hays
San Mateo, CA, USA

ISSN 0927-3042   ISSN 2509-8497 (electronic)
Cancer Treatment and Research
ISBN 978-3-030-96375-0   ISBN 978-3-030-96376-7 (eBook)
https://doi.org/10.1007/978-3-030-96376-7

This Springer imprint is published by the registered company Springer Nature Switzerland AG
The registered company address is: Gewerbestrasse 11, 6330 Cham, Switzerland

For Iyer and Jay
Two Awesome, Amazing, Loving Brothers

Preface

This volume was born out of an idea I had thought of through an email from the publisher Springer Nature calling for Editors for the Cancer Treatment and Research series. Having had experience in both industry and publishing on cancer immunotherapies, in particular immune checkpoint inhibitors and chimeric antigen T cell therapies, I thought I would approach the Editor at the time about my idea to produce an edited collection on Cancer Immunotherapies. It was well received by the Editor, and the proposal was passed along to the Series Editor for the Cancer Treatment and Research series, who ultimately approved. After a series of requests for chapters and further communication, the volume started to take shape, and finally we had ten contributors established. These contributors are eminent researchers and scholars from a variety of national and international locales and preeminent cancer centers, including the Fred Hutchinson Cancer Research Center at University of Washington, Duke University, Memorial Sloan Kettering Cancer Center, Oxford University, Indiana University School of Medicine, University of Calgary, The University of Hong Kong, Winship Cancer Institute at Emory University, and the National Yang-Ming Chiao Tung University in Taipei, Taiwan. The volume begins with an in-depth discussion on the development of cancer immunotherapies and covers immune checkpoint inhibitor therapies for solid tumors such as melanoma. Other contributions focus on chimeric antigen receptor T cell therapies, CAR-T cell therapy for glioblastoma multiforme, and comprehensive accounts on genitourinary malignancies and acute myeloid leukemia, as well as molecules in the tumor microenvironment such as Lag3, along with single-cell sequencing for enhancing cancer immunotherapy. We aim to provide state-of-art knowledge and expertise on the fast-paced field of cancer immunotherapies for the clinician, scientist, and interested stakeholders in this collection that the audience will find enlightening.

San Mateo, CA, USA   Priya Hays

Acknowledgements

This edited collection, *Cancer Immunotherapies: Solid Tumors and Hematologic Malignancies* in the Cancer Treatment and Research Series published by Springer Nature, was the result of the copious efforts of many people and institutions. First, I would like to thank my editor Sydney Keen at Springer Nature who first proposed it to the Cancer Treatment and Research Editor, Corinna Hauser, who then submitted it to the Series Editor Steven Rosen, who approved it. I have to extend my utmost gratitude to the Society for Immunotherapy of Cancer who permitted me to advertise the call for chapter contributors on their website. A most gracious thanks go especially to the brilliant set of scholars and researchers from all across the globe with whom I collaborated to write and publish this edited collection. They grace this publication with their generous insight and vision for the state of current advances in the cancer immunotherapy field. Some are established Springer Nature contributors, while others are more junior in the field, but on their way to a wonderful future. I also thank Banu Dhayalan for her outstanding efforts as Production Editor in organizing this volume. They were all a delight to work with.

Introduction

Cancer Immunotherapies: Solid Tumors and Hematologic Malignancies contains the following chapters from national and international contributors, spanning the spectrum of the wide applications of immunotherapies for cancer malignancies of all types. The first chapter, "Development of Cancer Immunotherapies," by Diana DeLucia and John Lee fittingly discusses how cancer immunotherapies evolved as a "concept of harnessing the immune system" for the purposes of cancer therapy from its origins in "Coley's Toxins" to the production of cytokines and monoclonal antibodies to vaccines, immune checkpoint inhibitors and chimeric antigen receptor T cell therapy. The chapter "focuses on recent advances, current strategies and future outlook" for cancer immunotherapies. The second chapter "Melanoma" by Vishal Navani and colleagues offers a fascinating window into the role of immune therapies in non-cutaneous subtypes and a "review of the impact of underlying genomic, transcriptomic, epigenetic, proteomic and immunological correlates alongside their interaction with patient phenotypes" in understanding immune checkpoint inhibitor therapies for melanoma and the impact of immunotherapy response and resistance. Michael Brown in the third chapter "Engaging pattern recognition receptors in solid tumors to generate systemic antitumor immunity" reflects with great insight on how "malignant frequently exploit innate immunity to evade immune surveillance" and how these contexts are "determined in large part by pathogen recognition receptors" whose "activation induces the delivery of T cell priming cues from antigen-presenting cells." Brown discusses how this phenomenon influences the tumor microenvironment, "ultimately providing a personalized antitumor response against relevant tumor-associated antigens." Zaki Molvi and Richard O'Reilly provide their contribution for the fourth chapter, "Allogeneic tumor antigen-specific T cells for broadly applicable adoptive cell therapy of cancer" and explain how "tumor antigen-specific, donor-derived T cells are expected to be the mainstay in the cancer immunotherapy armamentarium" and "analyze clinical evidence that tumor antigen-specific donor-derived T cells can induce tumor regressions". They conclude on the applicability of this technology in pre-clinical and clinical settings. An excellent, innovative read. Sarwish Rafiq and Amitesh Verma provide "Chimeric Antigen Receptor (CAR) T Cell Therapy for Glioblastoma," a well-researched chapter on a topic that focuses on a malignancy that has great urgency for novel and efficacious treatments, glioblastoma multiforme. They remark that the remarkable clinical outcomes have been observed in hematologic malignancies and that similar outcomes in solid tumors such as glioblastoma have been challenging. They note that "recent data support the clinical efficacy and safety of CAR-T cell therapy" in glioblastoma, and conclude on "emerging techniques of optimizing CAR-T cell therapy for GBM." Francesca Aroldi and colleagues, in "Lag-3: From Bench to Bedside" discuss Lymphocyte-activation gene 3, a transmembrane protein involved in cytokine release and inhibitory signaling in T cells, as a target to "overcome the resistance, improve the activity and reduce the toxicity of checkpoint inhibitor therapy." The explain in excellent detail how LAG-3 "is a negative regulator of both CD4+ T cell and CD8+ T cell and the activity on CD8+ T cell is independent of CD4+ activation" and how the "blockade of LAG-3 has been tested in several combination therapies." In another valuable contribution to this volume. Kevin Lu and colleagues provide an in-depth analysis of immunotherapies in genitourinary malignancies in their chapter, "Immunotherapy in genitourinary malignancy: evolution in revolution or revolution in evolution." They discuss how immunotherapies for this tumor type have evolved from IL-2 for metastatic renal cell carcinoma to immune checkpoint inhibitors, which "demonstrate meaningful survival benefit and durable clinical response." They cite common hurdles that arise from "benefits limited to a minority of unselected patients due to the complexities of biomarker development" and "figuring out which patients best respond to immune checkpoint inhibitors and which patients won't respond to immune checkpoint inhibitors?" They conclude on common therapeutic strategies for genitourinary cancers for achieving health-related quality of life and efficacy. Fabiana Perna and colleagues provide a comprehensive, outstanding account of immunotherapies for acute myeloid leukemia, spanning from allogeneic stem cell transplantation, immune checkpoint inhibitors, chimeric antigen T cell therapies, and antibody-drug conjugates. They state that immune checkpoint inhibitors "have been used with limited success in relapsed/refractory acute myeloid leukemia" since "AML mutational burden is low" and that "identification of cell surface targets is critical for the development of other antibody-drug conjugates potentially useful in the induction and maintenance regimens." They "highlight active areas of research investigations toward fulfillment of the great promise of immunotherapy to AML." The two penultimate chapters, written by Ryohichi Sugimura and colleagues, "*Off-the-shelf* chimeric antigen receptor immune cells from human pluripotent stem cells" and "The single-cell level perspective of the tumor microenvironment and its remodeling by CAR-T cells" are meant to be read in sequence. They write that "autologous donors" for autologous CAR-T therapy face "technical challenges" and provide evidence for "the development of safe and efficient allogeneic CAR-T therapy." "Since the advent of the generation of immune cells from pluripotent stems cells, numerous studies focus on the off-the-shelf generation of CAR immune cells derived from PSCs," they write, and conclude that the "combination of PSCs-derived immune cells and CAR engineering pave the way for developing next-generation cancer immunotherapy." The second of the two-series chapters discusses "delineating the tumor microenvironment at a single-cell level", a very timely topic, that "will provide useful information for cancer treatment." They discuss the "cellular and molecular features that curb response to CAR-T cells, for example, high expression of immune checkpoint molecules (PD-1, LAG3) and anti-inflammatory cytokines (IL-4, TGFb) that block CAR-T cell function" They discuss how newly invented single-cell technologies would benefit the understanding of cancer immunotherapy. The final chapter is written by the Guest Editor (Priya Hays) upon invitation from the Editor, Corinna Hauser, and Series Editor, Steven Rosen, entitled "Clinical Development and Therapeutic Applications of Bispecific Antibodies for Hematologic Malignancies," focusing on the canonical bispecific antibody blinatumomab for acute lymphocytic leukemia. Bispecific antibodies (also discussed by the Perna and colleagues for AML) are "composed of two monoclonal antibodies that are designed to target tumor cells by directing T cells to the antigens on these cells. They recognize and bind to two distinct antigens. The majority of bispecific antigens fall into the category of bispecific T cell engagers or BiTEs." Blinatumomab is the FDA-approved agent in the BiTE class with CD19 and CD3 epitopes. This chapter discusses the mechanism of action of BiTEs, their clinical development and efficacy, and adverse events associated with their use in treating hematologic malignancies.

In total, these chapters provide profound insight into our understanding of cancer immunotherapies. I speak for all of the chapter contributors and editors who worked diligently to produce an edited collection that advances the field forward and promotes a greater awareness of immunotherapies for all interested stakeholders for solid tumors and hematologic malignancies and beyond.

San Mateo, CA, USA   Priya Hays

Contents

Development of Cancer Immunotherapies

1

Diana C. DeLucia and John K. Lee

1.1 Cancer Immunotherapy—Precision Medicine

Cancer is a leading cause of death globally that is second only to cardiovascu-lar diseases such as ischemic heart disease and stroke [1]. It is likely cancer will become the primary cause of death worldwide as treatments for cardiovascular disease and infectious diseases continue to improve with modern medicine and the continued extension of average life expectancy. The development and imple-mentation of both localized and systemic therapeutic strategies, such as radiation therapy and chemotherapy, for the treatment of cancer have significantly improved the quality of life and survival rates for many individuals with cancer. Although radiation and systemic chemotherapy can successfully eliminate early disease and help manage disease, improving patient prognosis and overall survival, incomplete elimination of all cancerous tissue can occur alongside unwanted side effects due to off-target damage to healthy tissue [2, 3]. Targeted cancer treatment strategies utilize agents that block the progression of cancer by targeting molecules in the body that specifically promote the growth and spread of tumor cells. Variation in tumor-associated genetic alterations is responsible for a high level of molecular heterogeneity among targetable tumor molecules within patients as well as among different patients with the same or differing cancers. Such intra and interindivid-ual heterogeneity have formed the basis of "precision medicine" through which known characteristics of a patient's tumor are used to develop a personalized treatment regime. Despite the success of such strategies, millions of people con-tinue to die of cancer every year. Cancer immunotherapy is a form of precision

D. C. DeLucia · J. K. Lee (✉)
Fred Hutchinson Cancer Research Center and University of Washington, 1100 Fairview Avenue, Seattle, WA 98109, USA
e-mail: jklee5@fredhutch.org

P. Hays (ed.), *Cancer Immunotherapies*, Cancer Treatment and Research 183,
https://doi.org/10.1007/978-3-030-96376-7_1

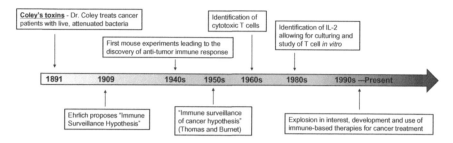

Fig. 1.1 Pioneering events that shaped the historical evolution of immune-oncology and immunotherapy

medicine whereby the genetic status or protein expression profile of an individual's tumor is used to formulate immune-based treatment agents that induce an anti-tumor immune response in patients. The goal of immunotherapy is to harness the antigen-specific nature and cytotoxicity of the human immune system to mount a potent anti-tumor immune response capable of eliminating all tumor cells without harming healthy tissue. Over the past century, an improved understanding of how the human immune system interacts with cancerous cells in the body has provided a foundation upon which immunotherapies are developed and studied today (Fig. 1.1).

1.2 Immune Surveillance and Recognition of Cancer

Efforts to treat cancer with immune-based therapies originated over a century ago, prior to the identification of tumor antigens and without an appreciation of the many components of the human immune system and anti-tumor immune response that we have today. Some of the first reported associations of immunity and cancer occurred in the late nineteenth century by German physicians Wilhelm Busch and Friedrich Fehleisen who observed that some of their cancer patients with concurrent erysipelas, a common streptococcal infection of the skin, experienced tumor regression or complete remission [4]. Similar findings and strategies were also observed and implemented by the American physician, Dr. William Coley, in the 1890s. Upon a vast review of patient medical records in search of information that might aid in the treatment of his own patients, Coley identified a number of cases where patients with otherwise incurable tumors entered into remission following a diagnosis of erysipelas. Coley subsequently used intratumoral inoculations of a cocktail of heat-killed strains of *Streptococcus* to treat his patients with soft tissue sarcomas with unforeseeable success [5, 6]. The streptococcal cocktail was later coined "Coley's Toxins". Despite the unrivaled success of Coley's toxins at the time, Coley's colleagues heavily debated and were hesitant to adapt the therapeutic technique due to a lack of understanding of the mechanism by which the bacterial inoculation induced tumor regression [5]. Consequently, Coley's achievements went unnoticed for the better part of the following 50 years while surgery and radiotherapy became the standard of care for cancer patients in the early twentieth century.

1.3 Cancer Immunoediting

In 1909, Paul Ehrlich was the first to propose the "immune surveillance hypothesis" which postulated that it was not unreasonable to think that the immune system can identify transformed cells, eliminate them from the body, and prevent primary tumor formation [7]. At the time, Ehrlich lacked the technical means to generate experimental evidence in support of his hypothesis. In the 1950s, the conversation of immune surveillance was resurrected by two scientists Lewis Thomas and Frank McFarlane Burnet. Soon after, the existence and immune recognition of tumor-specific antigens was empirically demonstrated [8] thereby providing evidenced-based support for Erlich's immune surveillance hypothesis. Despite these groundbreaking findings, it remained clear that many people continued to develop tumors and succumb to cancer, suggesting anti-tumor immune response did not always effectively prevent tumor formation. In the 2000s, Robert Schreiber proposed the "three E's theory" to describe three stages of cancer immunoediting: (1) Elimination phase, previously the immune surveillance hypothesis, (2) Equilibrium phase, and (3) Escape phase [9]. During the elimination phase, immune cells effectively identify and eliminate tumor cells in otherwise healthy tissues. The immune pressure applied to the tumor during the elimination phase drives some cancerous cells to undergo changes that mediate immune evasion resulting in a subset of variant tumor cells that are resistant to immune recognition and/or killing resulting in an equilibrium phase of immune-tumor interaction. Lastly, tumors enter the escape phase once all immunogenic tumor cells have been eliminated and only variant tumor cells remain to spread unhindered due to their acquired immune resistance. Additionally, tumor cells that have undergone Epithelial Mesenchymal Transition (EMT) often acquire improved metastatic potential through enhanced mobility and resistance to both apoptosis and immune-mediated killing [10, 11]. Therapeutic strategies targeting EMT have potential to improve the efficacy of immune-based therapies.

Acquisition of evidence directly demonstrating cancer immunoediting has historically been challenging due to the need for large patient numbers, the heavy demand for long-term study follow-up, and significant biological sample collection. The first collective body of data in support of human immune surveillance of cancer consisted primarily of studies comparing cancer incidence and progression rates in individuals with robust versus compromised immune systems. By the late 1980s, it was well documented that immunocompromised individuals, such as transplant patients, non-transplant patients receiving immunosuppressive treatment, established cancer patients receiving immunosuppressive chemotherapy, individuals with genetic immunodeficiencies, and individuals with acquired immunodeficiency syndrome, have a higher incidence of developing certain types of primary or secondary cancer [12]. A recent study of 907 immunocompromised heart and/or lung transplant patients between 1989 and 2004 found that the transplant patients were seven times more likely to develop cancer compared to the non-transplant population [13]. While the clear association between immune deficiency and cancer incidence supported an important role for the immune system in cancer development, mechanisms of cancer immune surveillance and the usefulness of immunotherapies became more evident as the scientific community's discovery and understanding of human tumor antigens advanced.

1.4 Tumor Antigens

Throughout much of the twentieth century, it was unclear whether the immune system could detect malignant cells or develop a protective immune response against tumors. The development of inbred mice strains and the discovery that tumors could be induced with carcinogens and transplanted between mice provided the tools necessary for the first essential studies that demonstrated immune recognition and elimination of tumors [14–17]. Mice injected with tumor cells rendered replication-incompetent through irradiation, were often protected against subsequent injection of growth competent cells of the same tumor type, but not of a different tumor type [18]. These experiments helped validate immune recognition and protective immunity against tumors and established the concept that different tumor types have unique immunogenic antigens, known as tumor recognition antigens.

1.4.1 Tumor-Specific Antigens

Tumor recognition antigens are categorized into two subsets: Tumor-specific antigens (TSAs) and tumor-associated antigens (TAAs). TSAs are "self" protein-derived peptides strictly found in tumor cells and not in nonmalignant cells that are presented on major histocompatibility complexes (MHC). TSAs are typically caused by genomic mutations or gene rearrangements leading to amino acid changes in proteins of cancer cells from which unique peptides can be generated, known as neoantigens or neopeptides [19]. Neoantigens form when the coding sequence of a peptide that is normally presented is altered leading to the de novo presentation of a novel peptide on MHC. For example, specific mutations in *CDK4* in melanoma cells result in presentation of highly immunogenic CDK4-derived neopeptides [20, 21]. Common mutations in tumor-suppressor genes such as *TP53* in cancer also represent immune-reactive TSAs [22, 23].

1.4.2 Tumor-Associated Antigens

TAAs are proteins or peptides that are expressed at a higher level in cancer cells while also being expressed in healthy cells but at a much lower level. Cancer-testis antigens, such as melanoma-associate antigens (MAGE) and New York esophageal squamous cell carcinoma 1 (NY-ESO-1), are well-studied examples of TAAs because they are only expressed in healthy male germ cells which lack MHC expression. TAAs also include differentiation antigens which are only expressed in select tissues such as CD19 on B lymphocytes and B cell lymphomas and MART-1 in melanoma [24–26]. Expression of TAAs can be upregulated in malignant cells through abnormal gene expression as well as post-transcriptional and post-translational modifications [26, 27]. Oncoviral proteins, foreign proteins expressed by cancer cells following infection and transformation by cancer-causing viruses, also serve as TAAs. The E6 and E7 proteins encoded by human papillomavirus type 16, one of the most common etiological agents of cervical cancer, and the K8.1 glycoproteins expressed by Kaposi's sarcoma virus are examples of immunogenic oncoviral proteins [28, 29].

1.5 Methods of Tumor Antigen Identification

The identification of cytotoxic T lymphocytes (T cells) in the 1960s [30] and the development of methods to propagate and study antigen-specific T cells in vitro in the 1970s and 80s [31–33] provided the foundational knowledge and technical methods required to identify tumor antigens. Additional technological advancements over the past 30 years have led to a sharp increase in the number of known tumor antigens, many of which are now utilized for cancer diagnostics and therapeutic targeting. A large range of molecular strategies continues to be explored to improve the identification of tumor antigens.

1.5.1 Gene Expression Profiling

Safety concerns when selecting tumor antigen targets are a primary focus for preclinical and clinical studies evaluating precision cancer therapies. To avoid off-tumor toxicity and undesired side effects, ideal tumor antigens for therapeutic targeting should be expressed exclusively or preferentially by tumor cells compared to healthy cells in the body. Tumor antigen expression should also be present across all or most malignant cells within tumors to maximize efficacy of the antigen targeting agent or drug and to minimize therapeutic resistance due to antigen escape. A common strategy to identify tumor-restricted antigens involves comparing the gene expression profiles of malignant and healthy cell lines or tissues using a variety of technologies such as microarrays, RNA-sequencing, direct digital mRNA counting, and real-time PCR [34–39]. Proteins encoded by genes widely expressed or overexpressed by tumor cells compared to healthy tissue can be investigated as potential tumor rejection antigens. However, analysis of gene expression alone does not always mirror protein expression profiles and is less effective for the discrimination of intracellular versus cell surface protein expression. An alternative approach is to integrate the analysis of transcriptomics and proteomics. We have recently used this method to specifically study differential cell surface protein expression to identify candidate tumor antigen targets for CAR T cell therapy in molecular subsets of prostate cancer [40, 41].

1.5.2 Serological Screening for Antibody-Targeting Antigens

Studies demonstrating humoral immune responses against tumors as well as a continued increase in the use of antibodies for immunological studies sparked interest in the identification of human antibody-binding tumor antigens. The first antigen discovery methodology of this type, serological screening of recombinant cDNA expression libraries (SEREX) was developed in 1995 [42]. SEREX utilized reverse proteomics whereby cancer patient sera were used to screen a phage cDNA library derived from various human malignancies for potential tumor antigens recognized by patient tumor-specific antibodies. Several novel tumor antigens have been identified using traditional SEREX strategies, including the cancer-testis antigen NY-ESO-1 [43]. The use of traditional, modified, and high-throughput optimized SEREX technology has aided in the identification of thousands of additional tumor antigens. The serological and proteomic evaluation of antibody responses (SPEAR) method combines the use of 2-D polyacrylamide gel electrophoresis of patient tumor samples, immunoblotting with patient sera, and mass spectrometric protein analysis to rapidly identify tumor antigens. SPEAR, as well as similar methods, has been used successfully in the setting of renal cell carcinoma for which identification of tumor antigens has been historically challenging [44, 45].

1.5.3 Recognition by Reactive Patient Lymphocytes

The use of tumor-reactive lymphocytes, specifically T cells, from patients as a screening tool has been one of the most widely used strategies for tumor antigen identification and continues to be used today. Tumor-infiltrating lymphocytes or lymphocytes isolated from patient peripheral blood are co-cultured with human leukocyte antigen (HLA)-matched cells expressing cDNA libraries corresponding to protein expression profiles derived from specific tumor types. Reactivity of the T cells is used to screen for immunogenic tumor antigens. The first tumor-associated antigen, melanoma-associated antigen MAGE-1 [46], was identified using this method and many others have followed [47–54]. While advancements in tumor antigen identification have led to the discovery of an abundance of potential targets for immunotherapy, antigen selection remains a challenging and essential step in the development of all targeted immunotherapeutic modalities.

1.6 Tumor Immune Evasion

Tumors employ various mechanisms to avoid activation of and detection by the host immune system such as decreasing overall immunogenicity through modulation of antigen expression, inducing T cell tolerance of tumor antigens, and creating a suppressive tumor microenvironment (TME). Some cancer types referred to as "cold tumors", inherently have poor immunogenicity whereby the majority of the tumor cells, as well as nonmalignant cells found in the TME, do not express tumor rejection antigens or other molecules that mediate immune cell infiltration and activation. Such tumors can (1) alter adhesion molecule expression that would otherwise mediate contact growth inhibition as well as proinflammatory signaling, (2) express few neoantigens due to a low tumor mutation burden, or (3) exhibit aberrant MHC expression due to somatic mutations in antigen-presenting genes or downregulation of MHC gene expression [55–57].

Tumor cells can also undergo antigen escape due to genetic instability resulting in altered amino acid sequences of tumor antigens or downregulation of tumor antigen expression to evade immune detection. Some tumor cells have been found to downregulate expression of T cell costimulatory molecules such as CD80 and CD86 which are essential for antigen-specific T cell activation and anti-tumor cytotoxicity [58, 59]. T cells that interact with tumor antigen presented on MHC on the tumor cell surface in the absence of these costimulatory signals are more likely to become tolerized rather than mediate killing of the tumor cell [60]. Additionally, cancer cells can induce immune suppression and promote a "pro-tumor" microenvironment by acting directly on cytotoxic T cells through the secretion of factors that directly suppress T cells such as tumor growth factor-beta (TGF-β) and indoleamine 2,3-dioxygenase (IDO) or induce tumor cell surface expression of T cell inhibitory molecules such as PD-L1 [61]. Some of the same factors can attract or drive the differentiation of suppressive immune cells such as regulatory T cells (Treg) and myeloid-derived suppressor cells (MDSC) to and within the tumor [61]. The production of collagen by tumor cells often functions to create an extracellular matrix through which immune cells struggle to pass. This physical barrier can create a tumor-induced privileged site that effectively blocks the entrance of any component of the cellular immune response.

Despite the multitude of mechanisms used by tumors to evade elimination by the immune system, a large body of research conducted over the past several decades has improved our understanding of the ways in which the immune system and cancer cells interact, thereby better positioning scientists to develop strategies for effective immunotherapy against cancer.

1.7 Passive Immunotherapy

The overall goal of cancer immunotherapy is to harness the antigen-specific and cytotoxic potency of the immune system to eliminate cancer cells from the body at both local and metastatic disease sites. An extensive range of modalities utilizing modified components of both the humoral and cellular immune response is under investigation for the generalized treatment of a broad spectrum of cancers. Immunotherapies are grouped into two main categories: passive immunotherapies and active immunotherapies. Passive forms of immunotherapy involve the introduction of donated or ex vivo-derived immune system components into a cancer patient to specifically target that patient's tumor. The most common forms of passive immunotherapy currently under investigation are tumor-specific antibodies, immune checkpoint inhibitors, and adoptive T cell therapy.

1.7.1 Tumor-Specific Antibodies

1.7.1.1 Monoclonal Antibodies

Monoclonal antibodies (mAb) have been routinely used for the clinical treatment of cancer for the past decade and represent some of the earliest immunotherapies tested in clinical trials and approved for clinical use due to their improved efficacy over current standard of care and versatility. Unlike polyclonal antibodies, which are a collection of antibodies that recognize different epitopes within the same antigen due to their isolation from a pool of B cells in an antigen immunized organism, mAb are isolated from clonal selection of antigen-specific hybridomas and are therefore identical to one another [62]. Hybridomas generated from the fusion of B cell clones from antigen immunized mice with immortalized myeloma cells lacking antibody production were first used to generate mAb [63]. The ability to produce large amounts of antibodies targeting a single epitope made monoclonal antibodies ideal for clinical use. Strong human anti-mouse antibody responses in patients have been demonstrated against some mouse monoclonal antibodies leading to decreased efficacy of the therapeutic antibody [64]. Molecular techniques have since been developed to humanize or generate chimeric versions of mouse mAb in order to eliminate patient immune reactivity without sacrificing the desired antigen specificity of the original antibody [65]. The variable regions of therapeutic mAb recognize and bind to protein antigens expressed on the surface of tumor cells effectively tagging the tumor cells for recognition by other immune cells which bind the constant region (Fc) of the antibody. Fc receptor binding facilitates direct killing of tumor cells through complement-dependent cytotoxicity, or indirectly through antibody-dependent cellular cytotoxicity (ADCC), usually mediated by natural killer (NK) cells [66] or phagocytosis of the tumor cell-mediated by myeloid cells such as macrophages [67]. The CD20-targeting chimeric mAb, Rituximab, was shown in a phase II clinical trial of 37 patients with relapsed non-Hodgkin's lymphoma following chemotherapy to provide improved patients outcomes and a better safety profile compared to chemotherapy alone [68]. Subsequently, in 1997 Rituximab became the first mAb to receive approval from the US Food and Drug Administration (FDA). Between 1997 and 2020, 16 mAb targeting 10 separate tumor antigens received FDA approval for the treatment of 13 types of cancer, including both hematologic and solid malignancies [69].

The primary efficacy-limiting challenges associated with mAb therapy are the half-lives of the antibodies in vivo, poor penetration of the therapeutic product into solid tumor tissue, and dependency on homogenous tumor antigen expression. Orthotopic genetically engineered mouse models (GEMMs) have allowed for the recapitulation of human tumor molecular heterogeneity and microenvironments in mouse tumors allowing for better assessment of mAb penetration and efficacy compared to subcutaneous or orthotopic syngeneic models [70, 71]. An alternative strategy by which mAb encoding DNA plasmids are introduced to patients through electroporation-enhanced intramuscular injection has been demonstrated with the PSMA-targeting mAb in preclinical models of prostate cancer [72]. Enhancement of tumor penetration with nanoparticle delivery systems that take advantage of the Enhanced Permeability and Retention effect to allow for greater accumulation of nanoparticles in newly vascularized tumor tissue by passive diffusion through endothelial gaps generated during extravasation has been demonstrated in a mouse sarcoma model [73, 74] and is worth exploring in other contexts. mAb-directed therapies, whereby mAbs are linked to drugs with toxic properties, promote antigen-guided delivery of genotoxic or cytotoxic payloads more specifically to tumor cells and enhance the anti-tumor effects of naked mAb therapies as discussed next.

1.7.1.2 Antibody-Targeted Radiotherapy

The localized delivery of high doses of ionizing radiation to cancerous regions of the body, or radiotherapy, is commonly used for the treatment of cancer. Radiotherapy is used for both curative and palliative intent in the treatment of cancer patients [75]. Although radiotherapy can be localized to known tumor regions within the body, damage to nearby healthy tissue and toxicity-associated side effects are commonly observed following radiation with significant impacts on quality of life [76]. Modernized methods to improve anti-tumor efficacy of external beam radiation therapy include improved tumor target delineation with different imaging strategies, fractionated dosing, and conformal techniques [77]. Yet it remains challenging to deliver a consistent, controlled dose of radiation to tumor sites using an external source. Antibody-targeted radiotherapy has been explored as a strategy to deliver a toxic dose of radioisotopes directly to tumor cells inside the body resulting in improved efficacy and reduced off-target toxicity [78]. The technology involves linking tumor antigen-targeting mAbs to radioisotopes that kill the cancer cell once the antibody is bound and internalized, thereby concentrating the genotoxicity of the radiotherapy within tumor cells and sparing surrounding healthy tissue. In the early 2000s, the first antibody-targeted radiotherapy drugs, yttrium-90 (90Y)-ibritumomab tiuxetan and iodine-131 (131I)–tositumomab, were

FDA-approved, both targeting CD20 for the treatment of non-Hodgkin's lymphoma [69]. Unfortunately, this form of radioimmunotherapy has not been widely used due to reports of severe systemic toxicities associated with treatment and poor efficacy in solid tumors [76]. Both high toxicity and poor solid tumor mass penetration have been linked to the size of the antibody isotype. High rates of bone marrow toxicity have been observed with IgG mAb-based drugs, which is believe to be a result of the long half-life and high rate of hepatic uptake compared to smaller antibody isotypes [79]. Smaller molecules, such as single-chain variable fragments (scFv) have demonstrated significantly higher rates of systemic clearance [80] and likely represent less toxic alternatives for the delivery of tumor toxic drugs.

Antibody-directed radiotherapy has demonstrated reasonable efficacy against solid tumors, including immunologically cold tumors, such as prostate cancer. Targeting prostate-specific membrane antigen (PSMA) has been heavily studied for the treatment of prostate cancer. PSMA is expressed on the tumor cell surface in the majority of prostate cancers, with increased expression observed in association with late-stage disease progression, making it a good candidate for antibody-targeting therapies. The PSMA-targeted mAb, J591, conjugated to ^{177}Lu or ^{90}Y radionuclides has demonstrated targetability of both soft tissue and bone metastases in prostate cancer and promising clinical efficacy across four clinical trials [81–84]. Based on the observed benefit of ^{177}Lu-PSMA-617 treatment of prostate cancer [85–87], the ongoing international phase III VISION trial aims to assess survival benefit of ^{68Ga}-PSMA-11 and ^{177}Lu-PSMA-617 in men with PSMA-expressing metastatic castrations-resistant prostate cancer [88]. Targeted radiotherapy for solid tumors such as prostate cancer could be also advanced with the use of ultrasmall gold nanoclusters, which have been shown to be internalized by prostate cancer cells and rapidly cleared from the body [89]. Extensive clinical trial data indicate that antibody-mediated delivery of radioisotopes to cancer cells improves the tolerability of radiotherapy in patients and supports the use of antibody-directed immunotherapies to improve the efficacy and safety of anti-cancer agents.

1.7.1.3 Antibody–Drug Conjugates

Antibody–drug conjugates (ADCs) are another form of antibody-targeted therapy that has gained great interest in the cancer immunotherapy field. ADCs consist of a therapeutic mAb connected by a chemical linker to a bioactive cytotoxic drug that is designed to be internalized by and kill tumor cells. To date, seven ADCs have been FDA-approved targeting a variety of tumor antigens, such as CD33, HER2, CD20, and recently TROP2, for the treatment of a range of both hematologic and solid tumors [69].

Over the past decade, three generations of ADCs have evolved during which the features of all three ADC components (antibody, linker, and cytotoxic payload) have been finely tuned. Considerations for the selection of targetable antigens for therapeutic mAb identification discussed earlier in this chapter are translatable to ADCs. Favorable characteristics of antibody–drug linkers include (1) appropriate stability to minimize drug release from the ADC prior to reaching target tumor cells and (2) effective cleavability for drug release and activity once internalized by tumor cells. Both cleavable and non-cleavable linkers have been investigated and utilized in FDA-approved ADCs. The current cleavable linkers used in clinical ADCs or in development utilize hydrolysis, enzymatic activity, or reduction by glutathione as cleavage mechanisms [90]. Examples include the valine–citrulline dipeptide linker, utilized in the ADC brentuximal vedotin for the treatment of CD30[+] Hodgkin lymphoma, which is sensitive to and cleaved by cathepsin B, a protease found in lysosomes [91]. CLA2, the linker in sacituzumab govitecan, is hydrolyzed by a decrease in pH following entry into the lysosomal pathway of tumor cells [92]. Recently, the CEACAM5-targeting ADC labetuzumab govitecan was demonstrated to have remarkable anti-tumor activity in preclinical studies against multiple CEACAM5-expressing prostate cancers, including highly lethal neuroendocrine prostate cancer with varying levels of and heterogeneous CEACAM5 expression [93]. The moderate stability of the CLA2 linker used in labetuzumab govitecan was hypothesized to contribute to a bystander effect allowing for complete clearance of tumors despite antigen heterogeneity. Linkers with poor stability, as were utilized with many first-generation ADCs, have a greater potential for unwanted toxicities in patients [94]. This suggests that although strong linkers are commonly preferred for safety reasons, there is also a clinical space for the use of moderately stable linkers as the efficacy of ADC linkers with differing stability is likely dependent on-tumor context.

Optimal cytotoxic payloads have high potency for cell killing, low hydrophobicity, and an adequate structural feature to which the antibody linker can bind. Although there are many toxic payloads being studied for use with ADCs, the majority fall into two categories: (1) DNA-targeting agents or (2) microtubule-targeting agents. DNA-targeting payloads, such as irinotecan, Doxorubicin, Pyrrolobenzodiazepine, and Duocarmycin function by creating DNA damage through topoisomerase inhibition or interfering with DNA processing in tumor cells leading to cell death. Such drugs function against both proliferating as well as non-proliferating cells [94]. Microtubule-targeting payloads, such as auristatins and maytansinoids provide anti-tumor activity by blocking tubulin assembly in tumor cells and therefore function against only proliferating cells [94]. Insufficient intracellular delivery of the payload to tumor cells is considered the primary reason for poor ADC efficacy and ultimate failure of ADCs in clinical settings. Identifying payloads with high potency/low IC_{50} values and reliable linkers have been a primary focus of second- and third-generation ADCs to promote the delivery of higher payload concentrations to target cells even in low antigen settings [94, 95]. ADC payload hydrophobicity has been directly linked to poor drug stability, as well as suboptimal clearance and efficacy in vivo [96, 97]. Because some linked payloads require some level of hydrophobicity to retain their biological function, the addition of polyethylene glycol or PEGylation of the linker or the payload has been used in many ADCs to help mask the hydrophobicity of the payload as well as improve ADC solubility [96]. Optimization of the ADC structure and components have expanded the therapeutic window of modern, third-generation ADCs compared to mAb therapy, chemotherapy, or earlier generation ADCs [98, 99].

1.7.1.4 Bi-specific T Cell Engagers

Bi-specific T cell engager antibodies (BiTEs) are composed of two conjugated single-chain antibodies, one of which is specific for the CD3 component of the T cell receptor (TCR) and the other which is specific for the tumor antigen of interest. The conjugated antibodies offer enhanced T cell activation and antigen-specific killing of tumor cells by cytotoxic T cells. Through engagement of the TCR, the CD3-specific antibody activates the cytotoxic effector functions of CD8 T cells while the tumor antigen-specific antibody binds to the surface of antigen-expressing tumor cells. The binding of both the T cell and tumor cell in proximity collectively redirects the cytotoxicity of the T cell toward the tumor cell, regardless of the cognate specificity of the T cell [100]. The only FDA-approved BiTE, blinatumomab, targets CD19 for the treatment of acute lymphoblastic leukemia [101]. Although blinatumomab conferred a significant improvement in overall survival compared to the chemotherapy standard of care, median survival in the blinatumomab group remained less than 8 months [102]. Several other BiTEs are currently under investigation, including a HER2 BiTE in the context of gastric and breast cancer and an EGFR BiTE in the context of non-small cell lung cancer (NSCLC) [69]. Due to the extremely short half-life of standard first-generation BiTEs which require frequent continuous infusion due to the absence of antibody Fc domains, extended half-life BiTEs with fused Fc domains that impart improved stability and bioavailability are favored for the design of clinical products [100].

1.7.2 Immune Checkpoint Inhibitors

The immune system has a set of checks and balances in place designed to minimize or prevent over-activation of cellular immune responses resulting in unwanted, indiscriminate killing of healthy or self-tissue during a response mounted against a foreign antigen in the body, such as a pathogen or a tumor-specific neoantigen. Immune checkpoint receptors are commonly upregulated by T cells following antigen-specific activation and cytotoxicity which serve as negative regulators of T cell function [103, 104]. The expression of immune checkpoints by T cells also influences positive selection of antigen-specific T cells in the thymus during T cell development and ultimately influences the T cell repertoire [105]. Tumor cells also upregulate immune checkpoint receptors and ligands to dampen anti-tumor T cell responses [106]. Immune checkpoint inhibitors (ICIs) are antibodies administered therapeutically that are designed to competitively bind to checkpoint receptors or ligands and block the inhibitory activity of checkpoint receptors to improve anti-tumor T cell responses. The two checkpoint receptors/ligand pairs that have been most commonly targeted are cytotoxic T-lymphocyte associated protein 4 (CTLA-4) binding to CD80 or CD86 and programmed cell death protein 1 (PD-1) binding to programmed death-ligand 1 (PD-L1). During T cell activation, CD80 and CD86 receptors on antigen-presenting cells (APC) bind to CD28 and provide a costimulatory signal following MHC peptide binding to the TCR [107]. Once the T cell has performed its effector function, CTLA-4 and PD-1 are upregulated on the T cell surface and serve as a T cell "off switch" which can competitively bind to CD80 or CD86 on APCs resulting in inhibition of TCR signaling to prevent any potential damage caused by a prolonged T cell response [108]. Following antigen-specific activation, T cells similarly upregulate PD-1 on the cell surface which can be engaged by PD-L1 expressed on the surface of anti-inflammatory immune cells, such as Tregs to suppress T cell activation and increase T cell apoptosis [109]. Tumor cell upregulation of PD-L1 and effective blocking of T cell activation upon tumor cell-T cell interaction is well documented [110].

The use of immune checkpoint blockade for the treatment of cancer was first proposed by James P. Allison in the early 1990s, however, significant doubt and lack of support was displayed from the cancer and immunology research communities for many years. In the 2000s, Allison and colleagues demonstrated that targeting of CTLA-4 with a fully human mAb led to durable responses in patients with melanoma [111, 112], which ultimately led to FDA approval of Ipilimumab for the treatment of melanoma [113]. Allison's perseverance and ultimate success in treating melanoma with ICI earned him the Nobel Prize in Physiology or Medicine in 2018. The approval of Ipilimumab in 2011 generated significant excitement in the cancer field and re-ignited a fierce interest in immunotherapy paradigms for the treatment in the scientific and clinical oncology communities. Six additional ICIs targeting, PD-1, PD-L1, and CTLA-4 have been approved for clinical use since Ipilimumab in 2011, the majority of which are for solid tumor indications.

Despite the success of ICIs in many patients targeting numerous tumor types, widespread use of ICIs has been hindered by relatively low response rates and severe immune-related adverse events (irAEs) in some patients [114, 115]. The presentation of irAEs can vary significantly with ICI and tumor type. ICI-associated irAEs most commonly affects the skin, but have been observed across many tissues, including vital organs, but can often be effectively treated with immune-suppressive drugs [115]. Neurotoxicity caused by neurological immune-related adverse events (NirAEs), is of particular concern with ICI treatment. Although NirAEs is a relatively rare side-effect of ICIs and more commonly affect the peripheral rather than the central nervous system, their potential severity makes them a significant clinical concern [116]. This has led to significant investigation into the identification of biomarkers associated with ICI response [106]. Specific genomic profiles across different cancers associated with tumor expression of neoantigens, antigen presentation, DNA repair, and oncogenic pathways have been correlated to patient response to ICIs [117]. Additionally, epigenetic changes in nonmalignant cells within the TME have been shown to contribute to solid tumor escape from immune surveillance and subsequent resistance to ICIs [114, 118].

Novel strategies to reduce the non-specific toxicities associated with ICIs are being actively investigated preclinically and clinically. Enhancing ICI targeting to tumor to reduce off-target toxicities has been demonstrated by leveraging exposed collagen in tumors. The fusion of ICI antibodies to collagen-binding domains has demonstrated strong on-tumor efficacy with improved safety over traditional ICIs [119]. ICIs used in combination with other ICIs or other approved cancer therapies, such as chemotherapy, radiotherapy, or tumor-specific therapies have demonstrated considerably enhanced tumor responses and patient benefit compared to ICI monotherapy [120]. Oncogenic kinases have been shown to increase immune suppression of solid tumor TMEs and kinase inhibition in combination with ICI has demonstrated clinical benefit for the treatment of breast cancer [121]. Approved kinase inhibitor monotherapies used for the treatment of other cancer types may also be strong candidates for combination with ICI therapy.

Although the research community has extensively explored mechanisms to improve and optimize antibody-based therapy design for tumor targeting, the majority of drugs tested in clinical trials do not meet the criteria for FDA approval. To improve the efficiency and speed at which novel therapies are translated into the clinical, improved models are required for preclinical assessment of drug effectiveness and safety. A well-known example is that of the CD28 agonist antibody TGN1412, a drug that performed well with an extremely high safety profile in animal models, but caused life-threatening toxicities, in the first patients tested in the first phase of clinical trials [122, 123]. In fact, the preclinical studies evaluating TGN1412 used cynomolgus and rhesus non-human primates rather than mice or rats because they demonstrated drug binding affinities similar to that of humans. While small changes in cytokine production were detected in the TGN1412 monkeys, the preclinical trials did not predict the life-threatening cytokine release syndrome observed in the first phase I clinical trial [122, 124, 125], indicating the potential for poor translation between seemingly representative preclinical models and in vivo human testing.

1.7.3 Adoptive T Cell Therapies

An alternative form of passive immunotherapy involves the adoptive transfer of cellular components of the immune system, most commonly T cells that naturally express or are engineered to express TCRs that recognize tumor antigens. The importance of T cells in tumor-specific immune responses has been apparent based on both mouse and human studies for several decades [126, 127]. The genetic engineering of patient T cells provides an avenue through which large numbers of T cells expressing high-affinity antigen-specific surface receptors can be produced in a lab generating a highly potent therapeutic product. Optimization of protocols for adoptive T cell product manufacturing has significantly reduced the time from peripheral blood collection to re-infusion of a therapeutic product. A variety of genome modification techniques have been employed to improve production of "off the shelf" chimeric antigen receptor (CAR) T and NK cell products for broad use in patients [128]. Here we discuss the development, use, and challenges associated with the three main types of adoptive T cell therapy products under investigation or in clinical use: tumor-infiltrating lymphocytes, TCR transgenic T cell, and CAR T cells.

1.7.3.1 Tumor-Infiltrating Lymphocytes

Tumor-associated immune cells can provide both pro- and anti-tumor functions from within the TME [129]. Due to their ability to penetrate the solid tumor environment and greater potential to recognize TSA/TAAs due to their intertumoral trafficking, tumor-infiltrating lymphocytes (TILs) have been widely studied for their prognostic value [130–134] and were also the first form of autologous T cells utilized for adoptive therapy against cancer [135]. Adoptive TIL therapy aims to improve T cell-mediated tumor elimination through the introduction of a large number of tumor-specific T cells in cancer patients. Most studies and clinical trials have focused on adoptive TIL therapy for melanoma, arguably the most immunogenic tumor type, and have demonstrated objective response rates of 50% and durable responses in approximately 10% of treated patients across studies performed at multiple treatment centers across the US [136]. TIL therapy has also demonstrated tolerability and sustained anti-tumor responses for at least 12 months against Epstein Barr virus (EBV)-driven nasopharyngeal carcinoma [137]. More recently, adoptive TIL therapy showed moderate anti-tumor activity but high levels of infused TIL exhaustion in a small pilot study of patients with metastatic ovarian cancer [138]. Interest in the therapeutic use of tumor-specific B cells has also recently emerged. The adoptive transfer of tumor-infiltrating B cells has demonstrated efficacy and patient benefit against lung cancer [139].

Studies have shown that ex vivo antigen-specific TIL expansion prior to infusion into patients is associated with therapeutic benefit in the context of nasopharyngeal carcinoma and metastatic melanoma [140, 141]. However, some tumor types, such as prostate cancer, exhibit minimal to no infiltration of lymphocytes. Prostate cancer is considered immunologically "cold" due to a lack of immune activation within the tumors. 3D models, such as tumor organoids or spheroids, have been developed to aid in studies investigating mechanisms to improve the infiltration of immune cells and immunotherapeutic drugs in vitro [142]. While researchers have reported methods to successfully isolate tumor-reactive TILs from prostate tumors [143], the reliable identification and isolation of TILs within prostate tumors and other cold tumors remains challenging.

1.7.3.2 TCR Transgenic T Cells

An alternative method to TIL therapy is the genetic engineering of transgenic T cells isolated from patient peripheral blood to express receptors that recognize TSA/TAAs. Transgenic TCR therapies are generated by isolating patient peripheral blood from which T cells are isolated and genetically modified to express a TCR specific to a tumor antigen. Once a small number of patient T cells are modified to express the therapeutic TCR, the transgenic cells are expanded in vitro to larger numbers that are then infused back into the patient [144]. Although there are no TCR transgenic T cell therapies currently approved for clinical use, several have demonstrated strong efficacy in clinical trials for the treatment of melanoma, myeloma, AML [145–147]. Other TAAs, such as mesothelin, HA-1, WT-1, alpha-fetoprotein, gp100, and numerous oncoviral proteins have demonstrated strong preclinical efficacy, many of which are currently being tested in clinical trials [148–150].

The development of TCR transgenic T cells involves antigen selection, the isolation of tumor-specific T cells, identification of TCR receptors recognizing tumor antigen-derived MHC-presented peptides from tumor-specific T cell clones, engineered expression of TCRs in patient T cells, and expansion of the resultant TCR transgenic T cells for infusion back into the patient. Optimized strategies for the design and manufacturing of transgenic TCR cell products for adoptive therapy have evolved over the past several decades.

Antigen Selection
While antibody-based therapies are restricted to targeting epitopes of cell surface antigens, TCR-targeted therapies can target a relatively broad range of tumor antigens localized within the cell or on the cell surface. Native TCRs recognize antigen in the form of peptides presented on the cell surface of various antigen-presenting cells including tumor cells in complex with MHC molecules. Proteins associated with a cell, including both self and foreign or neo-proteins, are routinely processed into peptides and transiently presented on the cell surface [151]. The diversity of MHC molecules encoded by the human genome is responsible for the significant diversity in antigen presentation and recognition by the immune system both within and among individuals. Human MHC molecules are highly polymorphic and encoded by HLA genes from which a single allele is inherited from each parent resulting in a total of 12 potential MHC molecules expressed by cells in the body [151]. Each T cell expresses a unique TCR that recognizes a specific HLA-type (i.e., MHC molecule) and will only become activated upon recognition of the proper cognate peptide antigen presented by the MHC molecules to which it is specified. Unique MHC molecules present different peptide antigens derived from the same protein based on the characteristics of the MHC peptide-binding groove and the properties of the amino acids encoded by the peptide. Some MHC types recognize similar or shared amino acid motifs and are often grouped into HLA supertypes, but the percentage of peptides within the peptidome capable of being presented by multiple MHC is believed to be relatively limited [152]. Therefore, a transgenic T cell utilizing a TCR that recognizes a peptide presented by one HLA-type, will likely only be effective in patients who have the same HLA allele. This HLA restriction limits the patient population eligible for individual TCR therapies. While most work has focused on antigens restricted by the common HLA-A*0201 allele, targeting shared antigens presented by multiple HLA types, as well as targeting multiple peptide antigens within a single protein would improve the reach of TCR therapies within HLA diverse patient populations.

Some of the earliest transgenic TCR therapies were targeted against melanoma-associated antigens such as the melanoma-associated antigen recognized by T cells

(MART-1) and melanoma-associated antigen 3 (MAGE-A3). In a clinical trial of 36 patients, a TCR developed against the MART-1$_{27-35}$ peptide epitope demonstrated tumor regression in 30% of melanoma patients and favorable persistence of adoptively transferred T cells in the peripheral blood. However, off-target toxicities were reported in 41% of patients receiving the adoptive therapy, primarily in the form of skin rash, anterior uveitis, and ototoxicity, due to the expression of MART-1 in healthy melanocytes found in the skin, eyes, and ear, respectively [145]. Similarly, another clinical trial in nine melanoma patients resulted in promising clinical responses in 56% of patients that received an adoptive TCR therapy targeting MAGE-A3$_{112-120}$ accompanied by a high rate of neurotoxicity associated with the expression of MAGE-A12 protein in the brain, within which the MAGE-A3$_{112-120}$ is conserved [153]. While these studies were instrumental in demonstrating the power of TCR therapies for targeting tumor cells based on specific tumor antigens, they also suggested that the potency of TCR therapy likely requires selection of target antigens that have minimal or no expression in healthy tissue to avoid unwanted therapeutic toxicities. A later study demonstrated an 80% response rate in a clinical trial of 20 myeloma patients treated with adoptively transferred T cells expressing a TCR targeting the NY-ESO-1$_{157-165}$ peptides [146]. The study reported minimal off-target toxicities likely owing to the restricted expression of NY-ESO-1 to myeloma cells. Adoptive TCR therapy targeting NY-ESO-1 has also demonstrated efficacy in preclinical models of neuroblastoma with no evidence of off-target toxicity [154]. More studies are now focusing on identifying and utilizing TCRs specific for neoantigens driven by mutations rather than TAAs [155]. Tumor types with high mutation rates, such as melanoma, and tumors from which patient-specific tumor mutations can be identified have been the primary focus of such studies [156–159]. However, the significant effort and cost associated with identification of patient-specific tumor mutations and the generation of a patient-customized therapeutic product is a current limitation of this approach.

Antigenic Peptide Identification
In silico-based predictive peptide-MHC binding algorithms are commonly used as a fast and inexpensive method of epitope discovery, however, experimentally identified peptides with high-affinity for selected MHC molecules have been shown to rank poorly by such predictive methods [160]. Furthermore, peptides selected by in silico methods must be experimentally validated for MHC binding affinity. Mild acid elution and immunoaffinity chromatography methods for the immunoprecipitation of cell surface MHC are alternatively used to identify MHC-presented peptides from cells [161–163], but commonly result in modest recovery of MHC which results in poor representation of the immunopeptidome. Additional methodologies designed to identify immunopeptides from a single protein antigen include antigen immunization of and peptide-MHC isolation from humanized HLA mice or the assessment of large overlapping peptide matrices representative of entire protein sequences [164]. As tumor antigens represent a small fraction of the total

peptidome of a tumor cell due to intracellular variation in target antigen expression and the presentation of multiple HLA molecules by each cell, analysis of the complete peptidome is quintessential for antigen discovery methods.

Antigen-Specific T Cell Isolation and TCR Identification

TCRs with high-affinity for selected tumor antigens have been identified from naturally occurring T cells isolated from patients with antigen-expressing tumors. As TILs are more likely to contain TCRs specific for tumor antigens, therapeutic TCR candidates have been identified from TILs isolated from tumors expressing the target antigen [156, 165, 166]. TCR candidates have also been identified from peripheral blood T cells isolated from similar patients [166]. The most common strategies used to identify antigen-specific T cells include labeling based on peptide-MHC (pMHC) multimer affinity and T cell cytokine production in response to antigen stimulation. pMHC multimers are complexes consisting of varying numbers of pMHC molecules typically conjugated to fluorochromes that are used to identify reactive T cells by flow cytometry. The interaction of individual pMHC with TCRs is relatively weak and short-lived, however pMHC multimers have higher TCR binding avidity due to the presence of multiple binding sites [167] which improves their functionality in antigen-specific T cell detection protocols. Traditionally, pMHC trimers and tetramers, formed by biotinylation of peptide-MHC molecules and streptavidin binding for multimerization, have been used for antigen-specific T cell detection. Single polypeptides consisting of MHC molecule and peptide joined by linkers, known as single-chain trimers, were later developed as a highly stable alternative to traditional multimers [168] and have been successfully used to detect antigen-specific T cells [169] and identify tumor antigen recognizing TCRs [170]. More recently, pMHC dextramers, containing a large number of fluorescently labeled pMHC covalently attached to a dextran backbone have been developed for T cell detection [171]. The high binding capacity of dextran glucose polymers allows for the formation of pMHC dextramers with up to 10 times more pMHC molecules than traditional tetramers which provides superior TCR binding avidity. In support of this, dextramers have an improved ability to identify lower affinity TCRs compared to traditional multimers [172, 173]. Alternatively, antigen-specific T cells can be detected based on expression of activation markers and/or cytokine production in response to stimulation with antigen in vitro. Detected antigen-specific cells, using multimer technology, activation status, or cytokine production can then be enriched by antibody-directed magnetic bead separation or fluorescence-activated cell sorting (FACS) [172, 174, 175].

Once T cells with demonstrated antigen specificity are identified, the sequences of the α and β chains of the TCR are determined to engineer their expression in patient samples. The two main methods currently used to identify TCR sequences are single-cell sequencing and the use of RNA bait libraries targeting TCR α and β chain loci for TCR gene capture [176, 177]. Sufficient genomic material for TCR sequencing from single cells following intracellular cytokine production analysis and FACS has until recently been unattainable. A new droplet-based polymerase chain reaction (PCR) technique called CLINT-seq utilized reversible cross-linking that allows for the isolation of genomic material following intracellular cytokine staining (ICS) that is comparable to the use of viable cells, unlike traditional methods of fixation and permeabilization for ICS [178].

Historically, TCR affinity for cognate antigen in complex with MHC has been directly associated with T cell activation and response to antigen [179, 180]. However, the absolute nature of this correlation has been challenged [181]. The induction of cytotoxic T cell effector function typically requires intracellular signaling mediated by the binding of TCR variable regions to cognate peptide, as well as CD8 interaction with the MHC molecule. However, the binding of high-affinity TCRs to peptides alone can be sufficient to elicit a cytotoxic T cell response [182]. Consequently, this allows for the generation of transgenic CD4 and CD8 T cells. Therapeutic adoptive T cell products consisting of TCR transgenic CD4 and CD8 T cells have been shown to have superior efficacy and greater therapeutic potential than products that contain transgenic CD8 T cells alone [183–185].

Identifying high-affinity TCRs targeting TAAs can be challenging as many TAAs are self-antigens against which strongly reactive T cells are eliminated during thymic selection. The MART-1-targeting TCR named DMF4 showed antigen reactivity in patients but was characterized as having low affinity and demonstrated an inability to target cells with lower levels of antigen expression. This was thought to be the major reason for poor efficacy of the TCR in a clinical trial of 31 melanoma patients, of which only 13% of patients exhibited an objective clinical response [186, 187]. One method to engineer high-affinity TCRs is through the humanization of mouse TCRs. Human HLA-matched transgenic mice immunized with human TAA of therapeutic interest produce antigen-reactive T cells from which high-affinity mouse anti-human TCRs can be isolated [145, 188, 189]. Genetic strategies to improve TCR interactions and affinity to peptide-MHC complexes by altering the physical properties of the TCR are also being explored, including mutations to the complementarity determining regions CDR1 and CDR2 to improve binding to MHC [190]. The enhancement of TCR antigen affinity must be carefully monitored for safety as high-affinity TCRs are more likely to surmount the activation threshold that is defined during thymic selection which can lead to undesired off-target toxicity and autoimmunity. Extensive safety studies are required prior to the clinical assessment of affinity-enhanced TCRs. While some affinity-enhanced TCRs have demonstrated high toxicity in vivo [191, 192], others appear to be safe [193]. Expression levels of antigen in healthy tissue and the essential nature of those healthy tissues remain the main factors associated with the severity of toxicities related to TCR therapies.

Eliminating Transgenic—Endogenous TCR Interactions
Transgenic expression of therapeutic TCR sequences into otherwise unaltered patient T cells poses the risk of TCR αβ chain mismatch pairing with endogenously expressed TCR α and β chains [194]. Unique pairing of transgenic and endogenous TCR chains can result in TCRs with unexpected specificities which have the potential to lead to off-target toxicity and lack of TAA specificity [195].

Replacement of the constant regions of the human TCR with mouse TCR constant regions has been used to prevent mismatch pairing while maintaining antigen specificity of the therapeutic TCR [196]. The introduction of specific point mutations into the constant region of the α and β chains to add cysteine residues has also been shown to improve pairing of the transgenic chains and limit mismatch pairing with endogenous chains [197]. An alternative method under heavy investigation is the disruption of endogenous TCR expression in patient T cells prior to introduction of the therapeutic TCR. This has been accomplished successfully by using zinc-finger nucleases, transcription activator-like effector nucleases, and meganucleases [198, 199] as well as through targeted insertion of the therapeutic TCR coding sequences into the endogenous TCR locus [200].

Chimeric Antigen Receptor T Cells

An additional type of genetically engineered T cell receptor under heavy clinical investigation for adoptive T cell therapy are CARs or chimeric antigen receptors. CARs are designed to combine the antigen specificity of a TAA/TSA-specific antibody with the cytotoxic signaling of a CD8 T cell receptor. The receptor consists of two primary components: (1) the light and heavy variable regions of the tumor antigen-specific antibody in the form of a single-chain variable fragment (scFv) which confers the antigen specificity of the receptor and (2) the transmembrane and cytoplasmic signaling domain of the CD8 T cell receptor, through which cytotoxic T cell effector functions are induced following scFv engagement with tumor antigen [201] (Fig. 1.2). Unlike TCRs, most CARs do not recognize peptide-MHC, therefore their specificity and activation are not HLA-dependent. Although multiple types of structures have been used as ligand-binding domains in CAR design, scFvs are the most common and recognize intact native protein or lipid epitopes present on the surface of cells [201]. Additionally, CARs are monomeric receptors, therefore the risk of mismatch pairing with endogenous TCR chains is not of concern for CAR T cell therapies.

The manufacturing of CAR T cell products is similar to TCR transgenic T cells. Autologous T cells isolated from the peripheral blood of patients are engineered to express an optimized tumor surface antigen-specific CAR through viral or non-viral transgenesis. CAR T cells are subsequently expanded to large numbers before infusion back into the patient following systemic lymphodepletion. Numerous strategies related to tumor antigen-specific antibody identification, scFv selection, spacer optimization, and the production of CAR T cell therapeutic products have been employed to optimize the efficacy of and minimize toxicities associated with CAR T therapy.

Selecting Antigens for CAR T Therapy

Several CAR T therapies have received FDA-approved, all for the treatment of B cell malignancies. Three CAR T cell adoptive therapies targeting CD19 were approved between 2017 and 2021, and a novel CAR targeting B cell maturation antigen (BCMA) was approved for the treatment of multiple myeloma in 2021. All four approved CAR T therapies generated impressive response rates and improved

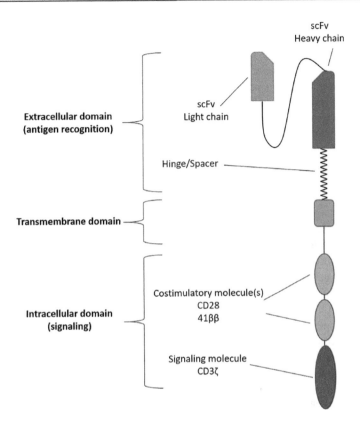

Fig. 1.2 CAR structure

patient outcomes in clinical trials [202]. However, long-term treatment targeting CD19 that is expressed on both healthy and malignant B cells results in chronic B cell aplasia and hypogammaglobulinemia in patients due to the on-target therapeutic elimination of healthy B cells [203–205]. While B cell aplasia is anticipated, tolerated, and managed clinically with immunoglobulin replacement therapy [206], the selection of non-self, TSAs is likely a safer option for future CAR T therapies. TSAs such as EGFRvIII, NY-ESO-1, and the glycolipid GD2 are being investigated clinically [207].

ScFv Antigen Affinity
The overall affinity and functional avidity of CARs are likely determined by several factors including scFv affinity for antigen, costimulatory signaling, and the level of antigen expression of target cells. Similar to TCRs, scFv with higher affinity for cognate antigen is often associated with better antigen recognition and more potent T cell responses [208, 209]. However, high-affinity can result in a greater potential for off-target toxicities. scFv with finely tuned antigen affinities have been demonstrated to better discriminate between higher densities of TAAs

on-tumor cells and lower antigen expression on healthy tissue [210]. In instances where TSAs with exclusive expression in tumors and absence in heathy tissue have not been identified, CARs with lower affinity scFv for the targeted TAA may exhibit improved anti-tumor efficacy and better safety profiles against healthy cells as has been demonstrated with CD19-targeting CAR T therapy [211]. This is most likely due to the ability of the CAR to bind tumor cells with higher surface antigen concentrations and inability to bind healthy cells with lower antigen density or binding insufficient for activation of cytotoxic T cell effector function.

Optimization of CAR Spacer Length

Unlike antibodies, which due to their small size and solubility have the ability to bind to a greater diversity of confirmational epitopes of surface antigens even with complex or large, 3-dimensional structures, the scFv of CARs can have greater difficulty accessing and binding antigens due to physical hindrances associated with being bound to the T cell surface [201]. Optimization of the CAR spacer length and isotype for specific scFv has been shown to improve the ability of the CAR to engage the targeted antigen. Optimal spacer length is often dependent on the location of the epitope within the extracellular antigen on the target cell surface that the CAR is required to bind. Longer spacers are usually better suited for binding epitopes near the cell membrane or with limited exposure, while shorter spacers are often sufficient for highly-exposed membrane distal epitopes [212]. However, this is a general observation, therefore the unique context of using a specific CAR to target a particular antigen often requires individual optimization. Spacer hinges derived from the IgG4 isotype have been shown to interact with Fc receptors of certain mouse cells leading to sequestering of therapeutic CAR T cells in mouse lungs and decreased efficacy in a preclinical study [213]. However, the introduction of specific mutations in the spacer domain led to diminished Fc receptor binding and improved tumor localization in both CD19- and PSCA-targeting CARs [213, 214].

Selection of T Cell Costimulatory Domains

First-generation CARs constructs consisted of an antigen-binding extracellular scFv and hinge, a transmembrane spanning domain, and an intracellular CD3ζ T cell receptor signaling domain linked to a co-receptor, such as CD4 or CD8. Such prototype chimeric receptors were first developed in the 1990s and were capable of targeting directing T cells to tumor cells expressing the antigen targeted by the scFvs. While the scFv and linkage of the intracellular CD3ζ domain to CD4 or CD8 provide "signal 1" to the T cells required for antigen-specific T cell activation and proliferation, such first-generation CARs lacked the ability to employ cytolytic functions against targeted tumor cells due to a lack of T cell costimulatory signaling (i.e., "Signal 2") [215]. The efficacy of most first-generation CARs was poor due to issues with CAR T cell persistence and therapeutic-associated toxicities in vivo [216–219]. A lack of costimulatory signaling domains and an effective signal 2 in first-generation CARs has also been associated with CAR T cell tumor antigen tolerance and elimination following antigen engagement [220].

Second- and third-generation CARs included single or multiple costimulatory domains, respectively in addition to the scFv and CD3ζ signaling domains (Fig. 1.2), which provided additional signaling to the T cell essential for effector function activation following antigen engagement [221, 222]. The type of costimulatory signal, as well as the order of the intracellular domains and their association with CAR function and efficacy, has been extensively studied. The main costimulatory molecule domains utilized to provide an activating "signal 2" to the T cells are CD28, 4-1BB, OX40, and ICOS [222]. In the context of CD19 CAR T therapy, clinical data show minimal to no difference in clinical response rates to CARs that use either CD28 or 4-1BB domains [223]. Preclinical studies have demonstrated CAR T cells with both CD28 and 4-1BB costimulatory domains induce greater cytokine production, anti-tumor activity in mice, and persistence of T cells compared to single costimulatory domain CAR T cells [224]. Additionally, a CD19 CAR that uses a CD28-derived hinge in addition to a 4-1BB intracellular costimulatory domain yielded a higher complete response rate compared to most prior studies using single costimulatory domain CARs [225]. In a preclinical study, CARs with dual ICOS and 4-1BB domains targeting mesothelin expressed by NSCLC tumors demonstrated greater anti-tumor efficacy and longer in vivo persistence but was dependent on the ICOS domain being positioned proximal to the plasma membrane [226]. A more recent fourth-generation of CARs has been created utilizing the addition of stimulated production of cytokines, such as IL-12, to single costimulatory domains to further augment tumor-targeted effector function of the T cells. It has been suggested that the presence of IL-12 is "signal 3" and required for full interferon-driven $CD8^+$ T cell effector function [227]. IL-12 is believed to improve CAR T efficacy via a multi-pronged approach including autocrine effects of the CAR T cells as well as manipulation of APCs within the tumor microenvironment to promote inflammation and T cell killing [228]. Such T cell products are commonly referred to as T cells redirected for universal cytokine-mediated killing or TRUCKs [229]. This concept of adding autocrine T cell stimulation with cytokines or the expression of activation and proinflammatory ligands is referred to as "armoring" T cells [228] and has focused primarily on the addition of IL-12, IL-15, CD40L, and 4-1BBL to second- or third-generation CAR constructs [228]. An extensive set of armored approaches is under investigation to continue the advancement of the efficacy and safety of future adoptive T cell therapy. Current studies focus on the efficacy of armored CARs and the identification of biological markers of response as well as therapeutic-associated toxicities. The development of humanized mouse models that allow for studying human cell lines or patient tumor-derived xenografts in immunodeficient mice or mice with autologous human immune systems by peripheral blood mononuclear cell or hematopoietic stem cell transplant has been essential for assessing adoptive T cell therapy efficacy in vivo [230]. Carcinogen-induced syngeneic models and GEMMs that recapitulate tumor molecular heterogeneity and complexity of the tumor microenvironment in immune-competent mice are being developed as better models for assessing toxicity and the broader effect of adoptive T cell therapies on anti-tumor immune responses preclinically [231, 232].

Reducing Toxicities Associated with CAR T Therapy

Toxicities associated with CAR T therapy are driven by two mechanisms: (1) excess on-target CAR T activity against tumor cells, referred to as on-target/on-tumor toxicity and (2) CAR T cytotoxic activity against non-tumor antigen-expressing cells, referred to as on-target/off-tumor toxicity. Toxicity due to on-target/on-tumor immune activation following infusion of a CAR T product occurs when CAR T cells engage the targeted antigen on-tumor cells and undergo activation resulting in a systemic proinflammatory immune response with massive cytokine production known as cytokine release syndrome (CRS) [233]. Similarly, on-target/off-tumor toxicity results from potent immune activation that occurs following unwanted CAR T cell recognition of cognate antigen on healthy cells. While many cytokines are produced during the cytokine storm responsible for CRS, studies in humanized mouse models have demonstrated that IL-6 is most commonly inked to the majority of tissue-damaging symptoms associated with the detrimental effects of CRS [234, 235]. CRS severity has a significant range and is clinically graded to aid with clinical management [236]. Symptoms include fever and varying degrees of hypoxia, hypotension, and organ damage [236], many of which can be mitigated by treatment with IL-6 blocking drugs [237] as well as other immune dampening medications such as corticosteroids [234]. Changes in immune monitoring of patients through temporal tumor biopsy and peripheral blood collection during clinical trials have promoted our understanding of the factors implicated in the development of CRS and neurotoxicity associated with CAR T therapy, as well as identification of markers of response to therapy [231, 238].

Several strategies are under investigation for the design of CAR T cells to help mitigate CAR T-related toxicities. While on-target/off-tumor toxicities can be eliminated or reduced by targeting tumor-restricted antigens or by optimizing scFv antigen affinity as previously discussed, on-target/on-tumor toxicities have been more challenging. If a tumor antigen is expressed on both tumor cells and healthy tissue, strategies that require dual antigen engagement to allow T cell killing can improve selective targeting [239]. A novel example is logic-gated CAR expression controlled by synthetic Notch (synNotch) receptors. Upon engagement of a T cell synNotch receptor with antigen #1, regulated intracellular signaling triggers expression of the CAR which will only lead to T cell-mediated killing of the tumor cell if the CAR-targeted antigen #2 is present [240, 241]. The use of suicide genes or safety switches is also being explored to regulate CAR T cells once infused into the patient. Such genes are included in the expression vectors of the CAR constructs and allow for selective elimination of transgenic T cells in patients [242]. This concept was adapted from techniques used to treat graft-versus-host disease (GVHD) following hematopoietic stem cell transplant (HSCT). Safety switches function through conditional expression of a gene in the transgenic T cells that encodes a product that converts a non-toxic pro-drug into a toxic payload that eliminates the T cell, thereby reducing T cell-mediated toxicity. The pro-drug is given upon observation of therapy-driven irAEs in a patient. Induction of Herpes simplex virus-tyrosine kinase (HSV-tk) with ganciclovir treatment has been extensively tested in and used to treat GVHD in leukemia patients receiving

HSCT [243] and has also been used to control expression of HSV-tk-expressing adoptive T cells targeting CD44 [244]. More recently an inducible Cas9 safety switch system has been shown to effectively mitigate CAR T toxicity in the context of patients with acute leukemia relapse receiving HSCT and adoptive T cell therapy [245].

Switch Receptors to Improve Efficacy of CAR T Against Solid Tumors
Unique genetic engineering of CAR constructs is being explored to counteract the immunosuppressive features of solid tumors that often restrict CAR T efficacy, such as expression of immune checkpoints. Chimeric switch receptors containing an extracellular domain of PD-1 linked to an intracellular CD28 T cell costimulatory domain equip therapeutic T cells with the ability to undergo activation upon engagement with PD-L1 rather than inactivation with a classic PD-1 receptor. This strategy has been shown to confer improved CAR T efficacy against several preclinical solid tumor models [246]. Dominant-negative TGF-βRII has been implemented as well to block the suppressive effect of TGFβ on T cells and was shown to enhance PSMA-directed CAR T therapies against multiple mouse models of prostate cancer [247]. However, there is some evidence in mice that use of a dominant-negative TGF-βRII can promote autoimmune disease and further exacerbate off-target toxicities [248]. While much of what we know about T cell therapy-associated toxicities have been collected from clinical studies, the development and use of orthotopic syngeneic or GEMMs in animals with functional immune systems are imperative for identifying the toxicity of T cell based therapies in preclinical setting.

1.8 Active Immunotherapy

The second overarching category of immunotherapies is active immunotherapy. Unlike passive immunity, which involves introducing exogenously produced tumor targeting agents or cells to a patient, active immunotherapy modalities are designed to stimulate or activate a patient's native immune system to mount a response to cancer. The most common types of active immunity currently under investigation are cancer vaccines and oncolytic viruses.

1.8.1 Prophylactic Cancer Vaccines

Some of the most successful cancer vaccines to date are preventative cancer vaccines that are prophylactically given to at-risk individuals and are designed based on successful vaccines against infectious agents. Such preventative cancer vaccines are designed to induce a long-term adaptive immune response against oncoviruses to reduce the risk of cancers that can be caused by the targeted viruses and most commonly function by eliciting a strong antibody response against virus-derived antigens. Several oncoviral-targeted cancer vaccines have been FDA-approved including Cervarix, Gardasil, and Gardasil-9 for the prevention of several reproductive cancers and head and neck cancers caused by human papilloma viruses (HPC) and HEPLISAV-B for the prevention of liver cancer due to hepatitis B virus (HBV). The HPV vaccines consist of virus-like particles (VLP) that lack the HPV genome and viral replication ability but retain the immunogenic HPV L1 protein on the VLP surface for immune recognition without the risk of vaccine-induced active viral infection [249]. HEPLISAV-B consists of a cocktail of immunogenic HBV surface proteins in recombinant form as well as the toll-like receptor 9 agonists, cytidine-phospho-guanosine, as an adjuvant [250]. Despite the success of cancer vaccines targeting etiological oncoviral antigens, such therapies are restricted to the few cancers that form due to viral infection-induced transformation and therefore have limited reach in the cancer patient population.

1.8.2 Therapeutic Cancer Vaccines

Similar to preventative cancer vaccines, therapeutic vaccines are designed to induce a long-term adaptive immune response but target non-viral TAAs or TSAs within a patient with a cancer diagnosis. Most therapeutic vaccines concentrate on inducing or improving the presentation of peptide-derived antigens in complex with MHC molecules for the activation of anti-tumor T cell responses. Types of therapeutic cancer vaccines that have been investigated include viral- and cellular-based cancer vaccines, as well as peptide, RNA, and DNA vaccines.

1.8.2.1 Viral Vector-Based Cancer Vaccines

Several viral-based cancer vaccines have been developed using strategies drawn from vaccines developed for infectious diseases. Both inactivated (i.e., replication-incompetent) and active, replication-competent viruses containing tumor antigens have been used to infect tumor cells and improve anti-tumor immune responses [251, 252]. PVSRIPO, a recombinant nonpathogenic poliovirus-based vaccine targeting the CD155 receptor often expressed on glioblastoma cells, has demonstrated improved survival in a study of 61 patients with recurrent, malignant glioblastoma [253]. Other viral vectors, such as vaccinia, adenovirus, and alphaviruses have been tested against a variety of cancer types, including prostate cancer, NSCLC, and colorectal cancer [254–257]. However, the development of neutralizing antibodies by patients against components of the native viral vectors limits the use and effectiveness of viral-based cancer vaccines [252].

1.8.2.2 Cellular-Based Cancer Vaccines
Autologous or Allogeneic Tumor Cells

The use of autologous tumor cells as a therapeutic product involves the use of replication-incompetent tumor cells or tumor cell fragments to stimulate an immune response to the antigens expressed by the tumor. Whole tumor cell vaccines are prepared in a variety of ways, most of which involve induction of tumor cell death through apoptosis or necrosis or by rendering the cells replication-incompetent. Ultraviolet B irradiation induces tumor cell apoptosis which ultimately results in uptake of tumor cell components by APCs both in vitro and in vivo thereby improving the likelihood of antigen presentation in patients [258]. Hypochlorous acid has been used in a similar manner to induce tumor cell necrosis for whole tumor vaccine production [259]. Lastly, the heating of tumor cells has been shown to not only increase MHC expression and antigen production by tumor cells but also increase heat shock protein expression and release from tumor cells which can induce proinflammatory immune signaling through engagement with toll-like receptors on immune cells [260].

Cancer cell lines and patient tumor cell lysates have also been used directly for vaccination of patients. In particular cancer settings, tumor lysates have been shown to perform as well or better than defined tumor antigens in clinical trials [261]. Peptide complexes derived from the heat shock-gp96 protein isolated from melanoma patient tumors have been used to successfully treat some patients following autologous vaccination [262]. Autologous tumor cell lysate has been used as a source of antigen for antigen loading for APC-based vaccines in several studies [263–265]. Autologous tumor cell RNA has also been transfected into APCs to allow for patient tumor-specific antigen expression and presentation for vaccination [266].

Expression of immune-activating cytokines by genetic modification of tumor cells for vaccination has been used to combat the immunosuppressive environments of many tumors. Granulocyte–macrophage colony-stimulating factor (GM-CSF) transgene expression is arguably the most comprehensively studied growth factor in the context of cancer vaccines. It is been well established for over two decades that GM-CSF expression in tumor cells can improve tumor antigen presentation and proinflammatory cytokine secretion of APC such as dendritic cells and macrophages and is associated with more productive anti-tumor immune responses [267]. Additionally, a phase I/II clinical trial of vaccination with autologous tumor cells expressing GM-CSF resulted in improved overall survival of NSCLC patients compared to vaccination with cells not modified to excrete GM-CSF [268]. GVAX, an allogeneic, tumor cell vaccine engineered to express GM-CSF has shown promise in clinical trials in various patient and cancer settings [269]. GVAX for prostate cancer, consisting of irradiated GM-CSF-secreting LNCaP and PC3 prostate cancer cell lines, was associated with a significant PSA decline in phase II clinical trials compared to chemotherapy [270, 271]. These findings led to two phases III clinical trials, VITAL-1, and VITAL-2, testing GVAX alone or in combination with chemotherapy compared to chemotherapy alone. However, both trials were discontinued due to futility and worse patient outcomes in the GVAX arm, respectively [272, 273]. A pancreatic GVAX vaccine has also been tested in the setting of resectable as well as unresectable metastatic pancreatic ductal adenocarcinoma. Phase II clinical trials have demonstrated that, while pancreatic GVAX monotherapy or in combination with cyclophosphamide, live, attenuated mesothelin-expressing Listeria (CRS-207), and nivolumab did not improve overall survival of patients compared to standard therapy, the combination led to improved immune cell infiltration into the TME [274, 275]. GVAX lung has been tested in several clinical trials of patients with NSCLC with modest clinical benefit in a small number of participants compared to standard, approved treatment regimens [268, 276, 277]. GVAX is also currently being tested in the neoadjuvant setting for melanoma and NSCLC [278]. Belagenpumatucel-L is an allogenic tumor cell vaccine consisting of a cocktail of NSCLC cell lines modified to express a TGFβ2 antisense gene which has been shown to improve immunogenicity of the tumor cells in early clinical trials [279, 280].

Dendritic Cell Vaccines

Dendritic cells (DC) are professional APCs well-known for their role in the induction of antigen-specific T cell responses. Once technical advances were developed that improved the ability of scientists to study and generate DC in vitro, DC-based vaccines have become the primary form of cell-based cancer vaccines under investigation, overshadowing the use of irradiated autologous tumor cells as a source of tumor antigen for immune activation. All DC lineages are derived from $CD34^+$ hematopoietic progenitor cells within the bone marrow [281]. The majority of DC are localized in tissues throughout the body where they survey the environment for pathogens and other foreign antigens that they are designed to take up, process, and present to T cells in complex with MHC within primary and secondary lymphoid tissues [281]. While many differentiated subsets of DC have been identified, blood-derived DC has been used the most in the context of cancer vaccines due to the ease with which they can be generated and manipulated in vitro. The earliest DC-based cancer vaccines were tested in the 1990s and early 2000s against well-known tumor antigens, including MART-1, gp100, and MAGE. Therapeutic DCs were derived in vitro from $CD14^+$ monocytes isolated from patients using GM-CSF and IL-4. Following differentiation, DV was loaded with established immunogenic tumor antigen peptides or patient tumor lysates and infused back into patients [282–284]. Recent advances in DC vaccine design have focused on targeting better antigens such as neoantigens arising from mutations. This approach has yielded positive results but requires labor-intensive strategies for personalized antigen discovery [285]. Preconditioning of the injection site with adjuvants such as tetanus toxoid prior to DC vaccination is also being explored [286]. DC vaccines have also been tested in clinical trials against acute myeloid leukemia (AML), and B cell lymphoma [287]. Despite evidence of immunogenicity and acceptable safety profiles, these DC vaccines showed inferior efficacy compared to other standard of care treatments.

The only FDA-approved DC vaccine to date is the prostatic acid phosphatase antigen (PAP) targeting vaccine, Sipuleucel-T (Provenge), for the treatment of prostate cancer. Sipuleucel-T contains DCs derived from autologous patient monocytes that presents an HLA-A*0201-restricted PAP-derived peptide [288]. The FDA approval of Sipuleucel-T in 2010 has generated significant interest in the utilization of DC vaccines for additional cancer types by the scientific community. Efforts to optimize development and improve efficacy of DC vaccines have been at the forefront of the field and have focused on identifying optimal in vitro culture conditions for DC generation and enhanced antigen-loading strategies.

Selecting a Source and Differentiation State for DC Generation

The majority of DC-based cancer vaccines have utilized autologous DC-derived in vitro from CD14$^+$ monocytes isolated from patient blood due to the ease of access to such cells. Several studies have used directly isolated patient CD34$^+$ hematopoietic stem cells for DC derivation. Other sources of DC include differentiated DC such as plasmacytoid DC or conventional DC; however, such DC primarily reside in tissues and are challenging to isolate from the periphery due to their low frequency in the blood [289, 290]. A more recent approach involving antigen loading and stimulation of DC within patients, such as the liposome-based vaccine-targeting DC, Lipovaxin-MM, has shown preclinical and safety in clinical trials for patients with melanoma while eliminating the need to produce a DC product ex vivo [291, 292].

In addition to DC source, the maturation status of DC used for vaccination has been identified as an important factor impacting vaccination efficacy. Most DC patrolling tissues for antigen are phenotypically immature and characterized by low expression of T cell costimulatory molecules. However, they are uniquely positioned for optimal antigen uptake, processing, and presentation [293]. Upon antigen uptake, immature DC undergoes maturation involving the upregulation of adhesion molecules that aid in DC transport to lymphoid tissues for antigen presentation to T cells [294]. Mature DC are better equipped to activate T cell responses to their presented antigen [295–297].

Antigen-Loading Strategies

Various antigen sources have been assessed for DC antigen loading including various lengths of MHC class I and II-restricted synthetic peptides, whole recombinant protein, killed autologous tumor cells, and tumor cell lysates, all of which have demonstrated immunogenicity in vitro [285]. Although MHC class I restricted peptide antigens may be most effective for CD8 cytotoxic T cell activation, whole protein antigen sources, such as recombinant, full-length protein, and autologous tumor cells, have demonstrated CD4 and CD8 T cell activation leading to a more potent cellular anti-tumor response. Intact protein antigens serve as a natural source of both MHC class I and II-restricted antigens that can be processed and presented by DC following exogenous protein uptake. This strategy also eliminates HLA-specific restrictions and may therefore be more suitable for the generation of autologous DC vaccines for patients across all HLA haplotypes. More recent strategies to develop efficacious DC vaccines have focused on the loading and targeting of tumor-specific neoantigens, as well as antigen presentation mediated by lentiviral antigen expression platforms [285]. Other strategies currently being explored to improve DC vaccine T cell activation include the use of immune-activating adjuvants, such as co-immunization with tetanus toxoid [286].

While DC vaccines have demonstrated significant clinical promise and DC remain a powerful tool for the initiation of anti-tumor T cell responses, the many factors involved in DC vaccines design and determinants of vaccine efficacy such as DC subset selection, DC maturation status, and antigen selection and loading have yet to be fully elucidated.

B Cell Vaccines

Although DC is considered the primary professional APC, both B cells and macrophages are heavily involved in antigen presentation and can induce antigen-specific T cell responses. Autologous B cells stimulated with CD40L and loaded with antigen in the form of peptides or tumor cell lysates have been shown to induce anti-tumor T cell responses in vitro against melanoma antigens [298]. B cell vaccines have also demonstrated antigen reactivity in models of bladder cancer and lymphoma [299, 300]. The ease of access to B cells in the periphery and their insensitivity to many immune-suppression mechanisms employed by tumors make them an exciting option for cell-based immunotherapy vaccines [301].

1.8.2.3 Nucleic Acid Vaccines

Direct vaccination of tumor antigen peptides not in association with autologous cells is an alternative strategy under investigation. Genetic material in the form of RNA or DNA encoding tumor antigen peptides or proteins has also been used for direct vaccination. Cells that receive the RNA or DNA transiently express tumor antigens for induction of an immune response [302]. Alternatively, viral and bacterial vectors have been used to introduce genes encoding the desired vaccine tumor antigen into patient cells. Vaccinia, adenovirus, and lentivirus-based vectors among others have all been tested for this purpose [303]. Vector delivery systems, while lacking the replication competency of live pathogens, enable efficient delivery of antigen genetic material to patient cells compared to direct injection or the use of gene guns. Viral and bacterial vector vaccines can generate both humoral and cell-mediated adaptive immune responses against target antigens [302]. Historically, non-formulated plasmid DNA vaccines have shown strong safety profiles but modest efficacy in patients. Modern DNA vaccines are focused on targeting better tumor antigens and employing combination strategies to counter immunosuppressive mechanisms. Most DNA vaccines currently under investigation encode a tumor antigen as well as other genes that aid in immune activation such as IL-2 (IL-2) or GM-CSF [304].

Delivery Methods

A variety of methods have been used to deliver plasmid-based vaccines including mucosal, subcutaneous, intradermal, and intramuscular administration. Methods that improve DNA entry into cells, such as gene guns and DNA tattooing are modern methods being explored as alternatives to standard injections [305, 306]. Plasmid DNA detected within the cell can induce non-specific innate immunity as double-stranded DNA (dsDNA) represents a pathogen-associated molecular pattern recognized by toll-like receptor 9 when outside of the nucleus. Once DNA has been successfully delivered to targeted APC, it is then transcribed and translated into protein using cellular machinery for subsequent MHC presentation to antigen-specific T cells.

Improving Immunogenicity of DNA Vaccines

Despite significant improvements in DNA delivery technology, DNA vaccines continue to have limited immunogenicity and efficacy in both preclinical models of cancer and in clinical trials [304]. Codon optimization has been used to both improve expression of vaccine encoded genes and ultimately improve gene-based vaccine immunogenicity designed for the treatment of both infectious disease and cancer [307–311]. To improve the overall immunogenicity and efficacy of DNA vaccines, most modern DNA vaccine designs focus on targeting multiple tumor antigens or neoantigens and make use of adjuvants. Multi-epitope introduction and presentation can be easily implemented with gene-based vaccines and have demonstrated improved efficacy in the context of multiple myeloma, for example, targeting multiple epitopes across several myeloma-associated proteins [312]. One group recently demonstrated that a cocktail of MHC class I restricted synthetic neoantigens delivered by DNA vaccination elicited potent, CD8 T cell dominant, anti-tumor effects in preclinical models of several ICI-resistant tumor models, including lung and ovarian cancer [313]. Others have demonstrated immunogenicity and anti-tumor activity of multi-epitope DNA vaccines targeting patient tumor-derived peptides against preclinical models of breast cancer and pancreatic neuroendocrine cancer [314]. Transgenic mice with selective expression of tumor antigens, such as Her-2 and MUC-1 for breast and pancreatic cancer, respectively have been essential for testing vaccines targeting tumor antigens [315]. In addition to targeting multiple CD8-specific tumor antigens, the ability to induce both CD8 and CD4 T cell responses using both MHC class I and II-restricted antigens has been demonstrated to generate a more effective anti-tumor response [316]. Despite the compelling interest in targeting neoantigens, the processes of neoantigen discovery and subsequent vaccine design are laborious and time-consuming requiring several months from tumor biopsy to patient vaccination [304]. Novel strategies to improve the efficiency of both immunogenic antigen discovery and vaccine manufacturing are needed for practical, widespread use of personalized gene-based therapy.

1.8.2.4 Peptide Vaccines

Peptide-based vaccines injected into the dermis and/or muscle are designed to be taken up by resident antigen-presenting cells such as DC and subsequently presented to T cells. Most peptide vaccines consist of one or more short (i.e., 8–10 amino acids) MHC class I restricted peptides that are immunologically recognized by CD8 T cells. Such vaccines have demonstrated good tolerability and have been shown to activate cytotoxic CD8 T cell responses but have only translated to moderate anti-tumor efficacy in clinical trials [317], most likely due to a lack of CD4 T cell engagement and subsequently limited CD4-mediated "help" which improves activation and potentiation of CD8 T cell response [318, 319]. Longer peptides (i.e., 15–30 amino acids) designed to include CD4 and CD8 epitopes or combinations of CD4- and CD8-recognized peptides of various lengths are also being explored [320]. Although peptide-based vaccines are one of the most highly studied types of cancer vaccine, no peptide-based formulations have received FDA approval due to limited therapeutic responses thus far in clinical trials. Novel strategies are needed to overcome the numerous challenges faced by peptide cancer vaccines.

Improving Peptide-Vaccine Efficacy

Peptide-based vaccines are often hindered by immunological tumor characteristics and vaccine design including heterogeneity in HLA restriction and tumor antigen expression within the patient population. Personalized peptide vaccines (PPV) are designed to trigger quick anti-tumor cytotoxic T cell responses by using peptide antigens to which a patient's immune system has already been exposed through the activation of memory T cells. Most PPV that has been tested utilize a cocktail of peptides derived from numerous TAA proteins restricted or shared by multiple HLA types [321]. Noguchi and colleagues have collectively studied 31 CD8-directed peptide candidates as multi-peptide cancer vaccines assessed against castration-resistant prostate cancer, glioblastoma multiforme, pancreatic cancer, several subtypes of lung cancer, colorectal cancer, and others [321] and concluded that immunological memory to vaccine-targeting epitopes correlates strongly with the magnitude and duration of response to a peptide vaccine. This highlights the importance of determining a patient's immunological memory to the vaccine-targeting epitopes patient screening and selection for treatment.

Peptide-vaccine efficacy has also been linked to vaccine design, most commonly due to the use of formulations that do not prevent peptide degradation in vivo. Naked peptide formulations are highly susceptible to degradation prior to reaching APC using standard delivery systems [320]. The use of nanoparticles to improve peptide to cell delivery is under heavy investigation. Numerous types of organic and inorganic nanoparticles are being explored in peptide-vaccine strategies. Manipulation of particle makeup, size, and formulation have improved the immunogenicity and efficacy of peptide-based vaccines [322]. Overall, selection of epitopes guided by patient immunity and prolongation of vaccine persistence in vivo are promising techniques to enhance peptide-vaccine efficacy, but new strategies are still needed to identify biomarkers of response for improved epitope and peptide selection.

1.9 Conclusion

Twenty years ago, the idea of utilizing immune-based therapies for the treatment of cancer was shrouded in doubt. The success of CAR T cell therapy and ICI against hematologic malignancies and solid tumors, respectively has provided a significant boost to the exploration and implementation of novel cancer immunotherapies. Recent advances elucidating interactions between the immune system and tumor have uncovered novel opportunities for the use of immune-based therapeutic strategies against a broad range of cancer types. Over the past decade, multiple antibodies- and immune cell-based cancer therapies have been approved for clinical use and clinical trials are increasingly showing efficacy of a broad range of passive and active immunotherapies. Overall, to improve the clinical benefit of cancer immunotherapies, current research efforts are focused on tailoring antigen selection, optimizing drug development and manufacturing, developing improved preclinical models, identifying reliable biomarkers of response, and using combinatorial treatment strategies to overcome immune evasion and antigen heterogeneity within and among the patient population.

References

1. Mathers C, Stevens G, Hogan D, Mahanani WR, Ho J (2017) Global and regional causes of death: patterns and trends, 2000–15. In: Jamison DT, Gelband H, Horton S, Jha P, Laxminarayan R et al (eds) Disease control priorities: improving health and reducing poverty. Washington (DC)
2. Schirrmacher V (2019) From chemotherapy to biological therapy: a review of novel concepts to reduce the side effects of systemic cancer treatment (Review). Int J Oncol 54(2):407–419
3. Schaue D, McBride WH (2015) Opportunities and challenges of radiotherapy for treating cancer. Nat Rev Clin Oncol 12(9):527–540
4. Modlin RL (2012) Innate immunity: ignored for decades, but not forgotten. J Invest Dermatol 132(3 Pt 2):882–886
5. Decker WK, Safdar A (2009) Bioimmunoadjuvants for the treatment of neoplastic and infectious disease: Coley's legacy revisited. Cytokine Growth Factor Rev 20(4):271–281
6. Coley WB (1991) The treatment of malignant tumors by repeated inoculations of erysipelas. With a report of ten original cases. 1893. Clin Orthop Relat Res (262):3–11
7. P. E. Ueber den jetzigen Stand der Karzinomforschung. Ned Tijdschr Genneeskd. 1909(5):273–90
8. Old LJ, Boyse EA (1964) Immunology of experimental tumors. Annu Rev Med 15:167–186
9. Dunn GP, Old LJ, Schreiber RD (2004) The three Es of cancer immunoediting. Annu Rev Immunol 22:329–360
10. Mittal V (2018) Epithelial mesenchymal transition in tumor metastasis. Annu Rev Pathol 13:395–412
11. Terry S, Chouaib S (2015) EMT in immuno-resistance. Oncoscience. 2(10):841–842
12. Penn I (1988) Tumors of the immunocompromised patient. Annu Rev Med 39:63–73
13. Roithmaier S, Haydon AM, Loi S, Esmore D, Griffiths A, Bergin P et al (2007) Incidence of malignancies in heart and/or lung transplant recipients: a single-institution experience. J Heart Lung Transplant 26(8):845–849
14. Jaffee EM, Pardoll DM (1996) Murine tumor antigens: is it worth the search? Curr Opin Immunol 8(5):622–627
15. Yamagiwa K, Ichikawa K (1918) Experimental study of the pathogenesis of Carcinoma 3(1):1–29
16. Baldwin RW (1955) Immunity to methylcholanthrene-induced tumours in inbred rats following atrophy and regression of the implanted tumours. Br J Cancer 9(4):652–657
17. Foley EJ (1953) Antigenic properties of methylcholanthrene-induced tumors in mice of the strain of origin. Cancer Res 13(12):835–837
18. Klein G, Sjogren HO, Klein E, Hellstrom KE (1960) Demonstration of resistance against methylcholanthrene-induced sarcomas in the primary autochthonous host. Cancer Res 20:1561–1572
19. Vigneron N (2015) Human tumor antigens and cancer immunotherapy. Biomed Res Int 2015:948501
20. Wolfel T, Hauer M, Schneider J, Serrano M, Wolfel C, Klehmann-Hieb E et al (1995) A p16INK4a-insensitive CDK4 mutant targeted by cytolytic T lymphocytes in a human melanoma. Science 269(5228):1281–1284

21. Soufir N, Avril MF, Chompret A, Demenais F, Bombled J, Spatz A et al (1998) Prevalence of p16 and CDK4 germline mutations in 48 melanoma-prone families in France. French Familial Melanoma Study Group Hum Mol Genet 7(2):209–216
22. Deniger DC, Pasetto A, Robbins PF, Gartner JJ, Prickett TD, Paria BC et al (2018) T-cell responses to TP53 "Hotspot" mutations and unique neoantigens expressed by human ovarian cancers. Clin Cancer Res 24(22):5562–5573
23. Malekzadeh P, Pasetto A, Robbins PF, Parkhurst MR, Paria BC, Jia L et al (2019) Neoantigen screening identifies broad TP53 mutant immunogenicity in patients with epithelial cancers. J Clin Invest 129(3):1109–1114
24. Gilboa E (1999) The makings of a tumor rejection antigen. Immunity 11(3):263–270
25. Scheuermann RH, Racila E (1995) CD19 antigen in leukemia and lymphoma diagnosis and immunotherapy. Leuk Lymphoma 18(5–6):385–397
26. Finn OJ (2017) Human tumor antigens yesterday, today, and tomorrow. Cancer Immunol Res 5(5):347–354
27. Brentville VA, Vankemmelbeke M, Metheringham RL, Durrant LG (2020) Post-translational modifications such as citrullination are excellent targets for cancer therapy. Semin Immunol 47:101393
28. Stanley M (2008) Immunobiology of HPV and HPV vaccines. Gynecol Oncol 109(2 Suppl):S15-21
29. Chandran B, Bloomer C, Chan SR, Zhu L, Goldstein E, Horvat R (1998) Human herpesvirus-8 ORF K8.1 gene encodes immunogenic glycoproteins generated by spliced transcripts. Virology 249(1):140–149
30. Rosenau W, Moon HD (1961) Lysis of homologous cells by sensitized lymphocytes in tissue culture. J Natl Cancer Inst 27:471–483
31. Watson J, Gillis S, Marbrook J, Mochizuki D, Smith KA (1979) Biochemical and biological characterization of lymphocyte regulatory molecules. I. Purification of a class of murine lymphokines. J Exp Med 150(4):849–861
32. Gillis S, Baker PE, Ruscetti FW, Smith KA (1978) Long-term culture of human antigen-specific cytotoxic T-cell lines. J Exp Med 148(4):1093–1098
33. Watson J, Mochizuki D (1980) IL-2: a class of T cell growth factors. Immunol Rev 51:257–278
34. Beard RE, Abate-Daga D, Rosati SF, Zheng Z, Wunderlich JR, Rosenberg SA et al (2013) Gene expression profiling using nanostring digital RNA counting to identify potential target antigens for melanoma immunotherapy. Clin Cancer Res 19(18):4941–4950
35. Condomines M, Hose D, Reme T, Requirand G, Hundemer M, Schoenhals M et al (2009) Gene expression profiling and real-time PCR analyses identify novel potential cancer-testis antigens in multiple myeloma. J Immunol 183(2):832–840
36. Forghanifard MM, Gholamin M, Farshchian M, Moaven O, Memar B, Forghani MN et al (2011) Cancer-testis gene expression profiling in esophageal squamous cell carcinoma: identification of specific tumor marker and potential targets for immunotherapy. Cancer Biol Ther 12(3):191–197
37. Kuppers R, Klein U, Schwering I, Distler V, Brauninger A, Cattoretti G et al (2003) Identification of Hodgkin and Reed-Sternberg cell-specific genes by gene expression profiling. J Clin Invest 111(4):529–537
38. Maia S, Haining WN, Ansen S, Xia Z, Armstrong SA, Seth NP et al (2005) Gene expression profiling identifies BAX-delta as a novel tumor antigen in acute lymphoblastic leukemia. Cancer Res 65(21):10050–10058
39. Mathiassen S, Lauemoller SL, Ruhwald M, Claesson MH, Buus S (2001) Tumor-associated antigens identified by mRNA expression profiling induce protective anti-tumor immunity. Eur J Immunol 31(4):1239–1246
40. Lee JK, Bangayan NJ, Chai T, Smith BA, Pariva TE, Yun S et al (2018) Systemic surfaceome profiling identifies target antigens for immune-based therapy in subtypes of advanced prostate cancer. Proc Natl Acad Sci U S A 115(19):E4473–E4482

41. DeLucia DC, Lee JK (2020) Identification of cell surface targets for CAR T cell immunotherapy. Methods Mol Biol 2097:45–54

42. Sahin U, Tureci O, Schmitt H, Cochlovius B, Johannes T, Schmits R et al (1995) Human neoplasms elicit multiple specific immune responses in the autologous host. Proc Natl Acad Sci U S A 92(25):11810–11813

43. Chen YT, Scanlan MJ, Sahin U, Tureci O, Gure AO, Tsang S et al (1997) A testicular antigen aberrantly expressed in human cancers detected by autologous antibody screening. Proc Natl Acad Sci U S A 94(5):1914–1918

44. Unwin RD, Harnden P, Pappin D, Rahman D, Whelan P, Craven RA et al (2003) Serological and proteomic evaluation of antibody responses in the identification of tumor antigens in renal cell carcinoma. Proteomics 3(1):45–55

45. Klade CS, Voss T, Krystek E, Ahorn H, Zatloukal K, Pummer K et al (2001) Identification of tumor antigens in renal cell carcinoma by serological proteome analysis. Proteomics 1(7):890–898

46. Traversari C, van der Bruggen P, Van den Eynde B, Hainaut P, Lemoine C, Ohta N et al (1992) Transfection and expression of a gene coding for a human melanoma antigen recognized by autologous cytolytic T lymphocytes. Immunogenetics 35(3):145–152

47. Bakker AB, Schreurs MW, de Boer AJ, Kawakami Y, Rosenberg SA, Adema GJ et al (1994) Melanocyte lineage-specific antigen gp100 is recognized by melanoma-derived tumor-infiltrating lymphocytes. J Exp Med 179(3):1005–1009

48. Gee MH, Han A, Lofgren SM, Beausang JF, Mendoza JL, Birnbaum ME et al (2018) Antigen Identification for orphan T cell receptors expressed on tumor-infiltrating lymphocytes. Cell 172(3):549–563 e16

49. Kawakami Y, Eliyahu S, Delgado CH, Robbins PF, Rivoltini L, Topalian SL et al (1994) Cloning of the gene coding for a shared human melanoma antigen recognized by autologous T cells infiltrating into tumor. Proc Natl Acad Sci U S A 91(9):3515–3519

50. Kawakami Y, Eliyahu S, Sakaguchi K, Robbins PF, Rivoltini L, Yannelli JR et al (1994) Identification of the immunodominant peptides of the MART-1 human melanoma antigen recognized by the majority of HLA-A2-restricted tumor infiltrating lymphocytes. J Exp Med 180(1):347–352

51. Khong HT, Wang QJ, Rosenberg SA (2004) Identification of multiple antigens recognized by tumor-infiltrating lymphocytes from a single patient: tumor escape by antigen loss and loss of MHC expression. J Immunother 27(3):184–190

52. Lu YC, Yao X, Crystal JS, Li YF, El-Gamil M, Gross C et al (2014) Efficient identification of mutated cancer antigens recognized by T cells associated with durable tumor regressions. Clin Cancer Res 20(13):3401–3410

53. Robbins PF, El-Gamil M, Li YF, Kawakami Y, Loftus D, Appella E et al (1996) A mutated beta-catenin gene encodes a melanoma-specific antigen recognized by tumor infiltrating lymphocytes. J Exp Med 183(3):1185–1192

54. Wang RF, Robbins PF, Kawakami Y, Kang XQ, Rosenberg SA (1995) Identification of a gene encoding a melanoma tumor antigen recognized by HLA-A31-restricted tumor-infiltrating lymphocytes. J Exp Med 181(2):799–804

55. Harjunpaa H, Llort Asens M, Guenther C, Fagerholm SC (2019) Cell adhesion molecules and their roles and regulation in the immune and tumor microenvironment. Front Immunol 10:1078

56. Cornel AM, Mimpen IL, Nierkens S (2020) MHC Class I downregulation in cancer: underlying mechanisms and potential targets for cancer immunotherapy. Cancers (Basel) 12(7)

57. Ferrone S, Whiteside TL (2007) Tumor microenvironment and immune escape. Surg Oncol Clin N Am 16(4):755–774, viii

58. Feng XY, Lu L, Wang KF, Zhu BY, Wen XZ, Peng RQ et al (2019) Low expression of CD80 predicts for poor prognosis in patients with gastric adenocarcinoma. Future Oncol 15(5):473–483

59. Thomas GR, Chen Z, Leukinova E, Van Waes C, Wen J (2004) Cytokines IL-1 alpha, IL-6, and GM-CSF constitutively secreted by oral squamous carcinoma induce down-regulation of

CD80 costimulatory molecule expression: restoration by interferon gamma. Cancer Immunol Immunother 53(1):33–40

60. Nurieva R, Thomas S, Nguyen T, Martin-Orozco N, Wang Y, Kaja MK et al (2006) T-cell tolerance or function is determined by combinatorial costimulatory signals. EMBO J 25(11):2623–2633

61. Frey AB (2015) Suppression of T cell responses in the tumor microenvironment. Vaccine 33(51):7393–7400

62. Pelletier JPR, Mukhtar F (2020) Passive monoclonal and polyclonal antibody therapies. Immunol Concepts Transfusion Med 251–348

63. Parray HA, Shukla S, Samal S, Shrivastava T, Ahmed S, Sharma C et al (2020) Hybridoma technology a versatile method for isolation of monoclonal antibodies, its applicability across species, limitations, advancement and future perspectives. Int Immunopharmacol 85:106639

64. Liu JK (2014) The history of monoclonal antibody development - Progress, remaining challenges and future innovations. Ann Med Surg (Lond) 3(4):113–116

65. Almagro JC, Fransson J (2008) Humanization of antibodies. Front Biosci 13:1619–1633

66. Alderson KL, Sondel PM (2011) Clinical cancer therapy by NK cells via antibody-dependent cell-mediated cytotoxicity. J Biomed Biotechnol 2011:379123

67. Gul N, Babes L, Siegmund K, Korthouwer R, Bogels M, Braster R et al (2014) Macrophages eliminate circulating tumor cells after monoclonal antibody therapy. J Clin Invest 124(2):812–823

68. Maloney DG, Grillo-Lopez AJ, White CA, Bodkin D, Schilder RJ, Neidhart JA et al (1997) IDEC-C2B8 (Rituximab) anti-CD20 monoclonal antibody therapy in patients with relapsed low-grade non-Hodgkin's lymphoma. Blood 90(6):2188–2195

69. Zahavi D, Weiner L (2020) Monoclonal antibodies in cancer therapy. Antibodies (Basel) 9(3)

70. Maniati E, Berlato C, Gopinathan G, Heath O, Kotantaki P, Lakhani A et al (2020) Mouse ovarian cancer models recapitulate the human tumor microenvironment and patient response to treatment. Cell Rep 30(2):525–540 e7

71. Tanaka HY, Kano MR (2018) Stromal barriers to nanomedicine penetration in the pancreatic tumor microenvironment. Cancer Sci 109(7):2085–2092

72. Muthumani K, Marnin L, Kudchodkar SB, Perales-Puchalt A, Choi H, Agarwal S et al (2017) Novel prostate cancer immunotherapy with a DNA-encoded anti-prostate-specific membrane antigen monoclonal antibody. Cancer Immunol Immunother 66(12):1577–1588

73. Danhier F, Feron O, Preat V (2010) To exploit the tumor microenvironment: passive and active tumor targeting of nanocarriers for anti-cancer drug delivery. J Control Release 148(2):135–146

74. Matsumura Y, Maeda H (1986) A new concept for macromolecular therapeutics in cancer chemotherapy: mechanism of tumoritropic accumulation of proteins and the antitumor agent smancs. Cancer Res 46(12 Pt 1):6387–6392

75. Baskar R, Lee KA, Yeo R, Yeoh KW (2012) Cancer and radiation therapy: current advances and future directions. Int J Med Sci 9(3):193–199

76. De Ruysscher D, Niedermann G, Burnet NG, Siva S, Lee AWM, Hegi-Johnson F (2019) Radiotherapy toxicity. Nat Rev Dis Primers 5(1):13

77. Citrin DE (2017) Recent developments in radiotherapy. N Engl J Med 377(22):2200–2201

78. Koppe MJ, Postema EJ, Aarts F, Oyen WJ, Bleichrodt RP, Boerman OC (2005) Antibody-guided radiation therapy of cancer. Cancer Metastasis Rev 24(4):539–567

79. Steiner M, Neri D (2011) Antibody-radionuclide conjugates for cancer therapy: historical considerations and new trends. Clin Cancer Res 17(20):6406–6416

80. Borsi L, Balza E, Bestagno M, Castellani P, Carnemolla B, Biro A et al (2002) Selective targeting of tumoral vasculature: comparison of different formats of an antibody (L19) to the ED-B domain of fibronectin. Int J Cancer 102(1):75–85

81. Bander NH, Milowsky MI, Nanus DM, Kostakoglu L, Vallabhajosula S, Goldsmith SJ (2005) Phase I trial of 177lutetium-labeled J591, a monoclonal antibody to prostate-specific membrane antigen, in patients with androgen-independent prostate cancer. J Clin Oncol 23(21):4591–4601

82. Milowsky MI, Nanus DM, Kostakoglu L, Vallabhajosula S, Goldsmith SJ, Bander NH (2004) Phase I trial of yttrium-90-labeled anti-prostate-specific membrane antigen monoclonal antibody J591 for androgen-independent prostate cancer. J Clin Oncol 22(13):2522–2531

83. Tagawa ST, Beltran H, Vallabhajosula S, Goldsmith SJ, Osborne J, Matulich D et al (2010) Anti-prostate-specific membrane antigen-based radioimmunotherapy for prostate cancer. Cancer 116(4 Suppl):1075–1083

84. Tagawa ST, Milowsky MI, Morris M, Vallabhajosula S, Christos P, Akhtar NH et al (2013) Phase II study of Lutetium-177-labeled anti-prostate-specific membrane antigen monoclonal antibody J591 for metastatic castration-resistant prostate cancer. Clin Cancer Res 19(18):5182–5191

85. Rahbar K, Bode A, Weckesser M, Avramovic N, Claesener M, Stegger L et al (2016) Radioligand therapy with 177Lu-PSMA-617 as a novel therapeutic option in patients with metastatic castration resistant prostate cancer. Clin Nucl Med 41(7):522–528

86. Rahbar K, Schmidt M, Heinzel A, Eppard E, Bode A, Yordanova A et al (2016) Response and tolerability of a single dose of 177Lu-PSMA-617 in patients with metastatic castration-resistant prostate cancer: a multicenter retrospective analysis. J Nucl Med 57(9):1334–1338

87. Hofman MS, Violet J, Hicks RJ, Ferdinandus J, Thang SP, Akhurst T et al (2018) [(177)Lu]-PSMA-617 radionuclide treatment in patients with metastatic castration-resistant prostate cancer (LuPSMA trial): a single-centre, single-arm, phase 2 study. Lancet Oncol 19(6):825–833

88. Rahbar K, Bodei L, Morris MJ (2019) Is the vision of radioligand therapy for prostate cancer becoming a reality? An overview of the Phase III VISION trial and its importance for the future of theranostics. J Nucl Med 60(11):1504–1506

89. Luo D, Wang X, Zeng S, Ramamurthy G, Burda C, Basilion JP (2019) Targeted gold nanocluster-enhanced radiotherapy of prostate cancer. Small 15(34):e1900968

90. Bargh JD, Isidro-Llobet A, Parker JS, Spring DR (2019) Cleavable linkers in antibody-drug conjugates. Chem Soc Rev 48(16):4361–4374

91. Bradley AM, Devine M, DeRemer D (2013) Brentuximab vedotin: an anti-CD30 antibody-drug conjugate. Am J Health Syst Pharm 70(7):589–597

92. Fenn KM, Kalinsky K (2019) Sacituzumab govitecan: antibody-drug conjugate in triple-negative breast cancer and other solid tumors. Drugs Today (Barc) 55(9):575–585

93. DeLucia DC, Cardillo TM, Ang L, Labrecque MP, Zhang A, Hopkins JE et al (2021) Regulation of CEACAM5 and therapeutic efficacy of an anti-CEACAM5-SN38 antibody-drug conjugate in neuroendocrine prostate cancer. Clin Cancer Res 27(3):759–774

94. Beck A, Goetsch L, Dumontet C, Corvaia N (2017) Strategies and challenges for the next generation of antibody-drug conjugates. Nat Rev Drug Discov 16(5):315–337

95. Khongorzul P, Ling CJ, Khan FU, Ihsan AU, Zhang J (2020) Antibody-drug conjugates: a comprehensive review. Mol Cancer Res 18(1):3–19

96. Buecheler JW, Winzer M, Tonillo J, Weber C, Gieseler H (2018) Impact of payload hydrophobicity on the stability of antibody-drug conjugates. Mol Pharm 15(7):2656–2664

97. Lyon RP, Bovee TD, Doronina SO, Burke PJ, Hunter JH, Neff-LaFord HD et al (2015) Reducing hydrophobicity of homogeneous antibody-drug conjugates improves pharmacokinetics and therapeutic index. Nat Biotechnol 33(7):733–735

98. Donaghy H (2016) Effects of antibody, drug and linker on the preclinical and clinical toxicities of antibody-drug conjugates. MAbs 8(4):659–671

99. Panowski S, Bhakta S, Raab H, Polakis P, Junutula JR (2014) Site-specific antibody drug conjugates for cancer therapy. MAbs 6(1):34–45

100. Einsele H, Borghaei H, Orlowski RZ, Subklewe M, Roboz GJ, Zugmaier G et al (2020) The BiTE (bispecific T-cell engager) platform: development and future potential of a targeted immuno-oncology therapy across tumor types. Cancer 126(14):3192–3201

101. Goebeler ME, Bargou R (2016) Blinatumomab: a CD19/CD3 bispecific T cell engager (BiTE) with unique anti-tumor efficacy. Leuk Lymphoma 57(5):1021–1032

102. Kantarjian H, Stein A, Gokbuget N, Fielding AK, Schuh AC, Ribera JM et al (2017) Blina-tumomab versus chemotherapy for advanced acute lymphoblastic leukemia. N Engl J Med 376(9):836–847
103. Zha Y, Blank C, Gajewski TF (2004) Negative regulation of T-cell function by PD-1. Crit Rev Immunol 24(4):229–237
104. Teft WA, Kirchhof MG, Madrenas J (2006) A molecular perspective of CTLA-4 function. Annu Rev Immunol 24:65–97
105. Keir ME, Latchman YE, Freeman GJ, Sharpe AH (2005) Programmed death-1 (PD-1):PD-ligand 1 interactions inhibit TCR-mediated positive selection of thymocytes. J Immunol 175(11):7372–7379
106. Darvin P, Toor SM, Sasidharan Nair V, Elkord E (2018) Immune checkpoint inhibitors: recent progress and potential biomarkers. Exp Mol Med 50(12):1–11
107. Sharpe AH, Abbas AK (2006) T-cell costimulation–biology, therapeutic potential, and challenges. N Engl J Med 355(10):973–975
108. Riley JL (2009) PD-1 signaling in primary T cells. Immunol Rev 229(1):114–125
109. Cai J, Wang D, Zhang G, Guo X (2019) The role Of PD-1/PD-L1 axis in treg development and function: implications for cancer immunotherapy. Onco Targets Ther 12:8437–8445
110. Cha JH, Chan LC, Li CW, Hsu JL, Hung MC (2019) Mechanisms controlling PD-L1 expression in cancer. Mol Cell 76(3):359–370
111. Hodi FS, Mihm MC, Soiffer RJ, Haluska FG, Butler M, Seiden MV et al (2003) Biologic activity of cytotoxic T lymphocyte-associated antigen 4 antibody blockade in previously vaccinated metastatic melanoma and ovarian carcinoma patients. Proc Natl Acad Sci U S A 100(8):4712–4717
112. Phan GQ, Yang JC, Sherry RM, Hwu P, Topalian SL, Schwartzentruber DJ et al (2003) Cancer regression and autoimmunity induced by cytotoxic T lymphocyte-associated antigen 4 blockade in patients with metastatic melanoma. Proc Natl Acad Sci U S A 100(14):8372–8377
113. Lipson EJ, Drake CG (2011) Ipilimumab: an anti-CTLA-4 antibody for metastatic melanoma. Clin Cancer Res 17(22):6958–6962
114. Tran L, Theodorescu D (2020) Determinants of resistance to checkpoint inhibitors. Int J Mol Sci 21(5)
115. Marin-Acevedo JA, Chirila RM, Dronca RS (2019) Immune checkpoint inhibitor toxicities. Mayo Clin Proc 94(7):1321–1329
116. Vilarino N, Bruna J, Kalofonou F, Anastopoulou GG, Argyriou AA (2020) Immune-driven pathogenesis of neurotoxicity after exposure of cancer patients to immune checkpoint inhibitors. Int J Mol Sci 21(16)
117. Keenan TE, Burke KP, Van Allen EM (2019) Genomic correlates of response to immune checkpoint blockade. Nat Med 25(3):389–402
118. Perrier A, Didelot A, Laurent-Puig P, Blons H, Garinet S (2020) Epigenetic mechanisms of resistance to immune checkpoint inhibitors. Biomolecules 10(7)
119. Ishihara J, Ishihara A, Sasaki K, Lee SS, Williford JM, Yasui M et al (2019) Targeted antibody and cytokine cancer immunotherapies through collagen affinity. Sci Transl Med 11(487)
120. Patel SA, Minn AJ (2018) Combination cancer therapy with immune checkpoint blockade: mechanisms and strategies. Immunity 48(3):417–433
121. Ahn R, Ursini-Siegel J (2021) Clinical potential of Kinase inhibitors in combination with immune checkpoint inhibitors for the treatment of solid tumors. Int J Mol Sci 22(5)
122. Attarwala H (2010) TGN1412: from discovery to disaster. J Young Pharm 2(3):332–336
123. Suntharalingam G, Perry MR, Ward S, Brett SJ, Castello-Cortes A, Brunner MD et al (2006) Cytokine storm in a phase 1 trial of the anti-CD28 monoclonal antibody TGN1412. N Engl J Med 355(10):1018–1028
124. Hanke T (2006) Lessons from TGN1412. Lancet 368(9547):1569–1570; author reply 70
125. Dayan CM, Wraith DC (2008) Preparing for first-in-man studies: the challenges for translational immunology post-TGN1412. Clin Exp Immunol 151(2):231–234

126. Shankaran V, Ikeda H, Bruce AT, White JM, Swanson PE, Old LJ et al (2001) IFNgamma and lymphocytes prevent primary tumour development and shape tumour immunogenicity. Nature 410(6832):1107–1111

127. Halliday GM, Patel A, Hunt MJ, Tefany FJ, Barnetson RS (1995) Spontaneous regression of human melanoma/nonmelanoma skin cancer: association with infiltrating CD4+ T cells. World J Surg 19(3):352–358

128. Morgan MA, Buning H, Sauer M, Schambach A (2020) Use of cell and genome modification technologies to generate improved "Off-the-Shelf" CAR T and CAR NK cells. Front Immunol 11:1965

129. Zhang Y, Zhang Z (2020) The history and advances in cancer immunotherapy: understanding the characteristics of tumor-infiltrating immune cells and their therapeutic implications. Cell Mol Immunol 17(8):807–821

130. Kawata A, Une Y, Hosokawa M, Uchino J, Kobayashi H (1992) Tumor-infiltrating lymphocytes and prognosis of hepatocellular carcinoma. Jpn J Clin Oncol 22(4):256–263

131. Dirican A, Ekinci N, Avci A, Akyol M, Alacacioglu A, Kucukzeybek Y et al (2013) The effects of hematologic parameters and tumor-infiltrating lymphocytes on prognosis in patients with gastric cancer. Cancer Biomark 13(1):11–20

132. Fukunaga A, Miyamoto M, Cho Y, Murakami S, Kawarada Y, Oshikiri T et al (2004) CD8+ tumor-infiltrating lymphocytes together with CD4+ tumor-infiltrating lymphocytes and dendritic cells improve the prognosis of patients with pancreatic adenocarcinoma. Pancreas 28(1):e26-31

133. Loi S, Drubay D, Adams S, Pruneri G, Francis PA, Lacroix-Triki M et al (2019) Tumor-infiltrating lymphocytes and prognosis: a pooled individual patient analysis of early-stage triple-negative breast cancers. J Clin Oncol 37(7):559–569

134. Chen TH, Zhang YC, Tan YT, An X, Xue C, Deng YF et al (2017) Tumor-infiltrating lymphocytes predict prognosis of breast cancer patients treated with anti-Her-2 therapy. Oncotarget 8(3):5219–5232

135. Topalian SL, Solomon D, Avis FP, Chang AE, Freerksen DL, Linehan WM et al (1988) Immunotherapy of patients with advanced cancer using tumor-infiltrating lymphocytes and recombinant IL-2: a pilot study 6(5):839–853

136. van den Berg JH, Heemskerk B, van Rooij N, Gomez-Eerland R, Michels S, van Zon M et al (2020) Tumor infiltrating lymphocytes (TIL) therapy in metastatic melanoma: boosting of neoantigen-specific T cell reactivity and long-term follow-up. J Immunother Cancer 8(2)

137. Li J, Chen QY, He J, Li ZL, Tang XF, Chen SP et al (2015) Phase I trial of adoptively transferred tumor-infiltrating lymphocyte immunotherapy following concurrent chemoradiotherapy in patients with locoregionally advanced nasopharyngeal carcinoma. Oncoimmunology 4(2):e976507

138. Pedersen M, Westergaard MCW, Milne K, Nielsen M, Borch TH, Poulsen LG et al (2018) Adoptive cell therapy with tumor-infiltrating lymphocytes in patients with metastatic ovarian cancer: a pilot study. Oncoimmunology 7(12):e1502905

139. Wang SS, Liu W, Ly D, Xu H, Qu L, Zhang L (2019) Tumor-infiltrating B cells: their role and application in anti-tumor immunity in lung cancer. Cell Mol Immunol 16(1):6–18

140. He J, Tang XF, Chen QY, Mai HQ, Huang ZF, Li J et al (2012) Ex vivo expansion of tumor-infiltrating lymphocytes from nasopharyngeal carcinoma patients for adoptive immunotherapy. Chin J Cancer 31(6):287–294

141. Dudley ME, Wunderlich JR, Shelton TE, Even J, Rosenberg SA (2003) Generation of tumor-infiltrating lymphocyte cultures for use in adoptive transfer therapy for melanoma patients. J Immunother 26(4):332–342

142. Katt ME, Placone AL, Wong AD, Xu ZS, Searson PC (2016) In Vitro tumor models: advantages, disadvantages, variables, and selecting the right platform. Front Bioeng Biotechnol 4:12

143. Yunger S, Bar El A, Zeltzer LA, Fridman E, Raviv G, Laufer M et al (2019) Tumor-infiltrating lymphocytes from human prostate tumors reveal anti-tumor reactivity and potential for adoptive cell therapy. Oncoimmunology 8(12):e1672494

144. Rath JA, Arber C (2020) Engineering strategies to enhance TCR-based adoptive T cell therapy. Cells 9(6)

145. Johnson LA, Morgan RA, Dudley ME, Cassard L, Yang JC, Hughes MS et al (2009) Gene therapy with human and mouse T-cell receptors mediates cancer regression and targets normal tissues expressing cognate antigen. Blood 114(3):535–546

146. Rapoport AP, Stadtmauer EA, Binder-Scholl GK, Goloubeva O, Vogl DT, Lacey SF et al (2015) NY-ESO-1-specific TCR-engineered T cells mediate sustained antigen-specific antitumor effects in myeloma. Nat Med 21(8):914–921

147. Chapuis AG, Egan DN, Bar M, Schmitt TM, McAfee MS, Paulson KG et al (2019) T cell receptor gene therapy targeting WT1 prevents acute myeloid leukemia relapse posttransplant. Nat Med 25(7):1064–1072

148. Stromnes IM, Schmitt TM, Hulbert A, Brockenbrough JS, Nguyen H, Cuevas C et al (2015) T cells engineered against a native antigen can surmount immunologic and physical barriers to treat pancreatic ductal adenocarcinoma. Cancer Cell 28(5):638–652

149. Dossa RG, Cunningham T, Sommermeyer D, Medina-Rodriguez I, Biernacki MA, Foster K et al (2018) Development of T-cell immunotherapy for hematopoietic stem cell transplantation recipients at risk of leukemia relapse. Blood 131(1):108–120

150. Oppermans N, Kueberuwa G, Hawkins RE, Bridgeman JS (2020) Transgenic T-cell receptor immunotherapy for cancer: building on clinical success. Ther Adv Vaccines Immunother 8:2515135520933509

151. Wieczorek M, Abualrous ET, Sticht J, Alvaro-Benito M, Stolzenberg S, Noe F et al (2017) Major Histocompatibility Complex (MHC) Class I and MHC Class II proteins: conformational plasticity in antigen presentation. Front Immunol 8:292

152. Comber JD, Philip R (2014) MHC class I antigen presentation and implications for developing a new generation of therapeutic vaccines. Ther Adv Vaccines 2(3):77–89

153. Morgan RA, Chinnasamy N, Abate-Daga D, Gros A, Robbins PF, Zheng Z et al (2013) Cancer regression and neurological toxicity following anti-MAGE-A3 TCR gene therapy. J Immunother 36(2):133–151

154. Singh N, Kulikovskaya I, Barrett DM, Binder-Scholl G, Jakobsen B, Martinez D et al (2016) T cells targeting NY-ESO-1 demonstrate efficacy against disseminated neuroblastoma. Oncoimmunology 5(1):e1040216

155. Bonini C, Mondino A (2015) Adoptive T-cell therapy for cancer: the era of engineered T cells. Eur J Immunol 45(9):2457–2469

156. Parkhurst M, Gros A, Pasetto A, Prickett T, Crystal JS, Robbins P et al (2017) Isolation of T-cell receptors specifically reactive with mutated tumor-associated antigens from tumor-infiltrating lymphocytes based on CD137 expression. Clin Cancer Res 23(10):2491–2505

157. Pasetto A, Gros A, Robbins PF, Deniger DC, Prickett TD, Matus-Nicodemos R et al (2016) Tumor- and neoantigen-reactive T-cell receptors can be identified based on their frequency in fresh tumor. Cancer Immunol Res 4(9):734–743

158. Wang QJ, Yu Z, Griffith K, Hanada K, Restifo NP, Yang JC (2016) Identification of T-cell receptors targeting KRAS-mutated human tumors. Cancer Immunol Res 4(3):204–214

159. Lo W, Parkhurst M, Robbins PF, Tran E, Lu YC, Jia L et al (2019) Immunologic recognition of a shared p53 mutated neoantigen in a patient with metastatic colorectal cancer. Cancer Immunol Res 7(4):534–543

160. Wang P, Sidney J, Dow C, Mothe B, Sette A, Peters B (2008) A systematic assessment of MHC class II peptide binding predictions and evaluation of a consensus approach. PLoS Comput Biol 4(4):e1000048

161. Lanoix J, Durette C, Courcelles M, Cossette E, Comtois-Marotte S, Hardy MP et al (2018) Comparison of the MHC I immunopeptidome repertoire of B-cell lymphoblasts using two isolation methods. Proteomics 18(12):e1700251

162. Storkus WJ, Zeh HJ 3rd, Salter RD, Lotze MT (1993) Identification of T-cell epitopes: rapid isolation of class I-presented peptides from viable cells by mild acid elution. J Immunother Emphasis Tumor Immunol 14(2):94–103

163. Sturm T, Sautter B, Worner TP, Stevanovic S, Rammensee HG, Planz O et al (2021) Mild acid elution and MHC immunoaffinity chromatography reveal similar albeit not identical profiles of the HLA Class I immunopeptidome. J Proteome Res 20(1):289–304

164. Lu YC, Robbins PF (2016) Cancer immunotherapy targeting neoantigens. Semin Immunol 28(1):22–27

165. Shitaoka K, Hamana H, Kishi H, Hayakawa Y, Kobayashi E, Sukegawa K et al (2018) Identification of tumoricidal TCRs from tumor-infiltrating lymphocytes by single-cell analysis. Cancer Immunol Res 6(4):378–388

166. Li Q, Ding ZY (2020) The ways of isolating neoantigen-specific T cells. Front Oncol 10:1347

167. Wooldridge L, Lissina A, Cole DK, van den Berg HA, Price DA, Sewell AK (2009) Tricks with tetramers: how to get the most from multimeric peptide-MHC. Immunology 126(2):147–164

168. Yu YY, Netuschil N, Lybarger L, Connolly JM, Hansen TH (2002) Cutting edge: single-chain trimers of MHC class I molecules form stable structures that potently stimulate antigen-specific T cells and B cells. J Immunol 168(7):3145–3149

169. Chour W, Xu AM, Ng AHC, Choi J, Xie J, Yuan D et al (2020) Shared antigen-specific CD8$^+$ T cell responses against the SARS-COV-2 spike protein in HLA-A*02:01 COVID-19 participants. 2020:2020.05.04.20085779

170. Zhang H, Sun M, Wang J, Zeng B, Cao X, Han Y et al (2021) Identification of NY-ESO-1 157-165 specific murine T cell receptors with distinct recognition pattern for tumor immunotherapy. Front Immunol 12:644520

171. Batard P, Peterson DA, Devevre E, Guillaume P, Cerottini JC, Rimoldi D et al (2006) Dextramers: new generation of fluorescent MHC class I/peptide multimers for visualization of antigen-specific CD8+ T cells. J Immunol Methods 310(1–2):136–148

172. Dolton G, Lissina A, Skowera A, Ladell K, Tungatt K, Jones E et al (2014) Comparison of peptide-major histocompatibility complex tetramers and dextramers for the identification of antigen-specific T cells. Clin Exp Immunol 177(1):47–63

173. Bethune MT, Comin-Anduix B, Hwang Fu YH, Ribas A, Baltimore D (2017) Preparation of peptide-MHC and T-cell receptor dextramers by biotinylated dextran doping. Biotechniques 62(3):123–130

174. Luxembourg AT, Borrow P, Teyton L, Brunmark AB, Peterson PA, Jackson MR (1998) Biomagnetic isolation of antigen-specific CD8+ T cells usable in immunotherapy. Nat Biotechnol 16(3):281–285

175. Basu S, Campbell HM, Dittel BN, Ray A (2010) Purification of specific cell population by fluorescence activated cell sorting (FACS). J Vis Exp (41)

176. De Simone M, Rossetti G, Pagani M (2018) Single cell T cell receptor sequencing: techniques and future challenges. Front Immunol 9:1638

177. Linnemann C, Heemskerk B, Kvistborg P, Kluin RJ, Bolotin DA, Chen X et al (2013) High-throughput identification of antigen-specific TCRs by TCR gene capture. Nat Med 19(11):1534–1541

178. Nesterenko PA, McLaughlin J, Cheng D, Bangayan NJ, Burton Sojo G, Seet CS et al (2021) Droplet-based mRNA sequencing of fixed and permeabilized cells by CLInt-seq allows for antigen-specific TCR cloning. Proc Natl Acad Sci U S A 118(3)

179. Busch DH, Pamer EG (1999) T cell affinity maturation by selective expansion during infection. J Exp Med 189(4):701–710

180. Zehn D, Lee SY, Bevan MJ (2009) Complete but curtailed T-cell response to very low-affinity antigen. Nature 458(7235):211–214

181. Galvez J, Galvez JJ, Garcia-Penarrubia P (2019) Is TCR/pMHC affinity a good estimate of the T-cell response? An answer based on predictions from 12 phenotypic models. Front Immunol 10:349

182. Stone JD, Chervin AS, Kranz DM (2009) T-cell receptor binding affinities and kinetics: impact on T-cell activity and specificity. Immunology 126(2):165–176

183. Frankel TL, Burns WR, Peng PD, Yu Z, Chinnasamy D, Wargo JA et al (2010) Both CD4 and CD8 T cells mediate equally effective In Vivo tumor treatment when engineered with a highly avid TCR targeting tyrosinase 184(11):5988–5998
184. Corthay A, Skovseth DK, Lundin KU, Rosjo E, Omholt H, Hofgaard PO et al (2005) Primary antitumor immune response mediated by CD4+ T cells. Immunity 22(3):371–383
185. Perez-Diez A, Joncker NT, Choi K, Chan WF, Anderson CC, Lantz O et al (2007) CD4 cells can be more efficient at tumor rejection than CD8 cells. Blood 109(12):5346–5354
186. Hughes MS, Yu YY, Dudley ME, Zheng Z, Robbins PF, Li Y et al (2005) Transfer of a TCR gene derived from a patient with a marked antitumor response conveys highly active T-cell effector functions. Hum Gene Ther 16(4):457–472
187. Morgan RA, Dudley ME, Wunderlich JR, Hughes MS, Yang JC, Sherry RM et al (2006) Cancer regression in patients after transfer of genetically engineered lymphocytes. Science 314(5796):126–129
188. Li LP, Lampert JC, Chen X, Leitao C, Popovic J, Muller W et al (2010) Transgenic mice with a diverse human T cell antigen receptor repertoire. Nat Med 16(9):1029–1034
189. Obenaus M, Leitao C, Leisegang M, Chen X, Gavvovidis I, van der Bruggen P et al (2015) Identification of human T-cell receptors with optimal affinity to cancer antigens using antigen-negative humanized mice. Nat Biotechnol 33(4):402–407
190. Stromnes IM, Schmitt TM, Chapuis AG, Hingorani SR, Greenberg PD (2014) Re-adapting T cells for cancer therapy: from mouse models to clinical trials. Immunol Rev 257(1):145–164
191. Cameron BJ, Gerry AB, Dukes J, Harper JV, Kannan V, Bianchi FC et al (2013) Identification of a Titin-derived HLA-A1-presented peptide as a cross-reactive target for engineered MAGE A3-directed T cells. Sci Transl Med 5(197):197ra03
192. Linette GP, Stadtmauer EA, Maus MV, Rapoport AP, Levine BL, Emery L et al (2013) Cardiovascular toxicity and titin cross-reactivity of affinity-enhanced T cells in myeloma and melanoma. Blood 122(6):863–871
193. Schmitt TM, Aggen DH, Stromnes IM, Dossett ML, Richman SA, Kranz DM et al (2013) Enhanced-affinity murine T-cell receptors for tumor/self-antigens can be safe in gene therapy despite surpassing the threshold for thymic selection. Blood 122(3):348–356
194. Fernandez-Miguel G, Alarcon B, Iglesias A, Bluethmann H, Alvarez-Mon M, Sanz E et al (1999) Multivalent structure of an alphabetaT cell receptor. Proc Natl Acad Sci U S A 96(4):1547–1552
195. van Loenen MM, de Boer R, Amir AL, Hagedoorn RS, Volbeda GL, Willemze R et al (2010) Mixed T cell receptor dimers harbor potentially harmful neoreactivity. Proc Natl Acad Sci U S A 107(24):10972–10977
196. Cohen CJ, Zhao Y, Zheng Z, Rosenberg SA, Morgan RA (2006) Enhanced antitumor activity of murine-human hybrid T-cell receptor (TCR) in human lymphocytes is associated with improved pairing and TCR/CD3 stability. Cancer Res 66(17):8878–8886
197. Kuball J, Dossett ML, Wolfl M, Ho WY, Voss RH, Fowler C et al (2007) Facilitating matched pairing and expression of TCR chains introduced into human T cells. Blood 109(6):2331–2338
198. Osborn MJ, Webber BR, Knipping F, Lonetree CL, Tennis N, DeFeo AP et al (2016) Evaluation of TCR gene editing achieved by TALENs, CRISPR/Cas9, and megaTAL nucleases. Mol Ther 24(3):570–581
199. Provasi E, Genovese P, Lombardo A, Magnani Z, Liu PQ, Reik A et al (2012) Editing T cell specificity towards leukemia by zinc finger nucleases and lentiviral gene transfer. Nat Med 18(5):807–815
200. Legut M, Dolton G, Mian AA, Ottmann OG, Sewell AK (2018) CRISPR-mediated TCR replacement generates superior anticancer transgenic T cells. Blood 131(3):311–322
201. Jayaraman J, Mellody MP, Hou AJ, Desai RP, Fung AW, Pham AHT et al (2020) CAR-T design: elements and their synergistic function. EBioMedicine 58:102931
202. Sermer D, Brentjens R (2019) CAR T-cell therapy: full speed ahead. Hematol Oncol 37(Suppl 1):95–100

203. Nahas GR, Komanduri KV, Pereira D, Goodman M, Jimenez AM, Beitinjaneh A et al (2020) Incidence and risk factors associated with a syndrome of persistent cytopenias after CAR-T cell therapy (PCTT). Leuk Lymphoma 61(4):940–943

204. Fried S, Avigdor A, Bielorai B, Meir A, Besser MJ, Schachter J et al (2019) Early and late hematologic toxicity following CD19 CAR-T cells. Bone Marrow Transplant 54(10):1643–1650

205. Locke FL, Ghobadi A, Jacobson CA, Miklos DB, Lekakis LJ, Oluwole OO et al (2019) Long-term safety and activity of axicabtagene ciloleucel in refractory large B-cell lymphoma (ZUMA-1): a single-arm, multicentre, phase 1–2 trial. Lancet Oncol 20(1):31–42

206. Yakoub-Agha I, Chabannon C, Bader P, Basak GW, Bonig H, Ciceri F et al (2020) Management of adults and children undergoing chimeric antigen receptor T-cell therapy: best practice recommendations of the European Society for Blood and Marrow Transplantation (EBMT) and the Joint Accreditation Committee of ISCT and EBMT (JACIE). Haematologica 105(2):297–316

207. Arabi F, Torabi-Rahvar M, Shariati A, Ahmadbeigi N, Naderi M (2018) Antigenic targets of CAR T cell therapy. A retrospective view on clinical trials. Exp Cell Res 369(1):1–10

208. Hudecek M, Lupo-Stanghellini MT, Kosasih PL, Sommermeyer D, Jensen MC, Rader C et al (2013) Receptor affinity and extracellular domain modifications affect tumor recognition by ROR1-specific chimeric antigen receptor T cells. Clin Cancer Res 19(12):3153–3164

209. Lynn RC, Feng Y, Schutsky K, Poussin M, Kalota A, Dimitrov DS et al (2016) High-affinity FRbeta-specific CAR T cells eradicate AML and normal myeloid lineage without HSC toxicity. Leukemia 30(6):1355–1364

210. Stoiber S, Cadilha BL, Benmebarek MR, Lesch S, Endres S, Kobold S (2019) Limitations in the design of chimeric antigen receptors for cancer therapy. Cells 8(5)

211. Ghorashian S, Kramer AM, Onuoha S, Wright G, Bartram J, Richardson R et al (2019) Enhanced CAR T cell expansion and prolonged persistence in pediatric patients with ALL treated with a low-affinity CD19 CAR. Nat Med 25(9):1408–1414

212. Lindner SE, Johnson SM, Brown CE, Wang LD (2020) Chimeric antigen receptor signaling: Functional consequences and design implications. Sci Adv 6(21):eaaz3223

213. Hudecek M, Sommermeyer D, Kosasih PL, Silva-Benedict A, Liu L, Rader C et al (2015) The nonsignaling extracellular spacer domain of chimeric antigen receptors is decisive for in vivo antitumor activity. Cancer Immunol Res 3(2):125–135

214. Watanabe N, Bajgain P, Sukumaran S, Ansari S, Heslop HE, Rooney CM et al (2016) Fine-tuning the CAR spacer improves T-cell potency. Oncoimmunology 5(12):e1253656

215. Abken H (2021) Building on synthetic immunology and T cell engineering: a brief journey through the history of chimeric antigen receptors. Hum Gene Ther 32(19–20):1011–1028

216. Thistlethwaite FC, Gilham DE, Guest RD, Rothwell DG, Pillai M, Burt DJ et al (2017) The clinical efficacy of first-generation carcinoembryonic antigen (CEACAM5)-specific CAR T cells is limited by poor persistence and transient pre-conditioning-dependent respiratory toxicity. Cancer Immunol Immunother 66(11):1425–1436

217. Kershaw MH, Westwood JA, Parker LL, Wang G, Eshhar Z, Mavroukakis SA et al (2006) A phase I study on adoptive immunotherapy using gene-modified T cells for ovarian cancer. Clin Cancer Res 12(20 Pt 1):6106–6115

218. Lamers CH, Sleijfer S, Vulto AG, Kruit WH, Kliffen M, Debets R et al (2006) Treatment of metastatic renal cell carcinoma with autologous T-lymphocytes genetically retargeted against carbonic anhydrase IX: first clinical experience. J Clin Oncol 24(13):e20–e22

219. Pule MA, Savoldo B, Myers GD, Rossig C, Russell HV, Dotti G et al (2008) Virus-specific T cells engineered to coexpress tumor-specific receptors: persistence and antitumor activity in individuals with neuroblastoma. Nat Med 14(11):1264–1270

220. Weinkove R, George P, Dasyam N, McLellan AD (2019) Selecting costimulatory domains for chimeric antigen receptors: functional and clinical considerations. Clin Transl Immunol 8(5):e1049

221. van der Stegen SJ, Hamieh M, Sadelain M (2015) The pharmacology of second-generation chimeric antigen receptors. Nat Rev Drug Discov 14(7):499–509

222. Zhang C, Liu J, Zhong JF, Zhang X (2017) Engineering CAR-T cells. Biomark Res 5:22
223. Davey AS, Call ME, Call MJ (2020) The influence of chimeric antigen receptor structural domains on clinical outcomes and associated toxicities. Cancers (Basel) 13(1)
224. Zhao Z, Condomines M, van der Stegen SJC, Perna F, Kloss CC, Gunset G et al (2015) Structural design of engineered costimulation determines tumor rejection kinetics and persistence of CAR T cells. Cancer Cell 28(4):415–428
225. Ma F, Ho JY, Du H, Xuan F, Wu X, Wang Q et al (2019) Evidence of long-lasting anti-CD19 activity of engrafted CD19 chimeric antigen receptor-modified T cells in a phase I study targeting pediatrics with acute lymphoblastic leukemia. Hematol Oncol 37(5):601–608
226. Guedan S, Posey Jr AD, Shaw C, Wing A, Da T, Patel PR et al (2018) Enhancing CAR T cell persistence through ICOS and 4–1BB costimulation. JCI Insight 3(1)
227. Thomas R (2004) Signal 3 and its role in autoimmunity. Arthritis Res Ther 6(1):26–27
228. Yeku OO, Brentjens RJ (2016) Armored CAR T-cells: utilizing cytokines and pro-inflammatory ligands to enhance CAR T-cell anti-tumour efficacy. Biochem Soc Trans 44(2):412–418
229. Chmielewski M, Abken H (2015) TRUCKs: the fourth generation of CARs. Expert Opin Biol Ther 15(8):1145–1154
230. Allen TM, Brehm MA, Bridges S, Ferguson S, Kumar P, Mirochnitchenko O et al (2019) Humanized immune system mouse models: progress, challenges and opportunities. Nat Immunol 20(7):770–774
231. Srivastava S, Riddell SR (2018) Chimeric antigen receptor T cell therapy: challenges to bench-to-bedside efficacy. J Immunol 200(2):459–468
232. McFadden DG, Politi K, Bhutkar A, Chen FK, Song X, Pirun M et al (2016) Mutational landscape of EGFR-, MYC-, and Kras-driven genetically engineered mouse models of lung adenocarcinoma. Proc Natl Acad Sci U S A 113(42):E6409–E6417
233. Shimabukuro-Vornhagen A, Godel P, Subklewe M, Stemmler HJ, Schlosser HA, Schlaak M et al (2018) Cytokine release syndrome. J Immunother Cancer 6(1):56
234. Murthy H, Iqbal M, Chavez JC, Kharfan-Dabaja MA (2019) Cytokine release syndrome: current perspectives. Immunotargets Ther 8:43–52
235. Norelli M, Camisa B, Barbiera G, Falcone L, Purevdorj A, Genua M et al (2018) Monocyte-derived IL-1 and IL-6 are differentially required for cytokine-release syndrome and neurotoxicity due to CAR T cells. Nat Med 24(6):739–748
236. Lee DW, Santomasso BD, Locke FL, Ghobadi A, Turtle CJ, Brudno JN et al (2019) ASTCT consensus grading for cytokine release syndrome and neurologic toxicity associated with immune effector cells. Biol Blood Marrow Transplant 25(4):625–638
237. Si S, Teachey DT (2020) Spotlight on Tocilizumab in the treatment of CAR-T-cell-induced cytokine release syndrome: clinical evidence to date. Ther Clin Risk Manag 16:705–714
238. Hou B, Tang Y, Li W, Zeng Q, Chang D (2019) Efficiency of CAR-T therapy for treatment of solid tumor in clinical trials: a meta-analysis. Dis Markers 2019:3425291
239. Ebert LM, Yu W, Gargett T, Brown MP (2018) Logic-gated approaches to extend the utility of chimeric antigen receptor T-cell technology. Biochem Soc Trans 46(2):391–401
240. Srivastava S, Salter AI, Liggitt D, Yechan-Gunja S, Sarvothama M, Cooper K et al (2019) Logic-gated ROR1 chimeric antigen receptor expression rescues T cell-mediated toxicity to normal tissues and enables selective tumor targeting. Cancer Cell 35(3):489–503 e8
241. Roybal KT, Rupp LJ, Morsut L, Walker WJ, McNally KA, Park JS et al (2016) Precision tumor recognition by T cells with combinatorial antigen-sensing circuits. Cell 164(4):770–779
242. Casucci M, Bondanza A (2011) Suicide gene therapy to increase the safety of chimeric antigen receptor-redirected T lymphocytes. J Cancer 2:378–382
243. Bonini C, Ferrari G, Verzeletti S, Servida P, Zappone E, Ruggieri L et al (1997) HSV-TK gene transfer into donor lymphocytes for control of allogeneic graft-versus-leukemia. Science 276(5319):1719–1724

244. Casucci M, Falcone L, Camisa B, Norelli M, Porcellini S, Stornaiuolo A et al (2018) Extracellular NGFR spacers allow efficient tracking and enrichment of fully functional CAR-T cells co-expressing a suicide gene. Front Immunol 9:507
245. Di Stasi A, Tey SK, Dotti G, Fujita Y, Kennedy-Nasser A, Martinez C et al (2011) Inducible apoptosis as a safety switch for adoptive cell therapy. N Engl J Med 365(18):1673–1683
246. Liu X, Ranganathan R, Jiang S, Fang C, Sun J, Kim S et al (2016) A Chimeric switch-receptor targeting PD1 augments the efficacy of second-generation CAR T cells in advanced solid tumors. Cancer Res 76(6):1578–1590
247. Kloss CC, Lee J, Zhang A, Chen F, Melenhorst JJ, Lacey SF et al (2018) Dominant-negative TGF-beta receptor enhances PSMA-targeted human CAR T cell proliferation and augments prostate cancer eradication. Mol Ther 26(7):1855–1866
248. Yang GX, Lian ZX, Chuang YH, Moritoki Y, Lan RY, Wakabayashi K et al (2008) Adoptive transfer of CD8(+) T cells from transforming growth factor beta receptor type II (dominant negative form) induces autoimmune cholangitis in mice. Hepatology 47(6):1974–1982
249. Garbuglia AR, Lapa D, Sias C, Capobianchi MR, Del Porto P (2020) The use of both therapeutic and prophylactic vaccines in the therapy of papillomavirus disease 11(188)
250. Awad AM, Ntoso A, Connaire JJ, Hernandez GT, Dhillon K, Rich L et al (2021) An open-label, single-arm study evaluating the immunogenicity and safety of the hepatitis B vaccine HepB-CpG (HEPLISAV-B(R)) in adults receiving hemodialysis. Vaccine 39(25):3346–3352
251. Song Q, Zhang CD, Wu XH (2018) Therapeutic cancer vaccines: from initial findings to prospects. Immunol Lett 196:11–21
252. Larocca C, Schlom J (2011) Viral vector-based therapeutic cancer vaccines. Cancer J 17(5):359–371
253. Desjardins A, Gromeier M, Herndon JE 2nd, Beaubier N, Bolognesi DP, Friedman AH et al (2018) Recurrent glioblastoma treated with recombinant poliovirus. N Engl J Med 379(2):150–161
254. Yamano T, Kubo S, Fukumoto M, Yano A, Mawatari-Furukawa Y, Okamura H et al (2016) Whole cell vaccination using immunogenic cell death by an oncolytic adenovirus is effective against a colorectal cancer model. Mol Ther Oncolytics. 3:16031
255. Kantoff PW, Schuetz TJ, Blumenstein BA, Glode LM, Bilhartz DL, Wyand M et al (2010) Overall survival analysis of a phase II randomized controlled trial of a poxviral-based PSA-targeted immunotherapy in metastatic castration-resistant prostate cancer. J Clin Oncol 28(7):1099–1105
256. Harrop R, Chu F, Gabrail N, Srinivas S, Blount D, Ferrari A (2013) Vaccination of castration-resistant prostate cancer patients with TroVax (MVA-5T4) in combination with docetaxel: a randomized phase II trial. Cancer Immunol Immunother 62(9):1511–1520
257. Quoix E, Lena H, Losonczy G, Forget F, Chouaid C, Papai Z et al (2016) TG4010 immunotherapy and first-line chemotherapy for advanced non-small-cell lung cancer (TIME): results from the phase 2b part of a randomised, double-blind, placebo-controlled, phase 2b/3 trial. Lancet Oncol 17(2):212–223
258. Jenne L, Arrighi JF, Jonuleit H, Saurat JH, Hauser C (2000) Dendritic cells containing apoptotic melanoma cells prime human CD8+ T cells for efficient tumor cell lysis. Cancer Res 60(16):4446–4452
259. Chiang CL, Kandalaft LE, Tanyi J, Hagemann AR, Motz GT, Svoronos N et al (2013) A dendritic cell vaccine pulsed with autologous hypochlorous acid-oxidized ovarian cancer lysate primes effective broad antitumor immunity: from bench to bedside. Clin Cancer Res 19(17):4801–4815
260. Chen T, Guo J, Han C, Yang M, Cao X (2009) Heat shock protein 70, released from heat-stressed tumor cells, initiates antitumor immunity by inducing tumor cell chemokine production and activating dendritic cells via TLR4 pathway. J Immunol 182(3):1449–1459
261. Neller MA, Lopez JA, Schmidt CW (2008) Antigens for cancer immunotherapy. Semin Immunol 20(5):286–295

262. Belli F, Testori A, Rivoltini L, Maio M, Andreola G, Sertoli MR et al (2002) Vaccination of metastatic melanoma patients with autologous tumor-derived heat shock protein gp96-peptide complexes: clinical and immunologic findings. J Clin Oncol 20(20):4169–4180
263. Ridolfi L, Petrini M, Fiammenghi L, Granato AM, Ancarani V, Pancisi E et al (2010) Unexpected high response rate to traditional therapy after dendritic cell-based vaccine in advanced melanoma: update of clinical outcome and subgroup analysis. Clin Dev Immunol 2010:504979
264. O'Rourke MG, Johnson M, Lanagan C, See J, Yang J, Bell JR et al (2003) Durable complete clinical responses in a phase I/II trial using an autologous melanoma cell/dendritic cell vaccine. Cancer Immunol Immunother 52(6):387–395
265. O'Rourke MG, Johnson MK, Lanagan CM, See JL, O'Connor LE, Slater GJ et al (2007) Dendritic cell immunotherapy for stage IV melanoma. Melanoma Res 17(5):316–322
266. Nair SK, Morse M, Boczkowski D, Cumming RI, Vasovic L, Gilboa E et al (2002) Induction of tumor-specific cytotoxic T lymphocytes in cancer patients by autologous tumor RNA-transfected dendritic cells. Ann Surg 235(4):540–549
267. Dranoff G (2002) GM-CSF-based cancer vaccines. Immunol Rev 188:147–154
268. Nemunaitis J, Sterman D, Jablons D, Smith JW 2nd, Fox B, Maples P et al (2004) Granulocyte-macrophage colony-stimulating factor gene-modified autologous tumor vaccines in non-small-cell lung cancer. J Natl Cancer Inst 96(4):326–331
269. Nemunaitis J (2005) Vaccines in cancer: GVAX®, a GM-CSF gene vaccine. Expert Rev Vaccines 4(3):259–274
270. Higano CS, Corman JM, Smith DC, Centeno AS, Steidle CP, Gittleman M et al (2008) Phase 1/2 dose-escalation study of a GM-CSF-secreting, allogeneic, cellular immunotherapy for metastatic hormone-refractory prostate cancer. Cancer 113(5):975–984
271. Vuky J, Corman JM, Porter C, Olgac S, Auerbach E, Dahl K (2013) Phase II trial of neoadjuvant docetaxel and CG1940/CG8711 followed by radical prostatectomy in patients with high-risk clinically localized prostate cancer. Oncologist 18(6):687–688
272. Comiskey MC, Dallos MC, Drake CG (2018) Immunotherapy in prostate cancer: teaching an old dog new tricks. Curr Oncol Rep 20(9):75
273. Sonpavde G, Slawin KM, Spencer DM, Levitt JM (2010) Emerging vaccine therapy approaches for prostate cancer. Rev Urol 12(1):25–34
274. Tsujikawa T, Crocenzi T, Durham JN, Sugar EA, Wu AA, Onners B et al (2020) Evaluation of cyclophosphamide/GVAX pancreas followed by listeria-mesothelin (CRS-207) with or without nivolumab in patients with pancreatic cancer. Clin Cancer Res 26(14):3578–3588
275. Laheru D, Lutz E, Burke J, Biedrzycki B, Solt S, Onners B et al (2008) Allogeneic granulocyte macrophage colony-stimulating factor-secreting tumor immunotherapy alone or in sequence with cyclophosphamide for metastatic pancreatic cancer: a pilot study of safety, feasibility, and immune activation. Clin Cancer Res 14(5):1455–1463
276. Salgia R, Lynch T, Skarin A, Lucca J, Lynch C, Jung K et al (2003) Vaccination with irradiated autologous tumor cells engineered to secrete granulocyte-macrophage colony-stimulating factor augments antitumor immunity in some patients with metastatic non-small-cell lung carcinoma. J Clin Oncol 21(4):624–630
277. Nemunaitis J, Jahan T, Ross H, Sterman D, Richards D, Fox B et al (2006) Phase 1/2 trial of autologous tumor mixed with an allogeneic GVAX vaccine in advanced-stage non-small-cell lung cancer. Cancer Gene Ther 13(6):555–562
278. Lipson EJ, Sharfman WH, Chen S, McMiller TL, Pritchard TS, Salas JT et al (2015) Safety and immunologic correlates of Melanoma GVAX, a GM-CSF secreting allogeneic melanoma cell vaccine administered in the adjuvant setting. J Transl Med 13:214
279. Nemunaitis J, Dillman RO, Schwarzenberger PO, Senzer N, Cunningham C, Cutler J et al (2006) Phase II study of belagenpumatucel-L, a transforming growth factor beta-2 antisense gene-modified allogeneic tumor cell vaccine in non-small-cell lung cancer. J Clin Oncol 24(29):4721–4730

280. Nemunaitis J, Nemunaitis M, Senzer N, Snitz P, Bedell C, Kumar P et al (2009) Phase II trial of Belagenpumatucel-L, a TGF-beta2 antisense gene modified allogeneic tumor vaccine in advanced non small cell lung cancer (NSCLC) patients. Cancer Gene Ther 16(8):620–624

281. Granucci F, Foti M, Ricciardi-Castagnoli P (2005) Dendritic cell biology. Adv Immunol 88:193–233

282. Nestle FO, Alijagic S, Gilliet M, Sun Y, Grabbe S, Dummer R et al (1998) Vaccination of melanoma patients with peptide- or tumor lysate-pulsed dendritic cells. Nat Med 4(3):328–332

283. Mukherji B, Chakraborty NG, Yamasaki S, Okino T, Yamase H, Sporn JR et al (1995) Induction of antigen-specific cytolytic T cells in situ in human melanoma by immunization with synthetic peptide-pulsed autologous antigen presenting cells. Proc Natl Acad Sci U S A 92(17):8078–8082

284. Butterfield LH, Ribas A, Dissette VB, Amarnani SN, Vu HT, Oseguera D et al (2003) Determinant spreading associated with clinical response in dendritic cell-based immunotherapy for malignant melanoma. Clin Cancer Res 9(3):998–1008

285. Santos PM, Butterfield LH (2018) Dendritic cell-based cancer vaccines. J Immunol 200(2):443–449

286. Mitchell DA, Batich KA, Gunn MD, Huang MN, Sanchez-Perez L, Nair SK et al (2015) Tetanus toxoid and CCL3 improve dendritic cell vaccines in mice and glioblastoma patients. Nature 519(7543):366–369

287. Hsu JL, Bryant CE, Papadimitrious MS, Kong B, Gasiorowski RE, Orellana D et al (2018) A blood dendritic cell vaccine for acute myeloid leukemia expands anti-tumor T cell responses at remission. Oncoimmunology 7(4):e1419114

288. Plosker GL (2011) Sipuleucel-T. Drugs 71(1):101–108

289. Fu C, Zhou L, Mi QS, Jiang A (2020) DC-based vaccines for cancer immunotherapy. Vaccines (Basel) 8(4)

290. Bol KF, Schreibelt G, Rabold K, Wculek SK, Schwarze JK, Dzionek A et al (2019) The clinical application of cancer immunotherapy based on naturally circulating dendritic cells. J Immunother Cancer 7(1):109

291. van Broekhoven CL, Parish CR, Demangel C, Britton WJ, Altin JG (2004) Targeting dendritic cells with antigen-containing liposomes: a highly effective procedure for induction of antitumor immunity and for tumor immunotherapy. Cancer Res 64(12):4357–4365

292. Gargett T, Abbas MN, Rolan P, Price JD, Gosling KM, Ferrante A et al (2018) Phase I trial of Lipovaxin-MM, a novel dendritic cell-targeted liposomal vaccine for malignant melanoma. Cancer Immunol Immunother 67(9):1461–1472

293. Wang Y, Xiang Y, Xin VW, Wang XW, Peng XC, Liu XQ et al (2020) Dendritic cell biology and its role in tumor immunotherapy. J Hematol Oncol 13(1):107

294. Randolph GJ, Angeli V, Swartz MA (2005) Dendritic-cell trafficking to lymph nodes through lymphatic vessels. Nat Rev Immunol 5(8):617–628

295. Steinman RM, Witmer MD (1978) Lymphoid dendritic cells are potent stimulators of the primary mixed leukocyte reaction in mice. Proc Natl Acad Sci U S A 75(10):5132–5136

296. Reis e Sousa C (2006) Dendritic cells in a mature age. Nat Rev Immunol 6(6):476–483

297. Fujii S, Liu K, Smith C, Bonito AJ, Steinman RM (2004) The linkage of innate to adaptive immunity via maturing dendritic cells in vivo requires CD40 ligation in addition to antigen presentation and CD80/86 costimulation. J Exp Med 199(12):1607–1618

298. Lapointe R, Bellemare-Pelletier A, Housseau F, Thibodeau J, Hwu P (2003) CD40-stimulated B lymphocytes pulsed with tumor antigens are effective antigen-presenting cells that can generate specific T cells. Cancer Res 63(11):2836–2843

299. Leko V, McDuffie LA, Zheng Z, Gartner JJ, Prickett TD, Apolo AB et al (2019) Identification of neoantigen-reactive tumor-infiltrating lymphocytes in primary bladder cancer. J Immunol 202(12):3458–3467

300. Wennhold K, Weber TM, Klein-Gonzalez N, Thelen M, Garcia-Marquez M, Chakupurakal G et al (2017) CD40-activated B cells induce anti-tumor immunity in vivo. Oncotarget 8(17):27740–27753

301. Wennhold K, Shimabukuro-Vornhagen A, von Bergwelt-Baildon M (2019) B cell-based cancer immunotherapy. Transfus Med Hemother 46(1):36–46
302. Li L, Petrovsky N (2016) Molecular mechanisms for enhanced DNA vaccine immunogenicity. Expert Rev Vaccines 15(3):313–329
303. Lundstrom K (2018) Viral vectors in gene therapy. Diseases 6(2)
304. Lopes A, Vandermeulen G, Preat V (2019) Cancer DNA vaccines: current preclinical and clinical developments and future perspectives. J Exp Clin Cancer Res 38(1):146
305. van den Berg JH, Oosterhuis K, Schumacher TN, Haanen JB, Bins AD (2014) Intradermal vaccination by DNA tattooing. Methods Mol Biol 1143:131–140
306. Bergmann-Leitner ES, Leitner WW (2015) Vaccination using gene-gun technology. Methods Mol Biol 1325:289–302
307. Ternette N, Tippler B, Uberla K, Grunwald T (2007) Immunogenicity and efficacy of codon optimized DNA vaccines encoding the F-protein of respiratory syncytial virus. Vaccine 25(41):7271–7279
308. Narum DL, Kumar S, Rogers WO, Fuhrmann SR, Liang H, Oakley M et al (2001) Codon optimization of gene fragments encoding Plasmodium falciparum merzoite proteins enhances DNA vaccine protein expression and immunogenicity in mice. Infect Immun 69(12):7250–7253
309. Stachyra A, Redkiewicz P, Kosson P, Protasiuk A, Gora-Sochacka A, Kudla G et al (2016) Codon optimization of antigen coding sequences improves the immune potential of DNA vaccines against avian influenza virus H5N1 in mice and chickens. Virol J 13(1):143
310. Lopes A, Vanvarenberg K, Preat V, Vandermeulen G (2017) Codon-Optimized P1A-encoding DNA vaccine: toward a therapeutic vaccination against P815 mastocytoma. Mol Ther Nucleic Acids 8:404–415
311. Lin CT, Tsai YC, He L, Calizo R, Chou HH, Chang TC et al (2006) A DNA vaccine encoding a codon-optimized human papillomavirus type 16 E6 gene enhances CTL response and anti-tumor activity. J Biomed Sci 13(4):481–488
312. Bae J, Prabhala R, Voskertchian A, Brown A, Maguire C, Richardson P et al (2015) A multiepitope of XBP1, CD138 and CS1 peptides induces myeloma-specific cytotoxic T lymphocytes in T cells of smoldering myeloma patients. Leukemia 29(1):218–229
313. Duperret EK, Perales-Puchalt A, Stoltz R, G HH, Mandloi N, Barlow J et al (2019) A synthetic DNA, multi-neoantigen vaccine drives predominately MHC Class I CD8(+) T-cell responses, impacting tumor challenge. Cancer Immunol Res 7(2):174–182
314. Li L, Zhang X, Wang X, Kim SW, Herndon JM, Becker-Hapak MK et al (2021) Optimized polyepitope neoantigen DNA vaccines elicit neoantigen-specific immune responses in preclinical models and in clinical translation. Genome Med 13(1):56
315. Finn OJ, Forni G (2002) Prophylactic cancer vaccines. Curr Opin Immunol 14(2):172–177
316. Ostroumov D, Fekete-Drimusz N, Saborowski M, Kuhnel F, Woller N (2018) CD4 and CD8 T lymphocyte interplay in controlling tumor growth. Cell Mol Life Sci 75(4):689–713
317. Aranda F, Vacchelli E, Eggermont A, Galon J, Sautes-Fridman C, Tartour E et al (2013) Trial watch: peptide vaccines in cancer therapy. Oncoimmunology 2(12):e26621
318. Kennedy R, Celis E (2006) T helper lymphocytes rescue CTL from activation-induced cell death. J Immunol 177(5):2862–2872
319. Kumai T, Lee S, Cho HI, Sultan H, Kobayashi H, Harabuchi Y et al (2017) Optimization of peptide vaccines to induce robust antitumor CD4 T-cell responses. Cancer Immunol Res 5(1):72–83
320. Slingluff CL Jr (2011) The present and future of peptide vaccines for cancer: single or multiple, long or short, alone or in combination? Cancer J 17(5):343–350
321. Noguchi M, Sasada T, Itoh K (2013) Personalized peptide vaccination: a new approach for advanced cancer as therapeutic cancer vaccine. Cancer Immunol Immunother 62(5):919–929
322. Tornesello AL, Tagliamonte M, Tornesello ML, Buonaguro FM, Buonaguro L (2020) Nanoparticles to improve the efficacy of peptide-based cancer vaccines. Cancers (Basel) 12(4)

Melanoma: An immunotherapy journey from bench to bedside

2

Vishal Navani, Moira C. Graves, Hiren Mandaliya,
Martin Hong, Andre van der Westhuizen, Jennifer Martin,
and Nikola A. Bowden

Acronyms

Confidence Intervals (CI) are 95% unless otherwise stated

AE	Adverse event
APC	Antigen-presenting cell

V. Navani (✉)
Tom Baker Cancer Centre, Calgary, AB, Canada
e-mail: Vishal.navani@albertahealthservices.ca

M. C. Graves · A. van der Westhuizen · J. Martin · N. A. Bowden
Centre for Drug Repurposing and Medicines Research, University of Newcastle and Hunter
Medical Research Institute, University Dr, Callaghan, NSW 2308, Australia
e-mail: Moira.graves@newcastle.edu.au

A. van der Westhuizen
e-mail: Andre.vanderwesthuizen@calvarymater.org.au

J. Martin
e-mail: Jenniferh.martin@newcastle.edu.au

N. A. Bowden
e-mail: nikola.bowden@newcastle.edu.au

H. Mandaliya · M. Hong · A. van der Westhuizen
Calvary Mater Hospital Newcastle, Edith St, Waratah, NSW 2298, Australia
e-mail: Hiren.mandaliya@calvarymater.org.au

M. Hong
e-mail: Martin.hong@calvarymater.org.au

J. Martin
John Hunter Hospital, Newcastle, NSW, Australia

AUROC	Area under receiver operating characteristic
B2M	Beta 2 microglobulin
BOR	Best overall response
BRAFi	BRAF inhibitor
BSLD	Baseline sum of long diameters
CD4	Cluster of differentiation 4
CI	Confidence interval
CRS	Cytokine release syndrome
ctDNA	Circulating tumor DNA
CTL	Cytotoxic T lymphocyte
CTLA-4	Cytotoxic T lymphocyte associated antigen - 4
DCs	Dendritic cells
DMFS	Distant-metastases free survival
DNMT1	DNA methytransferase 1
DoR	Duration of response
EGF	Epidermal growth factor
FACS	Flow cytometry
FDA	United States Food & Drug Administration
FGFR	Fibroblast growth factor receptor
G3	Grade 3
G4	Grade 4
GEP	Gene expression profile
PTEN	Phosphatase and tensin homolog
MDSCs	Myeloid-derived suppressor cells
TCGA	The cancer genome atlas
LOH	Loss of heterozygosity
TMB	Tumor mutational burden
NGS	Next-generation sequencing
GM-CSF	Granulocyte–macrophage colony-stimulating factor
HDAC(S)	Histone deacetylase (s)
HLA	Human leukocyte antigen
HR	Hazard ratio
ICB	Immune checkpoint blockade
IC	Investigator's choice
ICC	Investigator's choice chemotherapy
ICOS	Inducible T-cell co-stimulatory
IFN-α	Interferon-alpha
IGF-I	Insulin-like growth factor 1
IHC	Immunohistochemistry
IPRES	Innate anti-PD-1 resistance signatures
IL-1	Interleukin 1
IL-10	Interleukin 10
IL-2	Interleukin 2
IL-2Rα	Interleukin 2 alpha receptor
IL-6	Interleukin 6

IL-8	Interleukin 8
ILC2	Innate lymphoid cells
irRECIST	Immune-related response evaluation criteria in solid tumors
L-35	Interleukin 35
LAG-3	Lymphocyte activation gene 3
LDH	Lactate dehydrogenase
M1	Macrophage 1
M2	Macrophage 2
mAB	Monoclonal antibody
MDS	Myelodysplastic syndrome
MDSC	Myeloid-derived suppressor cells
MEKi	MEK inhibitor
MGSA/GRO	Melanoma growth-stimulatory activity
MHC	Major histocompatability complex
ORR	Objective response rate
MHC	Major histocompatibility complex
mNSCLC	Metastatic non-small cell lung cancer
mOS	Median overall survival
mPFS	Median progression-free survival
MTD	Maximum tolerated dose
NDA	New drug application
NGF	Nerve growth factor
NK	Natural killer cells
NR	Not reached
RECIST	Response evaluation criteria in solid tumors
NSCLC	Non-small cell lung cancer
ORR	Objective response rate
OS	Overall survival
SOC	Standard of care
TCR	T-cell receptor
pCR	Pathological complete response
PD	Pharmacodynamics
PFS	Progression-free survival
PD	Progressive disease
SD	Stable disease
PR	Partial response
CR	Complete response
PD-1	Programmed cell death protein 1
PD-L1	Programmed death ligand 1
PD-L2	Programmed death ligand 2
PDGF	Platelet derived growth factor
PDGF-A -	Platelet derived growth factor A,
PDGFRα	Platelet derived growth factor receptor alpha
PK	Pharmacokinetcs
PTM	Post-translational modifications

RFS	Recurrence-free survival
T $_{regs}$	Regulatory T-cells
TCR	T-cell receptor
TFS	Treatment-free Survival
TGF-β	Transforming growth factor beta
Th1	T helper type 1
Th2	T helper type 2
TIGIT	T-cell immunoreceptor with Ig and ITIM domains
TIL	Tumor infiltrating lymphocytes
TIM3	T-cell immunoglobulin and mucin domain-containing protein 3
TMD	Therapeutic drug monitoring
TME	Tumor microenvironment
TAM	Tumor-associated macrophage
UM	Uveal melanoma
TME	Tumor microenvironment
TT	Targeted therapy
ULN	Upper limit of normal
UV	Ultraviolet
VEGF	Vascular endothelial growth factor
VEGFR	Vascular endothelial growth factor receptor
βFGF	Basic fibroblast growth factor

2.1   Introduction

Melanoma gave science a window into the role immune evasion plays in the development of malignancy. The entire spectrum of immune focused anti-cancer therapies has been subjected to clinical trials in this disease, with limited success until the immune checkpoint blockade era. That revolution launched first in melanoma, heralded a landscape change throughout cancer that continues to reverberate today.

We aim, in this chapter, to outline the pre-clinical developments that permitted the development of immune checkpoint-focused therapies and subsequently their role across all stages of the disease. We describe the practice changing pivotal data that accompanied these drugs and the rapidly evolving regulatory and reimbursement environment that was required to facilitate their quick deployment from bench to bedside.

This is followed by a focus on the role of immune therapies in non-cutaneous subtypes and our evolving understanding of the factors driving response and resistance to immune checkpoint blockade. We review the impact of underlying genomic, transcriptomic, epigenetic, proteomic and immunological correlates alongside their interaction with diverse patient phenotypes in order to understand the variability in outcomes seen with immune checkpoint blockade. There is a focus on the underlying pharmacology of these agents and the resultant impact on current dosing regimens. We explore the role of predictive and prognostic biomarkers in this space.

We conclude with insights into immune-based therapies beyond the checkpoint, including harnessing the innate immune system, combinations that target aberrant oncogenic signaling pathways, DNA-damaging agents, epigenetic approaches, and cellular therapies.

2.2 Immune Hallmarks of Melanoma

A combination of the observed phenomena of spontaneous remission [1], a brisk lymphocytic infiltrate in primary and metastatic disease alongside a UV-induced tumoral mutational burden have long highlighted melanoma as an immunogenic malignancy. Clemente [2] and colleagues demonstrated over 25 years ago that TILs were a positive prognostic factor for melanoma. It is now well documented that density, localization and composition of TILs in melanoma influence risk of metastasis and survival. Despite this, lymph node metastasis and Breslow thickness remain the most utilized prognostic factors for resectable melanoma, with "M" stage and LDH remaining critical in advanced disease [3–5].

Melanoma development and metastasis occur when tumorigenic cells interact with components of the immune system that normally prevent cancer progression [6]. Melanoma cells invade localized and distant tissue by interacting with the TME and suppressing the tumor immune response. Melanoma cells overcome the carefully controlled production and release of growth-promoting signals, which normally ensure homeostasis and maintenance of normal tissue structure and function, to direct their own cellular destinies [7]. These signals are mostly conducted by growth factors (typically containing intracellular tyrosine kinase domains) which bind to cell surface receptors. These tyrosine kinase growth factors regulate cell cycle progression, cell growth as well as other cell properties such as cell survival and energy metabolism. Autocrine growth factors (bFGF, MGSA/GRO, IL-8 and sometimes IL-6, PDGF-A, IL-10) produced by melanoma cells stimulate proliferation while paracrine growth factors (e.g., PDGF, EGF, TGF-β, IL-1, GM-CSF, IGF-I, NGF, VEGF) modulate the microenvironment to the benefit of tumor growth and invasion [8]. Melanoma cells also send PDGF to the TME to potentiate growth and invasion [9].

The melanoma microenvironment contains many immune-suppressive immune cells: T_{regs}, MDSC and TAM. CD4 + regulatory T-cells function suppresses the over-reactive immune response to prevent it from damaging the host [10]. In melanoma, the number of T_{regs} is increased both in the peripheral blood, lymph nodes, and within the tumor. MDSC are formed by cancer cells transforming myeloid cells (from the bone marrow) and are critical in cancer progression. Their function is to reduce T-cell function in the tumor microenvironment by disrupting T-cell metabolic pathways, resulting in T-cell apoptosis [11]. MDSC can differentiate into TAMs and the TAM phenotype is dependent upon the tumor microenvironment. There are two TAM phenotypes, M1 and M2. The M1 phenotype occurs in normoxic conditions and is associated with anti-tumor effects. Whereas, the M2 phenotype TAMs are hypoxic driven and observed as melanoma progresses.

There are two subclasses of helper T-cells, Th1 and Th2, which are responsible for the M1 and M2 TAM polarization, respectively [12]. In melanoma, the Th2 subclass is dominant with systemic chronic inflammation and supports melanoma progression [13] by driving TAMs into an M2 phenotype. As the disease progresses, there is a further shift from M1 toward M2 phenotype supporting tumor growth and tumor immune evasion [14].

Melanoma also evades the immune system by establishing a process of immune editing. This occurs by selection of subclones based on their capability to evade immune detection—a process known as elimination, equilibrium, and escape [15]. During melanoma progression, there is a progressive loss of antigen presentation capacity to cytotoxic T-cells by dendritic cells [16]. In the elimination phase, the dendritic cells detect immunogenic melanoma clones and capture these melanoma antigens. The melanoma antigens are then processed by the dendritic cells and presented by the MHC II on the outside of dendritic cells to naïve T-cells in the lymph nodes. Cytotoxic CD8 + T-cells, which are melanoma-specific, are produced from this antigen exposure and are able to eliminate transformed malignant melanoma cells. During the next equilibrium step, non-immunogenic clonal populations achieve a selection advantage and co-exist with the immune response in a dynamic equilibrium [17]. This phase can last for a significant period of time, with tumor cells undergoing a constant state of editing to achieve mmune-quiescence. However, while the immune system may eliminate the majority of the highly immunogenic melanoma clones, some that are irrelevant in the eyes of the adaptive immune response escape, rapidly proliferating and disseminating throughout the body, termed the escape phase.

The final immune feature of melanoma is the high TMB driven predominantly by UV exposure [18, 19]. Both intermittent and chronic sun exposure increases the UV-induced mutation burden and leads to local and systemic immune system suppression. UV-induced mutations are presented as neo-antigens by major histocompatibility complex MHC proteins to T-cells. To evade immune eradication, melanoma uses immune checkpoints to dampen T-cell reactivity.

2.2.1 Pre-clinical Rationale for Checkpoint Blockade

Melanoma can evade the body's defenses by manipulating intrinsic and extrinsic biological pathways. These evasive manipulations were classified by Hanahan and Weinberg [20] into eight biological components: sustaining proliferative signaling, evading growth suppressors, resisting cell death, enabling replicative immortality, inducing angiogenesis, activating invasion and metastasis, reprogramming of energy metabolism and evading immune destruction.

Checkpoint immune therapies target the molecules involved in the regulation of T-cells to block melanoma immune evasion. The PD-1/PD-L1 axis regulates the immune response and protects the host against autoimmunity [21, 22]. Cells in healthy tissue express PD-L1, which interacts with invading cytotoxic T-cells to prevent further T-cell activation and avoid cell damage from reactivity against self-antigens. The PD-1/PD-L1 axis is hijacked by melanoma cells to escape the immune system. Melanoma upregulates the expression of PD-L1 and PD-1 expressing T-cells interact with the PD-L1 tumor cell leading to T-cell exhaustion [23]. PD-L2 is the second ligand for PD-1 and is expressed by melanoma and antigen-presenting cells and is a negative regulator of cytotoxic T-cell activity [24]. In human melanomas, PD-L2 is more abundant than PD-L1, with PD-L2 having a greater affinity for PD-1. This suggests a differential contribution by PD-L1/PD-L2 in regulating immune response [25]. Clinical efficacy of PD-1 pathway inhibitors relies on the reactivation of endogenous tumor-specific immunity and on the priming of distinct TCR repertoire in lymphoid organs [26]. Yost [26] showed that pre-existing tumor-specific T-cells may have limited reinvigoration capacity and the T-cell response to checkpoint blockade derives from a distinct repertoire of T-cell clones that may have recently entered the tumor.

CTLA-4 is a negative regulator of the immune system. CTLA-4 inhibits T-cell activation by outcompeting CD28 for the ligands CD80/CD86. T-cell activation is inhibited via a tolerance mechanism where the lymphocyte is intrinsically functionally inactivated but remains alive for an extended period in a hyporesponsive state [27]. Within the lymph nodes, two signals are required for adequate T-cell activation—TCR interaction with the MHC/peptide complex (on dendritic cells) and the secondary signal of CD28 on T-cells, binding with CD80/CD86 on dendritic cells [28]. CTLA4 outcompetes CD28 for CD80/CD86 binding, inhibiting the downstream TCR signaling and impeding anti-tumor CD8+ T-cell function.

Anti-CTLA4 (ipilimumab) overcomes T-cell anergy within lymph nodes and allows anti-tumor T-cell cytotoxicity [28]. Ipilimumab binds to the CTLA4 on T-cells, preventing CTLA4 binding to CD80/CD86, allowing the expansion of antigen-specific, anti-tumor cytotoxic CD8 + T-cells and CD4 + T-cells [29, 30]. The role of PD-1, PD-L1 and CTLA-4 not only includes negative regulators of the immune system but might be responsible for tumor intrinsic cell proliferation, survival [31], and metastatic signals [32].

2.3 Immunotherapy in Metastatic Cutaneous Melanoma

2.3.1 The Development of Immunotherapy in Advanced Cutaneous Melanoma

2.3.1.1 IL-2

An understanding of the immunogenicity of melanoma and the dense lymphocytic infiltrate often seen pathologically led to the T-cell growth cytokine IL-2, to be used in advanced disease. Durable responses were restricted to only the

6% (17/270) of patients that experienced CR [33]. Ubiquitous expression of the IL-2Rα on Treg cells paradoxically restrains the intended anti-tumoral immune response when exposed to exogenous IL-2, hampering efficacy in melanoma [34]. Restricted efficacy, concurrent with significant toxicities such as vascular leak syndrome and hypotension hampered the utility of this immunotherapy.

2.3.1.2 ICB

A decade of unprecedented advances in the treatment landscape across advanced solid organ tumors, the ICB era, was ushered in with the 2010, MDX-010–20 study. Patients with treatment-refractory advanced melanoma were randomized to four cycles (induction) of ipilimumab, an anti-CTLA-4 mAb ± peptide vaccine gp100 or gp100. A small improvement in mOS (10 vs. 6.4 months HR 0.68 p < 0.001) belied the real benefit, namely a prolonged OS in a proportion of patients that represented the tail of the survival curve, with 23.5% of patients alive at 2 years (cf 13.7% with gp100 alone) Fig. 2.1 [35].

KEYNOTE-001, a phase I study heralded the first FDA-approved anti-PD-1 mAb, pembrolizumab. The adaptive study design facilitated rapid clinical development, with numerous expansion cohorts across tumor streams gaining approval in metastatic melanoma and NSCLC. Durable responses across a wide dose range of 2 mg/kg Q3W–10 mg/kg Q3W provided an ORR of 26% [36], with mOS

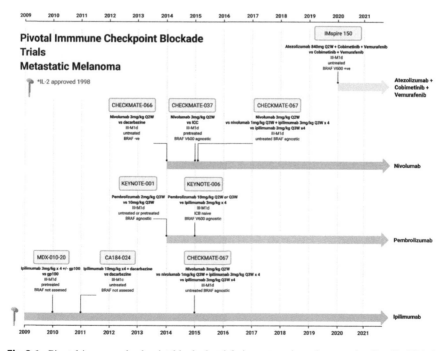

Fig. 2.1 Pivotal immune checkpoint blockade trials in metastatic melanoma timeline By Vishal Navani, made with BioRender

23.8 months in a cohort containing treatment naïve and ipilimumab experienced patients, with 34% achieving 5-year survival [37]. KEYNOTE-006 moved the use of pembrolizumab to the front-line setting, testing the higher dose 10 mg/kg against ipilimumab in ICB naïve patients. Overwhelming efficacy at the second interim analysis led the data and safety monitoring committee to unblind the study results and make pembrolizumab available to all patients with progression on ipilimumab, 1Y OS Q3W pembrolizumab 68.4% versus ipilimumab 58.2% HR 0.69 CI 0.52–0.90 $p = 0.0036$ [38]. In an exploratory subsequent analysis, these responses proved deep and durable with mOS in the pembrolizumab arm 32.7 vesus 15.9 months ipilimumab HR 0.79 $p = 0.00049$ [39].

Nivolumab, another anti-PD-1 mAb established efficacy at 3 mg/kg Q2W in randomized fashion against the previous standard of care treatment in melanoma, dacarbazine in CHECKMATE 066, ORR 40% versus 14% with overwhelming efficacy again prompting study unblinding early and crossover to ICB upon progression [40], mOS 37.5 months versus 11.2 (HR 0.46 $p < 0.01$) [41]. Traditional endpoints such as mPFS did not capture the benefit seen with nivolumab in CHECKMATE-037, which randomized patients that had progressed with ipilimumab to nivolumab 3 mg/kg Q2W (mPFS 4.7 months CI 2.3–6.5) versus ICC (mPFS 4.2 months CI 2.1–6.3 HR 0.82 descriptive) [42]. This was likely due to significant prognostic imbalances between groups favoring ICC such as 19 versus 14% M1d disease and 51 versus 35% baseline LDH > ULN, alongside recognition that tumor response kinetics with ICB could skew BOR toward pseudoprogression before exhibiting durable response. The impact of crossover and subsequent therapies (44% ICC arm receiving subsequent PD-1 vs. 11% nivolumab arm) on hard endpoints such as OS was also illustrated here, with similar mOS (15.7 months vs. 14.4 h 0.95 CI 0.73–1.23) influenced by increased drop-out in the ICC group when randomization informed (23% vs. 1%). Median duration of response was notably higher for nivolumab (32 vs. 13 months), keeping with the pattern seen in those responding to ICB [43].

Synergistic dual targeting of CTLA-4 and PD-1 with doublet ipilimumab and nivolumab induction, followed by maintenance nivolumab met co-primary endpoints confirming benefit in PFS over ipilimumab alone in CHECKMATE-067 (median PFS 11.5 CI 8.9–16.7 months vs. 2.9 CI 2.8–3.4 h 0.42 $p < 0.001$) [44]. Extended follow-up data met the pre-specified OS endpoint over ipilumumab, median NR CI 38.2-NR versus 19.9 months CI 16.8–24.6 h 0.52 $p < 0.001$ [45], with 52% of patients in the doublet arm alive at 5 years. Descriptive statistics, due to a lack of formal pre-specified power to compare the doublet ICB and single-agent nivolumab arm, found consistent improvement in mPFS (11.5 vs. 6.9 months HR 0.74 CI 0.6–0.92), with this direction of benefit maintained with mOS (72.1 CI 38.2-NR vs. 36.9 months CI 28.2-NR) [46]. Figure 2.2 outlines a cross-trial comparison of overall survival curves from the pivotal trials in metastatic melanoma.

Increased CD4+ & CD8+T lymphocyte infiltration [47] and upregulated melanoma antigen mRNA levels [48] during BRAFi/MEKi therapy gave translational merit to conducting a combination trial of anti-PD-L1 mAb atezolizumab

Overall Survival in Advanced Melanoma

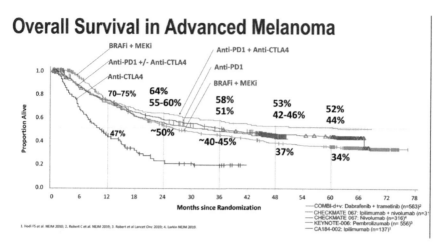

Fig. 2.2 Cross-trial comparison of overall survival. Reprinted by permission from Prof. Georgina Long: ASCO21 Educational Session—On the Shoulders of Giants, Journey to Success in Managing Melanoma, ASCO 2021

in combination with BRAFi vemurafenib/MEKi cobimetinib versus BRAFi/MEKi alone, IMspire 150. At the first published interim analysis, a mPFS benefit was noted (15.1 months vs. 10.6 h 0.78 CI 0.63–0.97 $p = 0.025$) albeit at a high cost of $\geq$ Grade 3 AE (79% vs. 73%), with similar ORR (66.3% vs. 65%) and no OS signal at this early follow-up (HR 0.83 CI 0.64–1.11 $p = 0.23$) [49]. PFS curves only began to separate after 7 months, invalidating traditional Cox-proportional hazards, but proving significant with a Flemington-Harrington approach, which adds weighting to late events more heavily. Similar studies evaluating triplet combination therapies have missed the PFS endpoint, such as COMBI-I, spartalizumab + dabrafenib/trametinib versus dabrafeninb/trametinib (HR 0.82 CI 0.655–1.027 $p = 0.042$), despite the numerically similar mPFS in the triplet of 16.2 months and improved landmark 24 months PFS of 50% [50] (35% estimated atezolizumab + cobimetinib/vemurafenib), suggesting statistical design, baseline prognostic characteristics and performance of the control arm as potentially impacting the positive result of Imspire 150.

2.3.1.3 The Evolving Supporting Landscape

Shortening of the clinical development time for ICB, from investigational NDA to market, was achieved by the breakthrough therapy designation for many ICB provided by regulatory authorities. Pembrolizumab was the first oncological therapy granted this status, allowing expedited development via rolling review, based on preliminary phase I evidence characterized by impactful ORR and DoR. Accelerated approval allowing a route to patients was subsequently granted, since meaningful clinical benefit was likely with these preliminary surrogate endpoints, prior to the confirmatory KEYNOTE-006.

Alongside regulatory innovation, novel imaging criteria were created to account for the nonclassical tumor kinetics seen with ICB response, such as pseudoprogression, delayed response, and mixed response prior to durable benefit [51]. Pseudoprogression, more commonly seen with ipilimumab, arises from CD4 + & CD8+ T-cell infiltration into sites of metastases [52] leading to tumor size increases often meeting traditional RECIST [53] defined progressive disease. Novel criteria, such as irRECIST, introducing the concept of unconfirmed progressive disease, requiring subsequent confirmation, facilitates treatment beyond progressing and an appreciation of novel patterns of response [54].

2.3.1.4 Advanced Cutaneous Melanoma in the Clinic
The advances in our understanding of the genomic basis of melanoma facilitated development of BRAFi/MEKi combinations as targeted therapy (TT) for patients with mutations in BRAF at the 600th codon (V600), with clear improvement in OS [55–57]. The ability to gain rapid, but limited control for the majority of eligible patients with targeted therapy contrasted with the delayed and durable benefit for a proportion receiving ICB. Prospective data comparing these approaches is best informed by SECOMBIT, which randomized patients to TT (encorafenib/binimetinib) versus doublet ICB versus TT induction followed by doublet ICB, with patients receiving the alternate regimen on progression. The trial showed similar 2-year PFS across all arms, with median 1st PFS mirroring trial data and OS results immature. Responses in both initial TT arms were limited, with only 30% durable to 2 years as opposed to the 70% seen with up front doublet ICB [58]. Real-world data identifying outcomes in patients with M1d disease progressing on TT found significantly poorer than expected outcomes with subsequent ICB therapy, with mOS of doublet ICB only 7.2 months (CI 5.2–9.1) [59]. Though this retrospective data in patients with poor prognosis disease should be taken with caution, it may inform practical treatment decision making. The randomized phase III DREAMseq is eagerly awaited [60]. If clinicians have the luxury of allowing early progression, as seen with inferior disease control rates with ICB, the propensity for long durable responses are attractive. Consensus guidelines advise consideration on ECOG, sites of disease, tumor progression kinetics, LDH, patient preference, and short versus long-term treatment goals in order to tailor treatment decision making [61].

The availability of long-term survival data has allowed inferences to be made regarding therapy choice despite the fraught nature of cross-trial comparisons. PFS curves, unadulterated by subsequent therapies may help define contribution to the picture revealed from OS analysis. In the BRAF +ve population of CHECKMATE-067, a PFS plateau at 3 years in the ipilimumab/nivolumab arm of 40% was maintained to 38% at 5 years, with a 32% to 29% plateau for nivolumab alone over the same timeframe [62].

No similar plateau is seen for dabrafenib/trametinib (3Y PFS 22%—5Y PFS 17%) [63]. Differences in baseline patient characteristics across these trials may contribute partly to these differences.

The potential for a persisting immune-mediated anti-tumor response from ICB is evidenced by a novel surrogate endpoint, TFS (treatment-free survival). This is defined by the area under the KM curve for two traditional endpoints, time to ICB cessation and time to next systemic therapy or death/censoring, for the entire randomized population. Analysis of this endpoint from 5 year CHECKMATE-067 follow-up data showed that patients initially randomized to doublet ICB alive at this timepoint, with the inherent immortal time bias of such a data set, spent 33% of that time free of any systemic treatment and alive (restricted mean TFS 19.7 months) (cf 17% with nivolumab restricted mean 9.9 months and 20% ipili-mumab restricted mean 11.9 months) [64]. Details of duration of a treatment-free interval for BRAF/MEKi pivotal trials are not available, with real-world case series suggesting a median time to next treatment ranging from only 2.5–6.6 months [65]. The undoubted patient and clinician preference for a clinical state alive without any systemic treatment are compelling.

Within ICB options, treatment selection between single-agent anti-PD-1 and doublet ICB approaches has to balance a higher immune-related toxicity and treat-ment discontinuation profile in the doublet arm, with an 8% overall survival benefit in absolute terms (52% vs. 44%) and the aforementioned absence of statistical power in the pivotal study to test for a difference between nivolumab contain-ing groups [45]. Favorable prognostic factors for a durable survival outcome with single-agent PD-1; M1b disease, normal LDH, and baseline sum of tumor dimen-sions under 102 mm [66, 67] may suggest nivolumab or pembrolizumab in this context. Alternatively, presence of disease in sites of immune privilege, e.g., active M1c/d disease, may prompt doublet ICB due to improved ORR and landmark OS outcomes: 63 vs. 51% 2Y OS [68].

Integrating the triplet atezolizumab + cobimetinib/vemurafenib approach into the current treatment paradigm is made challenging due to the rapidity of clinical development in this tumor and evolution of appropriate control arms. Arguably a suitable comparator now would involve anti-PD-1 ± anti-CTLA-4 and given the low likelihood of such a trial eventuating, cross-trial comparisons are again a necessity. A clinically meaningful OS signal, in the subgroup of patients likely to be most benefited by this all up-front approach, i.e., younger BRAF V600 + ve patients with M1c/d disease, high baseline LDH and numbers of metastatic sites, is necessary before incorporation into clinical practice routinely.

2.4   Adjuvant and Neoadjuvant Immune Checkpoint Inhibitor Therapy

2.4.1   Adjuvant Immune Checkpoint Blockade

With the transformational outcomes of anti-PD-1 and anti-CTLA4 treatment in the metastatic setting, efforts have been made to bring these treatments to earlier stages of cutaneous melanoma. In the setting of other tumor streams, adjuvant treatment is a commonly used strategy for high-risk resected disease to improve survival. Efforts to find effective treatments for high-risk, resected melanoma have been attempted for many years, with many strategies such as conventional chemotherapy, local therapy, isolated limb perfusion therapy, all not demonstrating any improvement in outcomes [69]. The first strategy to show some promise was high-dose IFN-α therapy, which had shown a recurrence-free and overall survival benefit at the time of the original paper, but has since shown to have only recurrence-free survival benefit [70, 71]. This loss of an overall survival benefit has been attributed to non-melanoma-related deaths in patients with the longer follow-up, given the lack of data on cause of death, a melanoma-specific survival could not be ascertained [71]. An individual patient-level data meta-analysis of all IFN-α trials showed only modest absolute RFS and overall survival (OS) benefit of approximately 3% at 10 years [72]. IFN-α therapy comes with significant toxicities, with 67% of patients in the trial suffering from Grade 3 toxicity, including hepatitis, myelosuppression, constitutional and neuropsychiatric symptoms, and in the E1684 trial, there were two deaths associated with hepatotoxicity. Thus, despite its FDA approval, it has not been considered standard of care.

Given the ongoing need for adjuvant treatment in the high-risk melanoma population, ICB was moved earlier into the treatment paradigm. Since 2015, there have been three positive pivotal trials in the adjuvant setting: EORTC 18,071, CheckMate 238 and KEYNOTE-054 looking at the use of ipilimumab, nivolumab, and pembrolizumab, respectively. EORTC 18,071 was the first trial looking at ICB treatment in the adjuvant setting. It was a double-blind placebo-controlled trial investigating the use of ipilimumab in patients with resected stage III cutaneous melanoma. Patients were randomly assigned to receive ipiliumumab 10 mg/kg or placebo three weekly for four cycles, then every 3 months for a maximum of 3 years. Between 2008 and 2011, 951 patients were assigned to receive ipilimumab or placebo, and the trial showed a statistically significant median RFS benefit of 26.1 months versus 17.1 months HR 0.75 (CI, 0.64–0.90; $p = 0.0013$) [73]. A subsequent 7-year follow-up results showed a persistent RFS benefit that translated into an OS benefit, with a hazard ratio of 0.73 (CI, 0.60–0.89; $p = 0.002$) and a 7-year OS rate of 60% versus 51.3% [74]. Furthermore, there was also a reduction in the distant-metastasis-free survival of 44.5% in the ipilimumab group versus 36.9% in the placebo group at 7 years. Notably, there was a significant proportion of people with toxicity from ipilimumab, with 54% of the patients who received ipilimumab experiencing Grade 3 or 4 toxicities. Similar to the metastatic setting, the most common Grade 3 or 4 immune-related toxicities were gastrointestinal, endocrine, and hepatic events.

CheckMate-238 was a randomized, active-controlled double-blind, phase 3 trial in patients with resected stage IIIB, IIIC, and IV which compared adjuvant nivolumab 3 mg/kg every 2 weeks to ipilimumab 10 mg/kg, which was given three weekly for four doses, then every 12 weeks for up to 1 year [75, 76]. 906 patients were randomized, and at a minimum follow-up of 48 months, the 4-year recurrence-free survival was 51.7% (CI 46.8–56.3) in the nivolumab group, compared to 41.2% (CI 36.4–45.9) in the ipilimumab group [76]. At the time of the final recurrence-free survival analysis, the overall survival data were immature but reported 4-year overall survival was 77.9% for nivolumab compared to 76.6% for ipilimumab. Furthermore, the 4-year distant-metastasis-free survival (DMFS) was 59.2% (CI 53.7–64.2) with nivolumab and 53.3% (CI 47.7–58.5) with ipilimumab. Importantly, the rates of toxicities were significantly better with nivolumab compared to ipilimumab, with rates of treatment-related Grade 3 or 4 toxicities occurring in 14.4% in patients receiving nivolumab, in contrast to 45.9% in the ipilimumab group [75].

Pembrolizumab was tested in the KEYNOTE-054 trial, which was a placebo-controlled, phase 3, double-blind trial of pembrolizumab 200 mg every 3 weeks for 18 doses or placebo for patients with resected stage IIIA, B, or C melanoma as per AJCC 7th edition [77–79]. 1019 patients were randomized 1:1 to receive pembrolizumab or placebo, and at a median follow-up of approximately 36 months in both groups, the 3.5-year recurrence-free survival was 59.8% (CI 55.3–64.1) in the pembrolizumab group in contrast to 41.4% (CI 37.0–45.8) in the placebo [79]. Similar to the CheckMate 238 trial, the overall survival data is immature. Pembrolizumab also reduced the occurrence of distant metastases, with a 3.5-year distant-metastasis-free survival of 65.3% (CI 60.9 – 69.5) versus 49.4% (CI 44.8 - 53.8) for pembrolizumab and placebo, respectively. Rates of Grade 3 or higher treatment-related adverse events similar to nivolumab seen in CheckMate 238.

The anti-PD-1 therapies have been approved for use in the adjuvant setting with the current information on the benefits on RFS and DMFS awaiting a confirmatory OS advantage. However, these therapies clearly demonstrate a reduction in the occurrence of distant metastases, which are a common first site of recurrence and can have a devastating impact on a person's quality and quantity of life. Furthermore, given the significance of patients who received subsequent ICB therapy in the control arms for both trials (57% in the ipilimumab group in CheckMate 238 and nearly 70% in the placebo arm in KEYNOTE-054), the overall survival data may not prove to be significant in the absence of information regarding sequencing of ICB therapy in melanoma. However, it is important to highlight that although RFS has been validated as a surrogate marker for OS in IFN-α treatment, it has not yet been validated in the ICB setting.

Combination ipilimumab and nivolumab have shown to be a highly effective strategy in the metastatic setting for high-risk patients, especially in those with brain metastases, showing greater response rates and overall survival than single-agent treatment. This approach was tested in the adjuvant setting with the CheckMate 915 trial. This trial randomly assigned 1844 patients with resected stage IIIB—IIID in a 1:1 ratio to receive either nivolumab 240 mg every 2 weeks with ipilimumab 1 mg/kg every 6 weeks versus nivolumab 480 mg every 4 weeks. The 2-year RFS in this study was 64.6% (CI, 61.3–67.7) compared to 63.2% (CI 59.9–66.4), and thus did not meet statistical significance [80]. Notably, those who received combination therapy had a shorter median duration of therapy compared to nivolumab alone, at 7.6 months compared to 11.1 months, which led to a lower cumulative nivolumab dose. Also, the incidence of Grade 3/4 toxicities was significant in the combination group, at 33% compared to 13% in the single-agent group. This study demonstrated that single-agent treatment is the standard of care in the adjuvant setting which balances the benefit of risk reduction against the toxicities.

2.4.2 Neoadjuvant Immune Checkpoint Blockade

Given the promising results of ICB in the adjuvant setting as described previously, efforts have been made to test ICB in the neoadjuvant setting. The rationale is that this will allow for assessment of the response to ICB of the disease, potentially reduce the tumor burden pre-operatively, and theoretically, the presence of the melanoma may enhance the T-cell responses to melanoma in the adjuvant setting given it has had the opportunity to respond to the melanoma while it is in situ [81]. 3 trials were published in 2018 looking at the safety and efficacy of ICB in the neoadjuvant setting for patients with resectable melanoma.

The first was from Amaria et al., who assessed the efficacy of either nivolumab (3 mg/kg fortnightly for four cycles) or the combination of ipilimumab/nivolumab (ipi 3 mg/kg; nivo 1 mg/kg three weekly for three cycles) in patients with AJCC 7th edition clinical stage III or oligometastatic, resect stage IV patients. All patients underwent surgery if there were no contraindications to surgery, then went on to receive 13 doses of nivolumab in the adjuvant setting. Twenty-three patients were randomized, 12 patients were randomized to the nivolumab alone group, and 11 patients received combination therapy. This trial was stopped early due to safety concerns regarding disease progression in single-agent treatment and the high rates of Grade 3 toxicities in the combination therapy. Treatment with nivolumab alone resulted in a pCR rate of 25%, compared to 45% in the combination ICB group [82]. Furthermore, although not statistically significant, there was improved survival in the combination ICB group compared to the single-agent group. Toxicity was significant with the combination group, with 73% of the patients having a Grade 3 toxicity, compared to only 8% in the single-agent group. The OpACIN trial was a Phase 1b trial assessing the feasibility of combination ICB in the adjuvant and neoadjuvant setting. Twenty patients with palpable stage III melanoma (i.e., IIIB and IIIC as per AJCC 7th) were randomized 1:1 to either receive 4 cycles of ipilimumab (3 mg/kg) or nivolumab (1 mg/kg) every 3 weeks all in an adjuvant fashion or two cycles neoadjuvant followed by two cycles of adjuvant. 78% of patients in the neoadjuvant arm had meaningful pathological responses to ICB (33% complete response, 33% near complete response, and 12% partial response) [81]. However, the median number of cycles received was two, with only one patient in each group receiving all four planned cycles. Most of the patients stopped treatment due to Grade 3/4 adverse events, the incidence of which was 90%, much greater than observed in the metastatic setting.

The OpACIN trial was followed up by the OpACIN-neo trial to assess whether a different regimen of anti-CTLA-4 and anti-PD-1 therapy may yield similar responses, but with reduced toxicity. Eighty-nine patients with histologically or cytologically confirmed stage III melanoma with macroscopic lymph node metastases were randomly assigned 1:1:1 to receive either: ipilimumab 3 mg/kg concurrently with nivolumab 1 mg/kg every 3 weeks (IPI3/NIVO1); ipilimumab 1 mg/kg concurrently with nivolumab 3 mg/kg every 3 weeks(IPI1/NIVO3); or two cycles ipilimumab 3 mg/kg followed by two cycles of nivolumab 1 mg/kg given every 2 weeks (second cycle of ipilimumab overlapped with the first cycle of nivolumab, sequentially). The trial had an early interim analysis due to safety concern of the sequential group, and the incidence of Grade 3/4 immune-related adverse events were 40% in the IPI3/NIVO1 group, and 50% in the sequential group, with one patient requiring a colectomy for colitis. This was in comparison to only 20% in the IPI1/NIVO3 group, which had no incidence of Grade 3/4 colitis. This reduction in toxicity did not affect the response rates, with 80% (CI, 61–92) in the IPI3/NIVO1 group having a pathological response, compared to 77% (CI, 58–90) of those who received IPI1/NIVO3 [83]. Notably, none of the patients who had a pathological response had relapsed at time of data analysis, with a minimum follow-up of 5.6 months. The results of this trial have sparked further efforts to assess viability of the neoadjuvant strategy in a larger group of patients in the phase III NADINA trial. In order to optimize surgical management, the PRADO trial looks to assess whether those who have had a pathological response to neoadjuvant therapy can be spared therapeutic lymph node dissection without detriment to survival and to give adjuvant therapy to those who have not had a response to neoadjuvant immunotherapy.

Adjuvant and now neoadjuvant immune checkpoint inhibitor therapy will likely change the paradigm of management of high-risk melanoma and keep more people free of metastatic disease. The key research area going forward, consistent with the metastatic setting, is to clearly define the patients who will benefit from ICB treatment in advance, in order to intensify or de-escalate approaches according to inherent risk.

2.5   Immunotherapy in Specific Contexts

2.5.1   Uveal Melanoma

UM arising from melanocytes in the pigmented components of the eye has been acknowledged to be immuno-quiescent. The molecular mechanisms underlying the immune-privileged site of the eye; upregulation of TGF-β induced CTLA-4 mediated APC tolerance [84], coupled with a CD8+ CTL inhibitory TME mediated by TAMs [85] at the liver (the most common site of metastatic disease) lead to these patients being excluded from most pivotal ICB trials. The low TMB of UM and resultant small number of neo-epitopes have limited efficacy of ICB, with an absence of robust randomized data for ICB use.

In this context, without a robust SOC, a bispecific artificial TCR fused to anti-CD3, tebentafusp, capable of redirecting T-cells independent of their native specificity, to melanocytes expressing target antigen gp100, showed improvement over IC therapy in the randomized phase III IMCgp100-202 trial in systemic therapy naïve patients Fig. 2.3. The control arm received primarily pembrolizumab

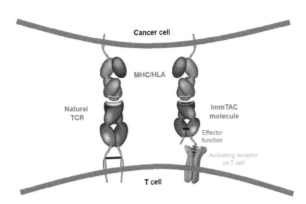

Fig. 2.3 ImmTAC (Immune mobilising monoclonal T-cell receptors against cancer) molecules are designed to mimic the natural immune synapse formed by the interaction of a TCR with a peptide human leukocyte antigen complex. The anti-CD3 effector function binds to CD3 (activating) receptors on T-cell surfaces, triggering T-cell-mediated cancer cell lysis. Reprinted by permission from **MDPI Basel** [Tebentafusp: T-cell Redirection for the Treatment of Metastatic Uveal Melanoma, Bertil E. Damato et al., Page 971, 2017].

or ipilimumab. Early separation of OS curves resulted in a stratified HR of 0.51 (CI 0.37–0.71) mOS 21.7 (CI 18.6–28.6) versus 16 (CI 9.7–18.4) $p < 0.0001$ in favor of the investigational agent. Consistent treatment effect was seen across all subgroups including those with poorer prognostic features such as M1c disease those with > ULN LDH. This benefit was seen without significant improvements in PFS and ORR, with OS benefit maintained in patients RECIST defined PD, with a landmark survival analysis of these patients at from 100 days revealing an HR of 0.40 (CI 0.248–0.642) [86]. This suggests the TCR therapeutic stabilizes disease after a period of initial ongoing tumor growth across metastatic sites, with many patients continuing on tebentafusp beyond conventional PD without additional toxicity, alongside a drop in their ctDNA. Unlike other T-cell-mediated approaches, CRS was limited to 1% G3/4 AE This TCR therapeutic is restricted to patient's expressing HLA-A*0201, limiting generalisability. Regulatory approval from the FDA has recently been granted in early 2022.

2.5.2 Mucosal Melanoma

Melanoma arising from mucosal surfaces, in the advanced setting, is characterized by higher LDH, BSLD, and fewer BRAF V600 mutations than metastatic disease with a cutaneous site of origin, all negative prognostic factors [87]. Robust randomized data for ICB in this tumor is absent, but exploratory analyses of mucosal subgroups ($n = 86$) from pooled ICB trials have shown a doublet approach over either PD-1 or CTLA-4 alone in monotherapy provided a prolonged PFS benefit once the curves separated at the 3-month mark (6.2 months HR 0.61 CI 0.39–0.96

$p = 0.116$ doublet vs. nivolumab, HR 0.42 CI 0.23–0.75 $p = 0.003$ doublet vs. ipilimumab), alongside a numerically larger ORR (37.1% vs. 23.3), with a median DoR NR (CI 7.6 to NR). This benefit was maintained across all clinically relevant subgroups, including M1c disease [88]. Given these were unplanned, post-hoc subgroups, no formal statistical comparison was undertaken with the outcomes seen in metastatic cutaneous melanoma, but the efficacy outcomes were poorer across all endpoints, likely representing variations in disease biology such as upregulated *KIT* expression, a lower somatic mutation rate and lack of immunogenic UV genetic mutation signature [89].

2.5.3   Melanoma with Brain Metastases

Brain metastases pose a significant unmet medical need in the management of metastatic melanoma, with 75% of patients having M1d disease at the time of their death [90]. Approaches such as surgical resection and sterotactic radiosurgery achieve effective local control rates, but have scant impact on development on other sites of intracranial disease and thus OS [68]. Metastasis to the brain, a sanctuary site due to the tight endothelium of the blood–brain barrier, compounded by its active drug efflux pumps [91], portends a poor prognosis. Multiple impactful clinical endpoints abound in this context, but intracranial PFS is particularly important as neurological compromise and a resultant drop in functional status must be avoided in order to countenance further systemic options. Prospective non-randomized data from COMBI-MB showed a 19% 12 month PFS (CI 10–31% $n = 76$), in patients with asymptomatic BRAF V600E +ve asymptomatic melanoma brain metastases without previous local therapy, using dabrafeninb/trametinib, with 50% of patients on corticosteroids. A cautious cross-trial comparison with a similar group without corticosteroid use permitted found a 59.5% (CI 47.9–69.3%) intracranial PFS [92] with doublet ICB. The co-operative group led ABC study, allowing pretreatment with BRAFi/MEKi and a maximum 10 mg of prednisolone for symptom control but maintaining the other aforementioned baseline characteristics found a 12 and 24 month intracranial PFS of 49%. A synergistic effect of combined checkpoint blockade is suggested given the poor single-agent nivolumab intracranial PFS of 20% and 15%, respectively [93]. This data suggests a role for systemic therapy with doublet ICB in the specific M1d context of asymptomatic metastases, under 40 mm maximally, agnostic to BRAF status, over a BRAFi/MEKi, single-agent PD-1 or local therapy approach. The role of targeted SABR alongside ICB is being outlined in the ABC-X study [94].

2.6   Pharmacology of Immune Checkpoint Blockade

Over the last ten years, there has been a significant jump in the understanding of the interaction between immune effector cells and tumor tissue leading to the development of drugs that enhance the body's immune surveillance ystems to promote apoptosis. These drugs include ICB mAbs that are directed at PD-1, PD-L1, or CTLA-4.

Despite the large number of jurisdictions that now have given full or provisional registration and the vast number of publications on the efficacy of these agents, few have critically outlined the key role of clinical pharmacology to guide optimal use for individual patients, preferring instead to recommend use as a "flat dose", or a dose per unit mass, a regimen known to be relatively unrelated to how the body handles most drugs in the adult setting. Yet drug dose, route of administration, entry to the "activity" compartment, the relationship of dose to exposure and exposure to efficacy and toxicity, dosing regimen, and dose in combination is proving key to our ability to predict safety and efficacy in the clinical setting.

It is notable that many of these therapies have "fast-tracked" regulatory approval in the FDA, and by other regulatory bodies without therefore the usual complete pharmacological work up or knowledge gleaned from an extensive clinical trial portfolio, which when available enables better predictions of likely drug efficacy and safety in real-world clinical practice. PK-PD modeling has helped with predictions, but even then the information on which the assumptions are made is sparse and does not include the complex patients treated. Specifically, "real-world" clinical practice often involves older patients with more comorbidities and complexities than the patients in clinical trials, an important fact as pharmacologically it has been observed that these drugs at similar doses have different effects in different population groups. An understanding of the clinical trial data alongside a knowledge of pharmacology helps predict alterations in doses and regimens, particularly for populations for which there is zero or limited clinical trial information [95].

Establishing the registered dose of mAbs and particularly of the ICBs is a challenge, with main criteria used for dose selection along with their limitations previously outlined [96, 97]. This work has identified the general features of the relationships between ICB pharmacokinetic parameters and treatment outcome (efficacy and/or toxicity) of the various drugs, which differ, sometimes significantly. Additionally, it is increasingly clear that the relationship between exposure and efficacy is complex, with variable relationships in exposure, age, and outcomes [97].

Understanding the pharmacology is also important for optimal use of these agents not just in monotherapy, but particularly in combination, where serious immune-related adverse events often have to be navigated.

What else is there that is known about the clinical pharmacology of these agents, or predictable based on a knowledge of the molecule and of clinical pharmacology principles? While summary of the predicted clinical pharmacology of these agents in special groups is available [51, 95] it may be helpful to revise some general aspects of the pharmacology of these agents. In summary, these ICBs are monoclonal antibodies, i.e., proteins, and thus share many of the clinical pharmacokinetic characteristics of mAb "targeted" therapies. These characteristics are different from our more traditional cytotoxic chemotherapies or the small-molecule targeted therapies such as tyrosine kinase inhibitors (e.g., "nibs".) However, it is known, for example, that inter-individual pharmacokinetic variability in clearance affects outcome [98]. Similarly, higher clearance values observed in some patients potentially result from characteristics associated with poor predictive factors of efficacy [98, 97] Finally, the schedule, and particularly its frequency of administration, in different population groups requires a consideration for dose individualization including regimen. The following section gives an overview of relevant pharmacokinetics of individual cancers and ICBs to illustrate such principles.

2.6.1 Pharmacokinetics of ICBs

The PK of all ICBs used for melanoma are relatively similar [99]. ICBs have low volumes of distribution (4.72–6.9 L), confinement to the vascular compartment [99], long half-lives, and receptor-mediated specific degradation and clearance mechanisms, ubiquitous throughout plasma and tissues [100]. This suggests drug interactions in the liver are unlikely but that renal impairment will affect clearance.

All dose ranges of ICBs showed linear PK characteristics with non-linear PK only occurring at lower doses of <0.1 mg/kg for pembrolizumab, <3 mg/kg for durvalumab and <10 mg/kg for avelumab [101]. MTD was not achieved in early phase trials and was therefore not available to guide dosing in subsequent Phase II and III studies as would occur in a traditional cancer therapeutic trial. MTD was not achieved due to absence of dose-limiting toxicities in these highly selected populations and when promising efficacy occurred, the trial design of rapid multiple expansion cohorts was adopted [102]. Subsequently, PK/PD modeling and simulation studies were utilized to determine dosing regimens and schedules for ICBs for fast track approvals [103]. The PK/PD simulation studies have estimated residual error ranging from 16 to 27% [99], albeit based on assumptions of the patients in the trials.

Serum pembrolizumab concentrations from three KEYNOTE studies (−001, −002, −006) utilized a model-based approach to justify the absence of dose adjustments in subpopulations. In these studies, those that had "non-reportable serum concentrations" were excluded [104], inclusion of non-detected serum concentrations may have provided valuable insight into inter-patient variability. Therefore, the range of plasma exposures seen in real-world context is yet to be determined.

2.6.2 Dose and Scheduling of ICBs

Loose and wide body weight-guided dosing was used across early phase trials of the first PD-1 inhibitors [99], followed by pivotal trials for pembrolizumab in melanoma using a variety of dosing regimens. Later trials of pembrolizumab used doses ranging from 50 to 400% of the phase I dose [105] contributing to multiple subsequent changes in dose and schedule. The limited early data on linearity of PK & PD and reliance on modeling to guide dosing for Phase II and III trials [106] also led to a number of adjustments to schedules and dosing of pembrolizumab, with a 200 mg flat Q3W dosing now FDA approved for melanoma [107].

Flat exposure–response and exposure-safety curves (from 0.5 to 10 mg/kg) and limited effect of body weight on PK were used as sponsors to support fixed-dose regimens for pembrolizumab [108] and nivolumab [109]. Population-based PK/PD modeling showed similarities of pembrolizumab exposure of 2 mg kg^{-1} Q3W and 200 mg Q3W, data from KEYNOTE-024 confirming these results with use of a fixed-dose regime [103]. A similar approach with nivolumab, initially based on modeling [109, 103] led to FDA approval of 240 mg Q2W and 480 mg Q4W [110] regimens across multiple tumor types. The 480 mg Q4W dosing regimen was later supported by an open-label extension of a phase III trial.

PD measured as receptor occupancy of PD-1 by nivolumab in circulating human T-cells saturates at 0.3 mg kg^{-1}, with flat exposure–response above 1 mg kg^{-1} Q2W [111]. Indeed, two neoadjuvant doses of nivolumab have led to a major pathological response in 45% ($n = 20$) of patients in an NSCLC trial [112]. Similarly, retrospective real-world evidence has found low-dose nivolumab (20–100 mg Q3W) had similar efficacy in a metastatic NSCLC cohort [113]. Taken together, this indicates that nivolumab could be a candidate for dose adjustment in the future [99].

Ipilimumab has been more extensively studied for appropriate dose selection. Phase III trial data from metastatic melanoma showed an improvement in mOS with 10 mg/kg Q3W versus 3 mg/kg Q3W dosing (15.7 vs. 11.5 months HR 0.84 $p = 0.04$). There was higher toxicity with $\geq$ Grade 3 adverse events in the higher dose group (34% vs. 18% $n = 362$) [114]. This is particularly relevant as patients with this toxicity often stop treatment and may not realize the full benefit of a potential increase in OS. As with other therapies a relationship between dose, survival benefit, and toxicity with ipilimumab, is not seen in the approved dosing ranges of anti-PD-1/PD-L1 PCIs [115]. The currently approved doses in the melanoma context for metastatic disease remain 3 mg/kg with the 10 mg/kg regimen approved, but largely unused, in the adjuvant setting.

The safest and most effective doses and schedules of ICBs may change as our appreciation of the underlying pharmacology grows, with dose individualization yet to be investigated, as it has for oral TKIs [116, 117].

2.6.3 Therapeutic Drug Monitoring

TDM is an approach used to individualize dosing with other systemic therapies and may play a role with ICB mAbs also [118]. TDM for any drug is most clinically relevant when there is large inter-patient variability, low intra-individual variability, a valid assay, and an exposure–response relationship [119]. PK modeling, prompted by concern from regulatory authorities [101], has suggested clearance of ICBs varies over time [120, 121]. Changes in BOR were associated with decreased maximal clearance change [122], despite a lack of trial data showing any relationship between plasma exposure (AUC steady state at 6 weeks) to pembrolizumab and overall survival, over a fivefold dose range (2 mg/kg and 10 mg/kg Q3W [123].

Prospective data from real-world practice has found a statistically and clinically significant negative clearance response relationship in patients with mNSCLC treated with nivolumab (PD mean clearance 0.244 L/day vs. PR/CR 0.172 L/day) [98]. A cohort of mNSCLC patients treated with an identical weight-based dosing of Q2W nivolumab found, within 10 weeks of ICB initiation, responders ($n = 15$) had 73% higher ($p = 0.002$) trough concentrations than non-responders and those with the highest trough concentrations had significantly longer mOS (NR vs. 306 days $p = 0.001$). Similarly, in a real-world practice study in metastatic melanoma treated with pembrolizumab, the high pembrolizumab exposure group (plasma trough concentration $= 55.9 \pm 25.6$ mcg/ml, $n = 14$) experienced meaningfully longer OS than the low exposure group (plasma trough concentration $= 104.2 \pm 8.1$ mcg/ml) with median OS not reached versus 48 months ($p = 0.014$). A similar positive exposure PFS relationship was found (median not reached vs. 48 months, $p = 0.045$) [124].

Higher baseline clearance has been associated with disease markers for cancer cachexia syndrome [125], suggesting higher malignancy-induced catabolic clearance and a lower chance of response. Given the number of confounders at play, both patient (altered catabolic state) and malignancy-related (histopathology, tumor burden, and receptor expression), it is difficult to determine whether higher drug exposure is the cause or effect of tumor response, especially given the aforementioned flat exposure–response curves [126, 127]. However, given the consistent data emerging out of real-world practice, further studies are required to identify whether a TDM for ICBs may identify patients more likely to need dose or therapeutic escalation early in their treatment course.

2.7   Response and Resistance to Immune Checkpoint Blockade

2.7.1   Somatic Genotype and Immune Phenotype

A framework to illustrate the underlying principles governing the immune–tumor interaction within the TME is outlined by the phenotypes in Fig. 2.4 [128]. The immune inflamed, immune excluded, and immune desert nomenclature [128], based on the characteristics of the immune infiltrate and local TME milieu prior to ICB identifies both threats and opportunities inherent in the immune response to cancer. Sites of metastatic disease in cutaneous melanoma have been shown, via a combined multiplex IHC & FACS approach, to primarily be immune inflamed, with a lack of ICB response associated with the excluded or desert phenotype [129]. This spatial approach to the TME has to be viewed orthogonally, through the lens of tumor genomic instability or transcriptional error leading to T-cell-specific tumor neoepitope recognition. Unrepaired ultraviolet radiation-induced DNA damage [130] with a resultant mutagenic signature characterized by a high mutational

Fig. 2.4 The Cancer-Immune Setpoint. Reprinted by permission from **Springer Nature**: **Nature** [Elements of cancer immunity and the cancer-immune set point, Daniel S. Chen et al., Page 324, 2017]

load is characteristic of melanoma. TMB is a composite measure of the total number of somatic mutations in the exome of a tumor [131]. Some of these mutations are immunogenic, capable of creating mutated proteins processed into neopeptides by the proteasome and subsequently competently presented as tumor-specific neoepitopes to CD8 + T-cell by MHC Class 1 [132]. Neoepitope recognition is uncommon, and thus somatic TMB is an imperfect predictive biomarker for ICB response. Pre-clinical work suggests only 10% of nonsynonymous point mutations leading to peptides presented with sufficient affinity by MHC Class 1, with even less capable of stimulating an immune response [128, 133]. These approaches are only a snapshot of the dynamic interaction between the immune system and sites of tumor, which may fluctuate between immune phenotype and genotypic status in periods of response and resistance.

The immune inflamed, or "hot" tumor consists of exhausted CD4 + and CD8+T lymphocytes in close proximity to tumor cells. Despite presence of type 1 IFN, IL-12, TNF-alpha, and other inflammatory cytokines [134], upregulated PD-L1 expression on tumor cells and the infiltrating immune cohort halts a pre-existing anti-tumor immune response in the TME [128], which may be re-invigorated by ICB [135]. A "cold" non-inflamed tumor, the immune-excluded phenotype is characterized by a dense stroma with an excluded T-cell infiltrate. This is likely mediated by upregulation of melanoma intrinsic oncogenic WNT/β catenin signaling, [136] preventing anti-tumor immunity even in the presence of ICB. Immunological ignorance or induction of tolerance with an absence of requisite CD8 + T-cells characterizes the "cold" immune desert, with limited tumor-specific neoepitopes to recognize, ineffective antigen presentation, and an immunosuppressive cytokine environment, poorly response to ICB.

2.7.2 Predictors of Response

2.7.2.1 Tumoral

TMB broadly correlates with response to anti-PD-1 therapy, (R^2 0.55 for difference in ORR across cancer types explained by TMB) [137], but despite the recent approval of anti-PD-1 therapy for tumor agnostic TMB-High patients [138], multiple factors contribute to this biomarker remaining imperfect for accurate predictive probability of ICB response. Defining TMB remains problematic, with variations in assay utilized genomic factors such as; breadth (WES vs. NGS vs. specific cancer gene panel), depth [132], mutation type (frameshift, indel, and RNA splicing mutations are more immunogenic than the nonsynonymous SNV that dominate TMB [139]) and germline testing. Patient technical factors such as inter and intra-tumoral genomic heterogeneity, site of biopsy, and specimen tumor content add to the limitations.

TMB makes no account of specific mutations within genes known to affect ICB response, e.g., JAK2 [140]. Concurrently equivalent weighting to all mutation locations and types is given, ignoring the significant heterogeneity in immunogenic quality of subsequent neoepitopes formed [51]. The substantial heterogeneity in the ability of various TMB cut-offs to predict response across various tumor sites of origin is clearly outlined with an AUROC for CR/PR in melanoma of only 0.63 (95% CI 0.53–0.74) with a 10 mut/MB cut off having a false positive rate of 55%, limiting confidence in using it exclusively for discriminating ability to predict objective response. The same cut-point in NSCLC generates a false positive rate of only 10% with similar sensitivity (80%), [141] enforcing that one size, and indeed one tumour site, does not fit all with this biomarker.

CheckMate 067 explored the utility of tumoral staining for PD-L1 as a predictive biomarker, with both doublet ICB and nivolumab alone approaches improving all clinical endpoints irrespective of PD-L1 cut-point. A Receiver operated characteristic curve for PD-L1 only showed an AUC of 0.53 (CI 0.46–0.60) again suggesting poor predictive accuracy. This is consistent with the high ORR (41%) seen in non-PD-L1 expressing tumors [45]. A hypothesis-generating, separation of survival curves in the PD-L1 < 1% cohort with doublet ICB over nivolumab after 9 months (post-hoc HR 0.69 CI 0.5–0.97, with overall HR 0.83 CI 0.66–1.03), absent in all other expression cohorts, suggests this expression level as a cut-point where the risk–benefit of doublet ICB should be weight carefully over single-agent PD-1 monotherapy. Similar to TMB, PD-L1 use is hampered by intratumoural heterogeneity of staining, fresh versus archival tissue, and significant discordance in assays and scoring [51].

Given that neither low PD-L1 expression nor low tumoral mutational load precludes a meaningful response to ICB [142], understanding the dynamic interaction of these immunological correlates within the TME is important in order to identify stratification tools for those most likely to respond to ICB. Transcriptomic avenues, such as GEPs create a molecular signature for mRNA expression across specific immune inflamed genes [143]. A T-cell inflamed GEP signature, containing IFN-γ responsive genes involved with antigen presentation, cytotoxicity, adaptive immune resistance, and chemokine upregulation [144] has been identified. This was combined with PD-L1, to reflect the extent of an inflamed TME, and assessed alongside TMB—a broad score for the inherent tumoural immunogenicity [145]. Low correlation between TMB and this T-cell inflamed GEP ($r = 0.252$), alongside retention of significance as independent predictors of objective response in a multivariate model allows stratification of meaningful ICB response along these axes. Moderate correlation between PD-L1 and GEP ($r = 0.53$) gives confidence that a PD-L1 IHC assay is an appropriate correlate for an immune inflamed TME. Objective responses were largely absent in the $GEP^{lo}TMB^{lo}$ melanoma cohort (1/89) with significantly longer PFS in the $^e GEP^{Hi}TMB^{Hi}$ versus $GEP^{lo}TMB^{lo}$ cohort (median 504 vs. 123 days HR 0.63 CI 0.36–1.09) [145].

Composite biomarker approaches, reflecting both TME and tumor immunogenicity, are necessary to accurately predict response in the new generation of ICB trials. This can only be achieved through a nuanced multi-omics methodology that can fully reflect the complex crosstalk between malignancy and the immune response stimulated by immune checkpoint blockade. Genomic and immunological characteristics such as increased T-cell clonality alongside CD4 + memory and CD8 + infiltrates in patients with responding tumors, alongside a B cell-enriched population, and development of tertiary lymphoid structures have all been associated with meaningful durable response [146, 147]. A multi-omic predictive model that includes immunoglobulin re-arrangement, T-cell receptor clonality, PD-L1 expression, and mutational burden has generated impressive predictive capacity for response in a small test cohort (AUROC 0.866 CI 0.72–0.99), but these approaches have yet to be translated to real-world practice [146].

2.7.2.2 Patient

Moving to patient phenotypic factors and their clinical assessment, most pivotal ICB trials did not enroll those with an ECOG ≥ 2. This cohort is heterogenous and the limited available trial data suggests an mOS of only 2.4 months post ipilimumab in ECOG 2 patients treated with nivolumab [148], with real-world retrospective data reflecting this finding (mOS 19.5 vs. 1.8 months HR 5.45 p < 0.0001 in ECOG < 2 vs. > 2)[149]. Identifying whether performance status is independently predictive of poor response or a prognostic clinical marker of advanced burden of disease is challenging. Cancer cachexia, a complex multi-system paraneoplastic disorder seen with worsening ECOG, is associated with a heightened immunosuppressive TME with upregulated IL-6, decreased T lymphocytes, and increased inhibitory MDSCs and Treg cells [150, 151], with resulting immune effector exclusion contributing to disease progression.

2.7.3 Mechanisms of Resistance

2.7.3.1 Primary/Innate and Adaptive

A consensus approach clinically defines resistance to ICB in the advanced setting into: primary—with best response as PD/SD < 6 months or secondary—with best response as CR/PR or SD > 6 months [152]. At a cellular level, a patient's tumor may never mount an immune response (clinically primary resistance), have an initial anti-tumor response rapidly halted by adaptive resistance mechanisms (clinically primary resistance), or benefit from an immune response that later progresses due to truly acquired resistance to ICB (clinically secondary resistance) [153]. Intrinsic cellular mechanisms of resistance are outlined in Fig. 2.5 [153]. In melanoma, low tumor immunogenicity due to a lack of tumor-specific antigens [154] or a defective antigen-presenting complex due to genetic loss of MHC expression (155) can lead to primary resistance. Upregulation of oncogenic signaling can occur early during the course of treatment with PD-1 mAbs, with loss of tumor suppressor PTEN protein expression resulting in an upregulated PI3K pathway. This leads to a drop in CD8+ T-cell tumor infiltration, a "cold" immune-excluded phenotype which translates clinically into lower incidence of objective response in individuals with PTEN loss, via a separate mechanism from the aforementioned WNT/β catenin pathway [156].

A combination of whole-exome and transcriptome sequencing from pre-ICB samples to identify mechanisms of resistance in melanoma that did not respond to ICB has identified increased expression of genes involved in wound healing, angiogenesis, ECM organization, and cell adhesion, with a transcriptomic signature, labeled IPRES (Innate anti-PD-1 Resistance), with amplified features of mesenchymal transition. This signature was seen more commonly in "cold" immune excluded, stromal dense, malignancies such as pancreatic adenocarcinoma [142].

Proximity of PD-1+ and PD-L1+ cells are required to facilitate the initial immune response post-anti-PD-(L)1 mAb dosing, with primary resistance mediated by a lack of these interactions in the tumor [157]. PD-L1 can be constitutively expressed on tumors or upregulated as a defense mechanism against T-cells upon IFN γ signaling, an adaptive resistance mechanism via the IFN receptor [158, 159].

2.7.3.2 Secondary/Acquired

Despite an initial response, up to 1/3rd of patients progress, despite ongoing ICB administration [160]. Clonal subpopulations of melanoma cells may acquire, epigenetic or phenotypic characteristics during the course of ICB that allow them to evade immune surveillance and thus gain a Darwinian advantage over other

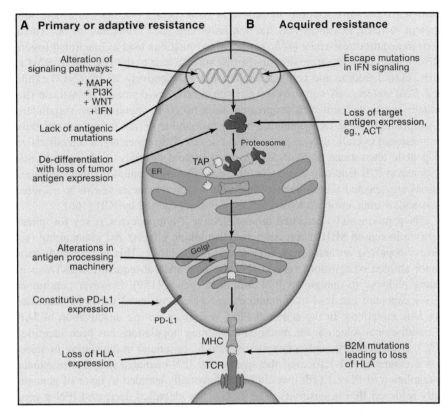

Fig. 2.5 Mechanisms of resistance. **a** Intrinsic factors that lead to primary or adaptive resistance include lack of antigenic mutations, loss of tumor antigen expression, loss of HLA expression, alterations in antigen processing machinery, alterations of several signaling pathways (MAPK, PI3K, WNT, IFN), and constitutive PD-L1 expression. **b** Intrinsic factors that are associated with acquired resistance of cancer, including loss of target antigen, HLA, and altered interferon signaling, as well as loss of T-cell functionality. Reprinted by permission from **Elsevier**: [Cell Primary, Adaptive, and Acquired Resistance to Cancer Immunotherapy, Padmanee Sharma et al., Page 710, 2017]

immune-sensitive populations [161]. Counterpoised against this immune surveillance may become ineffective over time due to immunological ignorance, as seen in transcriptomic work in PD-1 resistant biopsies that identified tumor MHC Class I downregulation associated with a de-differentiation phenotype in both primary and acquired PD-1 resistant samples [162]. ICB can re-invigorate exhausted T-cells, which have weak (though not absent) efficacy, but if the tumor burden remains high despite this then may render them permanently re-exhausted epigenetically, preventing an immune response [163, 164].

Paired baseline and recurrence biopsies in case studies of patients with acquired secondary ICB resistance (anti-CTLA-4 and anti-PD-1), by separate groups, have identified frameshift deletions in exon-1 of *B2M* (beta 2 microglobulin) and subsequent increase in cancer cell fraction harboring such mutations, to be critical steps in acquired resistance [165, 166]. These mutations lead to functional loss of MHC Class I protein expression on the cell surface, due to the key role of *B2M* in MHC class I folding and cell surface transport. Collectively 29.4% (5/17) exhibited *B2M* defects, all non-responders, with a 60% incidence (3/5) in those that initially responded and then progressed, with no *B2M* alterations in responders. Unsurprisingly *B2M* LOH/mutation has been found to enrich for poorer objective response and overall survival across all classes of ICB in melanoma. Identification of genetic aberrations in *B2M* also in pretreatment biopsies of those with little response to ICB lead to a composite inference that *B2M* mutated subclonal populations are selected via immunoediting early in tumorigenesis or can be acquired resistance events, enriched via selection pressures, exerted by ICB [166].

IFN-γ, produced by activated tumor-selective T lymphocytes, is key for initiating (via increased MHC expression and chemokine effects) and maintaining (via direct apoptosis) an anti-tumor immune response [167]. This is achieved, upon tumor antigen recognition, via the IFN receptor and subsequent JAK/STAT signaling pathway to upregulate IFN stimulated genes [159]. However continuous IFN-γ exposure can lead to immune escape [153], mediated partly by inactivating JAK mutations. In the acquired PD-1 resistance setting, inactivation of JAK 1/2 with clonal selection for mutation-containing populations has been identified [165]. Evolutionarily, such a mutation would be beneficial in the setting of selection pressure of PD-1 focused therapy, as the IFN-mediated adaptive resistance upregulation of PD-L1 [168] would be preferentially avoided in favor of genomically achieved IFN insensitivity. The TCGA has identified decreased IFN-γ gene expression as another mechanism by which PTEN loss contributes to immune resistance in melanoma.

Understanding this heterogenous and dynamic system is key to personalizing the approach to resistant disease. Identification of whether a TME has become immunologically cold, for example, could prompt intervention with an oncolytic virus-containing tumor-specific antigen to induce suitable T-cell activation. Informed approaches to resistant oncogenic signaling pathways may alternatively identify upregulated PI3K signaling with a resultant small-molecule inhibitor being more appropriate. Until introduction of adequate biomarkers to assess for such nuances are validated and used in practice, the problem of ICB resistance will remain unsolved.

2.8   Beyond the Checkpoint

2.8.1   Harnessing the Innate and Adaptive Immune Systems

Research in immune checkpoint blockade has largely been focused on checkpoint therapies intrinsically targeting the T cell which triggers the cancer-killing potential of the adaptive immune system. We are now realizing that much of the efficacy of the checkpoint therapies are attributable also to the innate immune system which is activated by direct and indirect pathways. Research is expanding into the role of the innate subsets of cells in the immune system and the tumor microenvironment. These cells include TAMs, DC, MDSCs, NK, and ILC2 [169]. Emerging checkpoints like LAG3, TIGIT, and TIM3 and additional co-stimulatory checkpoints (OX40, GIRR, ICOS) [170] are expressed in both the adaptive and innate immune systems. LAG3 for example has a role in regulating hematopoiesis [171] and contributes to T-cell exhaustion. By blocking LAG3 the TAMs can be shifted to an anti-tumor phenotype. Anti-TIGIT therapy can reverse NK cell dysfunction [172]. TIM3 is broadly expressed and directly regulates both innate and adaptive immunity. TIM3 blockade may increase DC activation and crosstalk between T-cells and DC cells, and increase antigen presentation [173].

These novel checkpoints may enhance the action of existing checkpoint therapies as demonstrated in the combination of nivolumab and relatlimab, a novel checkpoint inhibitor that targets LAG3. This demonstrated improved progression-free survival (PFS) compared to nivolumab monotherapy for patients with advanced melanoma, according to a PFS interim analysis of the global phase II/III RELATIVITY-047 trial. Median PFS in the relatlimab + nivolumab group (10.1 months CI, 6.4–15.7]) was significantly longer than in the nivolumab alone group (4.6 months CI, 3.4–5.6]; hazard ratio, 0.75 [CI, 0.6–0.9]; $p = 0.0055$). PFS rates at 12 months were 47.7% (CI, 41.8–53.2) and 36.0% (CI, 30.5–41.6), respectively. This is the first phase III trial to explore the clinical benefit of blocking LAG-3 [174].

2.8.2 Small Molecules

Small molecules that inhibit tyrosine kinases are able to shift the tumor microenvironment to an immune stimulatory state. Lenvatinib is an oral small-molecule inhibitor of VEGFR, FGFRs, PDGFRα, KIT, and RET proto-oncogene [175]. Several of these receptor tyrosine kinases are overexpressed in multiple cancers. The combination of lenvatinib and pembrolizumab showed immunomodulatory properties which included increased tumor infiltration of effector CD8 + T-cells and decreased monocytes and macrophages. The LEAP 004 study of lenvatinib and pembrolizumab combination therapy showed a 21.4% ORR in patients with advanced anti-PD1 resistant melanoma. The median duration of response was 8.3 months and 38.6% of responses were ongoing for more than 9 months. The majority of the participants had primary resistance (60.2%) in the metastatic setting and the ORR was comparable in patients with primary (22.6%) and those with secondary (22.7%) resistance [176].

2.8.3 IL-2

Exploration of IL-2 variants in melanoma and other tumor types continue in attempt to improve upon the safety profile of high-dose IL-2 while building upon the anti-tumor activity historically associated with high-dose IL-2 [177]. Nemvaleukin alfa ("nemvaleukin," ALKS 4230) is an engineered fusion protein composed of a circularly permuted IL-2 and IL-2 receptor-α (IL-2Rα) designed to selectively activate the intermediate-affinity IL-2R but not the high-affinity IL-2R. The high-affinity IL-2R is also expressed on vascular and pulmonary endothelial cells and may contribute to toxicity via capillary leak syndrome. Thus, selective activation of the intermediate-affinity IL-2R by nemvaleukin has the potential to enhance tumor killing as well as improve safety and tolerability. In vitro pharmacology data also supported the hypothesis that nemvaleukin signals through the intermediate-affinity IL-2R, leading to the expansion and activation of effector cells, while minimizing the expansion and activation of immunosuppressive CD4 + Tregs. Intravenous administration of nemvaleukin is currently being evaluated in ARTISTRY-1, a first in human, Phase 1/2 study (ALK4230-A101) in patients with advanced solid tumors after treatment failure of established therapies known to provide clinical benefit. Preliminary data show anti-tumor activity with IV nemvaleukin at the RP2D of 6 ug/kg/day given as an IV infusion daily for five consecutive days. The FDA granted fast track designation to nemvaleukin alfa for mucosal melanoma due to the positive responses seen in this anti-PD-1 resistant tumor . Enrollment has begun for the global phase 2 ARTISTRY-6 (NCT04830124) study designed to evaluate the anti-tumor activity of monotherapy nemvaleukin subcutaneously in advanced cutaneous melanoma and intravenously in advanced mucosal melanoma.

2.8.4 DNA-Damaging Agents

DNA damage repair agents are also showing early positive results in patients with advanced cutaneous, acral, and mucosal melanoma that have failed immunotherapy. ATR is a serine-treonine specific protein kinase important for DNA repair. Ceralasertib (AZD6738- AstraZeneca) has completed phase 1 trials and has demonstrated an ORR of 33%, a disease control rate of 60.6%, a medial progression-free survival of 3.6 months, and a median duration of response of 9.9 months in patients with melanoma when given with paclitaxel [178]. This agent is now being combined with anti-PD-L1 immunotherapy (durvalumab) in an ongoing phase 2 trial.

2.8.5 Adoptive Cell Therapy

Adoptive cell therapy or tumor-infiltrating lymphocyte therapy is prepared by extracting a fragment of tumor and expanding it ex vivo to remove the TILs from the hostile tumor microenvironment [179]. This reduces the immunosuppressive effects of intratumoural regulatory T-cells and rejuvenates the cells. Lifileucel (LN-144) was tested in a phase II open-label, single-arm, multicenter study in 66 patients with immunotherapy-resistant advanced melanoma and demonstrated durable responses. Patients received a nonmyeloablative lymphodepletion regimen, a single infusion of lifileucel with $> 1 \times 10^9$TIL cells, and up to six doses of high-dose IL-2. The objective response rate was 36% and the median duration of response was not reached at 18.7-month median study follow-up. The US Food and Drug Administration has granted lifileucel a Regenerative Medicine Advanced Therapy designation, Orphan Drug designation, and a Fast Track designation for advanced melanoma [180].

2.8.6 Epigenetics

Epigenetics are heritable alterations in gene expression without change in the DNA sequence. Epigenetics that has been successfully targeted for cancer treatment include DNA methylation and histone post-translational modifications. DNA methylation is the addition of a methyl group to the 5-carbon position of a cytosine nucleotide, almost exclusively in CpG dinucleotides. CpG dinucleotides cluster in "CpG islands"; located mostly in gene promoters and in regions of repetitive DNA. Abnormal methylation patterns are found frequently in cancer with a characteristic pattern of global DNA hypomethylation and promoter hypermethylation at sites including tumor suppressor genes [181]. The inactivation of tumor suppressor genes occurs as a consequence of promoter region hypermethylation and numerous studies have demonstrated a broad range of genes are silenced by DNA methylation across cancer types [182].

Anti-neoplastic agents that target DNA methylation in use for other cancer types include are decitabine and 5-azacitidine. Decitabine inhibits DNA methyltransferase 1 (DNMT1) after incorporating into DNA during S phase of the cell cycle. 5-azacitidine has a very similar mechanism of action but can be incorporated into both DNA and RNA. Decitabine and 5-azacitidine are both used for the treatment of MDS and AML. Low-dose 5-azacitidine in combination with anti-CTLA-4 therapy was shown to be more effective at controlling tumor growth in a mouse model of melanoma compared to each of the agents alone [183]. This provided the basis for clinical trials with immunotherapies such as ACTRN12618000053224: Early phase II study of Azacitidine and Carboplatin priming for Avelumab in patients with advanced melanoma who are resistant to immunotherapy. This seeks to understand if the combination of an epigenetic modifier and a DNA-damaging agent improves response to anti-PDL-1 immunotherapy in treatment-resistant metastatic melanoma [184].

References

1. Printz C (2001) Spontaneous regression of melanoma may offer insight into cancer immunology. J Natl Cancer Inst 93(14):1047–1048
2. Clemente CG, Mihm MC Jr, Bufalino R, Zurrida S, Collini P, Cascinelli N (1996) Prognostic value of tumor infiltrating lymphocytes in the vertical growth phase of primary cutaneous melanoma. Cancer 77(7):1303–1310
3. Keung EZ, Gershenwald JE (2018) The eighth edition American Joint Committee on Cancer (AJCC) melanoma staging system: implications for melanoma treatment and care. Expert Rev Anticancer Ther 18(8):775–84
4. Fridman WH, Pages F, Sautes-Fridman C, Galon J (2012) The immune contexture in human tumours: impact on clinical outcome. Nat Rev Cancer 12(4):298–306
5. Dudley ME, Wunderlich JR, Yang JC, Sherry RM, Topalian SL, Restifo NP et al (2005) Adoptive cell transfer therapy following non-myeloablative but lymphodepleting chemotherapy for the treatment of patients with refractory metastatic melanoma. J Clin Oncol Off J Am Soc Clin Oncol 23(10):2346–2357
6. Han D, Zager JS, Shyr Y, Chen H, Berry LD, Iyengar S et al (2013) Clinicopathologic predictors of sentinel lymph node metastasis in thin melanoma. J Clin Oncol 31(35):4387–4393
7. Piérard GE, Piérard-Franchimont C, Reginster M-A, Quatresooz P (2012) Smouldering malignant melanoma and metastatic dormancy: an update and review. Dermatol Res Pract 2012:461278
8. Lázár-Molnár E, Hegyesi H, Tóth S, Falus A (2000) Autocrine and paracrine regulation by cytokines and growth factors in melanoma. Cytokine 12(6):547–554
9. Furuhashi M, Sjöblom T, Abramsson A, Ellingsen J, Micke P, Li H et al (2004) Platelet-derived growth factor production by B16 melanoma cells leads to increased pericyte abundance in tumors and an associated increase in tumor growth rate. Cancer Res 64(8):2725–2733
10. Vignali DA, Collison LW, Workman CJ (2008) How regulatory T-cells work. Nat Rev Immunol 8(7):523–532
11. Gabrilovich DI (2017) Myeloid-derived suppressor cells. Cancer Immunol Res 5(1):3–8
12. Mills CD, Kincaid K, Alt JM, Heilman MJ, Hill AM (2000) M-1/M-2 macrophages and the Th1/Th2 paradigm. J Immunol 164(12):6166–6173
13. Nevala WK, Vachon CM, Leontovich AA, Scott CG, Thompson MA, Markovic SN (2009) Evidence of systemic Th2-driven chronic inflammation in patients with metastatic melanoma. Clin Cancer Res 15(6):1931–1939
14. Zhou J, Tang Z, Gao S, Li C, Feng Y, Zhou X (2020) Tumor-associated macrophages: recent insights and therapies. Front Oncol 10:188
15. Eddy K, Chen S (2020) Overcoming immune evasion in melanoma. Int J Mol Sci 21(23):8984
16. Tucci M, Passarelli A, Mannavola F, Felici C, Stucci LS, Cives M et al (2019) Immune system evasion as hallmark of melanoma progression: the role of dendritic cells. Front Oncol 9:1148
17. Cali B, Molon B, Viola A (2017) Tuning cancer fate: the unremitting role of host immunity. Open Biol 7(4)
18. Hayward NK, Wilmott JS, Waddell N, Johansson PA, Field MA, Nones K et al (2017) Whole-genome landscapes of major melanoma subtypes. Nature 545(7653):175–180
19. Cancer Genome Atlas N (2015) Genomic classification of cutaneous melanoma. Cell 161(7):1681–1696
20. Hanahan D, Weinberg RA (2011) Hallmarks of cancer: the next generation. Cell 144(5):646–674
21. Nishimura H, Nose M, Hiai H, Minato N, Honjo T (1999) Development of lupus-like autoimmune diseases by disruption of the PD-1 gene encoding an ITIM motif-carrying immunoreceptor. Immunity 11(2):141–151

22. Nishimura H, Okazaki T, Tanaka Y, Nakatani K, Hara M, Matsumori A et al (2001) Autoimmune Dilated Cardiomyopathy in PD-1 Receptor-Deficient Mice. Science 291(5502):319–322
23. Haanen JB (2013) Immunotherapy of melanoma. EJC Suppl 11(2):97–105
24. Latchman Y, Wood CR, Chernova T, Chaudhary D, Borde M, Chernova I et al (2001) PD-L2 is a second ligand for PD-1 and inhibits T-cell activation. Nat Immunol 2(3):261–268
25. Youngnak P, Kozono Y, Kozono H, Iwai H, Otsuki N, Jin H et al (2003) Differential binding properties of B7–H1 and B7-DC to programmed death-1. Biochem Biophys Res Commun 307(3):672–677
26. Yost KE, Satpathy AT, Wells DK, Qi Y, Wang C, Kageyama R et al (2019) Clonal replacement of tumor-specific T-cells following PD-1 blockade. Nat Med 25(8):1251–1259
27. Azuma M, Ito D, Yagita H, Okumura K, Phillips JH, Lanier LL et al (1993) B70 antigen is a second ligand for CTLA-4 and CD28. Nature 366(6450):76–79
28. Ribas A, Wolchok JD (2018) Cancer immunotherapy using checkpoint blockade. Science 359(6382):1350–1355
29. Kearney ER, Pape KA, Loh DY, Jenkins MK (1994) Pillars article: visualization of Peptide-specific T-cell immunity and peripheral tolerance induction in vivo. Immunity. 1:327–339; (2013) J Immunol 191(11):5327–5339
30. Kvistborg P, Philips D, Kelderman S, Hageman L, Ottensmeier C, Joseph-Pietras D et al (2014) Anti-CTLA-4 therapy broadens the melanoma-reactive CD8+ T-cell response. Sci Transl Med 6(254):254ra128
31. Clark CA, Gupta HB, Sareddy G, Pandeswara S, Lao S, Yuan B et al (2016) Tumor-Intrinsic PD-L1 signals regulate cell growth, pathogenesis, and autophagy in ovarian cancer and melanoma. Cancer Res 76(23):6964–6974
32. Contardi E, Palmisano GL, Tazzari PL, Martelli AM, Falà F, Fabbi M et al (2005) CTLA-4 is constitutively expressed on tumor cells and can trigger apoptosis upon ligand interaction. Int J Cancer 117(4):538–550
33. Atkins MB, Lotze MT, Dutcher JP, Fisher RI, Weiss G, Margolin K et al (1999) High-dose recombinant IL-2 therapy for patients with metastatic melanoma: analysis of 270 patients treated between 1985 and 1993. J Clin Oncol 17(7):2105–2116
34. Jiang T, Zhou C, Ren S (2016) Role of IL-2 in cancer immunotherapy. Oncoimmunology. 5(6):e1163462
35. Hodi FS, O'Day SJ, McDermott DF, Weber RW, Sosman JA, Haanen JB et al (2010) Improved survival with ipilimumab in patients with metastatic melanoma. N Engl J Med 363(8):711–723
36. Robert C, Ribas A, Wolchok JD, Hodi FS, Hamid O, Kefford R et al (2014) Anti-programmed-death-receptor-1 treatment with pembrolizumab in ipilimumab-refractory advanced melanoma: a randomised dose-comparison cohort of a phase 1 trial. Lancet 384(9948):1109–1117
37. Hamid O, Robert C, Daud A, Hodi FS, Hwu WJ, Kefford R et al (2019) Five-year survival outcomes for patients with advanced melanoma treated with pembrolizumab in KEYNOTE-001. Ann Oncol 30(4):582–588
38. Robert C, Schachter J, Long GV, Arance A, Grob JJ, Mortier L et al (2015) Pembrolizumab versus Ipilimumab in advanced melanoma. N Engl J Med 372(26):2521–2532
39. Robert C, Ribas A, Schachter J, Arance A, Grob JJ, Mortier L et al (2019) Pembrolizumab versus ipilimumab in advanced melanoma (KEYNOTE-006): post-hoc 5-year results from an open-label, multicentre, randomised, controlled, phase 3 study. Lancet Oncol
40. Robert C, Long GV, Brady B, Dutriaux C, Maio M, Mortier L et al (2015) Nivolumab in previously untreated melanoma without BRAF mutation. N Engl J Med 372(4):320–330
41. Ascierto PA, Long GV, Robert C, Brady B, Dutriaux C, Di Giacomo AM et al (2019) Survival outcomes in patients with previously untreated BRAF wild-type advanced melanoma treated with nivolumab therapy: three-year follow-up of a randomized phase 3 trial. JAMA Oncol 5(2):187–194

42. Weber JS, D'Angelo SP, Minor D, Hodi FS, Gutzmer R, Neyns B et al (2015) Nivolumab versus chemotherapy in patients with advanced melanoma who progressed after anti-CTLA-4 treatment (CheckMate 037): a randomised, controlled, open-label, phase 3 trial. Lancet Oncol 16(4):375–384
43. Larkin J, Minor D, D'Angelo S, Neyns B, Smylie M, Miller WH Jr et al (2018) Overall survival in patients with advanced melanoma who received nivolumab versus investigator's choice chemotherapy in CheckMate 037: a randomized, controlled, open-label phase III trial. J Clin Oncol 36(4):383–390
44. Larkin J, Chiarion-Sileni V, Gonzalez R, Grob JJ, Cowey CL, Lao CD et al (2015) Combined nivolumab and ipilimumab or monotherapy in untreated melanoma. N Engl J Med 373(1):23–34
45. Larkin J, Chiarion-Sileni V, Gonzalez R, Grob JJ, Rutkowski P, Lao CD et al (2019) Five-YEAR SURVIVAL WITH combined nivolumab and ipilimumab in advanced melanoma. N Engl J Med
46. Wolchok JD, Chiarion-Sileni V, Gonzalez R, Grob J-J, Rutkowski P, Lao CD et al (2021) (CheckMate 067: 6.5-year outcomes in patients (pts) with advanced melanoma. J. Clin. Oncol 39(15_suppl):9506
47. Wilmott JS, Long GV, Howle JR, Haydu LE, Sharma RN, Thompson JF et al (2012) Selective BRAF inhibitors induce marked T-cell infiltration into human metastatic melanoma. Clin Cancer Res 18(5):1386–1394
48. Frederick DT, Piris A, Cogdill AP, Cooper ZA, Lezcano C, Ferrone CR et al (2013) BRAF inhibition is associated with enhanced melanoma antigen expression and a more favorable tumor microenvironment in patients with metastatic melanoma. Clin Cancer Res 19(5):1225–1231
49. Gutzmer R, Stroyakovskiy D, Gogas H, Robert C, Lewis K, Protsenko S et al (2020) Atezolizumab, vemurafenib, and cobimetinib as first-line treatment for unresectable advanced BRAF(V600) mutation-positive melanoma (IMspire150): primary analysis of the randomised, double-blind, placebo-controlled, phase 3 trial. Lancet 395(10240):1835–1844
50. Nathan P, Dummer R, Long GV, Ascierto PA, Tawbi HA, Robert C et al (2020) LBA43 Spartalizumab plus dabrafenib and trametinib (Sparta-DabTram) in patients (pts) with previously untreated BRAF V600–mutant unresectable or metastatic melanoma: Results from the randomized part 3 of the phase III COMBI-i trial. Ann Oncol 31:S1172
51. Navani V, Graves MC, Bowden NA, Van Der Westhuizen A (2020) Immune checkpoint blockade in solid organ tumours: Choice, dose and predictors of response. Br J Clin Pharmacol
52. Borcoman E, Nandikolla A, Long G, Goel S, Le Tourneau (2018) Patterns of response and progression to immunotherapy. Am Soc Clin Oncol Educ Book 169–178
53. Eisenhauer EA, Therasse P, Bogaerts J, Schwartz LH, Sargent D, Ford R et al (2009) New response evaluation criteria in solid tumours: revised RECIST guideline (version 1.1). Eur J Cancer 45(2):228–247
54. Chiou VL, Burotto M (2015) Pseudoprogression and Immune-related response in solid tumors. J Clin Oncol 33(31):3541–3543
55. Larkin J, Ascierto PA, Dreno B, Atkinson V, Liszkay G, Maio M et al (2014) Combined vemurafenib and cobimetinib in BRAF-mutated melanoma. N Engl J Med 371(20):1867–1876
56. Robert C, Karaszewska B, Schachter J, Rutkowski P, Mackiewicz A, Stroiakovski D et al (2015) Improved overall survival in melanoma with combined dabrafenib and trametinib. N Engl J Med 372(1):30–39
57. Dummer R, Ascierto PA, Gogas HJ, Arance A, Mandala M, Liszkay G et al (2018) Encorafenib plus binimetinib versus vemurafenib or encorafenib in patients with BRAF-mutant melanoma (COLUMBUS): a multicentre, open-label, randomised phase 3 trial. Lancet Oncol 19(5):603–615

58. Ascierto PA, Mandala M, Ferrucci PF, Rutkowski P, Guidoboni M, Fernandez AMA et al (2020) LBA45 First report of efficacy and safety from the phase II study SECOMBIT (SEquential COMBo Immuno and Targeted therapy study). Ann Oncol 31:S1173–S1174

59. Kreft S, Gesierich A, Eigentler T, Franklin C, Valpione S, Ugurel S et al (2019) Efficacy of PD-1-based immunotherapy after radiologic progression on targeted therapy in stage IV melanoma. Eur J Cancer 116:207–215

60. ClinicalTrials. gov. DREAMseq (Doublet, Randomized Evaluation in Advanced Melanoma Sequencing) a Phase III Trial. [Available from: https://clinicaltrials.gov/ct2/show/NCT022 24781]

61. Keilholz U, Ascierto PA, Dummer R, Robert C, Lorigan P, van Akkooi A et al (2020) ESMO consensus conference recommendations on the management of metastatic melanoma: under the auspices of the ESMO guidelines committee. Ann Oncol 31(11):1435–1448

62. Michielin O, Atkins MB, Koon HB, Dummer R, Ascierto PA (2020) Evolving impact of long-term survival results on metastatic melanoma treatment. J Immunother Cancer 8(2)

63. Robert C, Grob JJ, Stroyakovskiy D, Karaszewska B, Hauschild A, Levchenko E et al (2019) Five-year outcomes with dabrafenib plus trametinib in metastatic melanoma. N Engl J Med 381(7):626–636

64. Regan MM, Mantia C, Werner L, Tarhini AA, Rao S, Moshyk A et al (2020) Estimating treatment-free survival (TFS) over extended follow-up in patients (pts) with advanced melanoma (MEL) treated with immune-checkpoint inhibitors (ICIs): five-year follow-up of CheckMate 067. J Clin Oncol 38(15_suppl):10043

65. Warburton L, Meniawy TM, Calapre L, Pereira M, McEvoy A, Ziman M et al (2020) Stopping targeted therapy for complete responders in advanced BRAF mutant melanoma. Sci Rep 10(1):18878

66. Weide B, Martens A, Hassel JC, Berking C, Postow MA, Bisschop K et al (2016) Baseline biomarkers for outcome of melanoma patients treated with pembrolizumab. Clin Cancer Res 22(22):5487–5496

67. Joseph RW, Elassaiss-Schaap J, Kefford R, Hwu WJ, Wolchok JD, Joshua AM et al (2018) Baseline tumor size is an independent prognostic factor for overall survival in patients with melanoma treated with pembrolizumab. Clin Cancer Res 24(20):4960–4967

68. Long GV, Atkinson V, Lo S, Sandhu S, Guminski AD, Brown MP et al (2018) Combination nivolumab and ipilimumab or nivolumab alone in melanoma brain metastases: a multicentre randomised phase 2 study. Lancet Oncol 19(5):672–681

69. Molife R, Hancock BW (2002) Adjuvant therapy of malignant melanoma. Crit Rev Oncol Hematol 44(1):81–102

70. Gray RJ, Pockaj BA, Kirkwood JM (2002) An update on adjuvant interferon for melanoma. Cancer Control 9(1):16–21

71. Najjar YG, Puligandla M, Lee SJ, Kirkwood JM (2019) An updated analysis of 4 randomized ECOG trials of high-dose interferon in the adjuvant treatment of melanoma. Cancer 125(17):3013–3024

72. Ives NJ, Suciu S, Eggermont AMM, Kirkwood J, Lorigan P, Markovic SN et al (2017) Adjuvant interferon-alpha for the treatment of high-risk melanoma: an individual patient data meta-analysis. Eur J Cancer 82:171–183

73. Eggermont AM, Chiarion-Sileni V, Grob JJ, Dummer R, Wolchok JD, Schmidt H et al (2015) Adjuvant ipilimumab versus placebo after complete resection of high-risk stage III melanoma (EORTC 18071): a randomised, double-blind, phase 3 trial. Lancet Oncol 16(5):522–530

74. Eggermont AMM, Chiarion-Sileni V, Grob JJ, Dummer R, Wolchok JD, Schmidt H et al (2019) Adjuvant ipilimumab versus placebo after complete resection of stage III melanoma: long-term follow-up results of the European organisation for research and treatment of cancer 18071 double-blind phase 3 randomised trial. Eur J Cancer 119:1–10

75. Weber J, Mandala M, Del Vecchio M, Gogas HJ, Arance AM, Cowey CL et al (2017) Adjuvant nivolumab versus Ipilimumab in resected stage III or IV Melanoma. N Engl J Med 377(19):1824–1835

76. Ascierto PA, Del Vecchio M, Mandala M, Gogas H, Arance AM, Dalle S et al (2020) Adjuvant nivolumab versus ipilimumab in resected stage IIIB-C and stage IV melanoma (Check-Mate 238): 4-year results from a multicentre, double-blind, randomised, controlled, phase 3 trial. Lancet Oncol 21(11):1465–1477

77. Eggermont AMM, Blank CU, Mandala M, Long GV, Atkinson V, Dalle S et al (2018) Adjuvant pembrolizumab versus placebo in resected stage III melanoma. N Engl J Med 378(19):1789–1801

78. Eggermont AMM, Blank CU, Mandala M, Long GV, Atkinson VG, Dalle S et al (2020) Longer follow-up confirms recurrence-free survival benefit of adjuvant pembrolizumab in high-risk stage III melanoma: updated results from the EORTC 1325-MG/KEYNOTE-054 trial. J Clin Oncol 38(33):3925–3936

79. Eggermont AMM, Blank CU, Mandala M, Long GV, Atkinson VG, Dalle S et al (2021) Adjuvant pembrolizumab versus placebo in resected stage III melanoma (EORTC 1325-MG/KEYNOTE-054): distant metastasis-free survival results from a double-blind, randomised, controlled, phase 3 trial. Lancet Oncol 22(5):643–654

80. Long GV, Schadendorf D, Vecchio MD, Larkin J, Atkinson V, Schenker M et al (2021) Abstract CT004: adjuvant therapy with nivolumab (NIVO) combined with ipilimumab (IPI) vs NIVO alone in patients (pts) with resected stage IIIB-D/IV melanoma (CheckMate 915). Cancer Res 81(13 Supplement):CT004-CT

81. Blank CU, Rozeman EA, Fanchi LF, Sikorska K, van de Wiel B, Kvistborg P et al (2018) Neoadjuvant versus adjuvant ipilimumab plus nivolumab in macroscopic stage III melanoma. Nat Med 24(11):1655–1661

82. Amaria RN, Reddy SM, Tawbi HA, Davies MA, Ross MI, Glitza IC et al (2018) Neoadjuvant immune checkpoint blockade in high-risk resectable melanoma. Nat Med 24(11):1649–1654

83. Rozeman EA, Menzies AM, van Akkooi ACJ, Adhikari C, Bierman C, van de Wiel BA et al (2019) Identification of the optimal combination dosing schedule of neoadjuvant ipilimumab plus nivolumab in macroscopic stage III melanoma (OpACIN-neo): a multicentre, phase 2, randomised, controlled trial. Lancet Oncol 20(7):948–960

84. Rossi E, Schinzari G, Zizzari IG, Maiorano BA, Pagliara MM, Sammarco MG et al (2019) Immunological backbone of uveal melanoma: is there a rationale for immunotherapy? Cancers (Basel) 11(8)

85. Tosi A, Cappellesso R, Dei Tos AP, Rossi V, Aliberti C, Pigozzo J et al (2021) The immune cell landscape of metastatic uveal melanoma correlates with overall survival. J Exp Clin Cancer Res 40(1):154

86. Piperno-Neumann S, Hassel J, Rutkowski P, Baurain JF, Butler MO, Schlaak M et al (2021) Phase 3 randomized trial comparing tebentafusp with investigator's choice in first line metastatic uveal melanoma. AACR Ann Meeting

87. Hamid O, Robert C, Ribas A, Hodi FS, Walpole E, Daud A et al (2018) Antitumour activity of pembrolizumab in advanced mucosal melanoma: a post-hoc analysis of KEYNOTE-001, 002, 006. Br J Cancer 119(6):670–674

88. D'Angelo SP, Larkin J, Sosman JA, Lebbe C, Brady B, Neyns B et al (2017) Efficacy and safety of nivolumab alone or in combination with ipilimumab in patients with mucosal melanoma: a pooled analysis. J Clin Oncol 35(2):226–235

89. Furney SJ, Turajlic S, Stamp G, Nohadani M, Carlisle A, Thomas JM et al (2013) Genome sequencing of mucosal melanomas reveals that they are driven by distinct mechanisms from cutaneous melanoma. J Pathol 230(3):261–269

90. Sloan AE, Nock CJ, Einstein DB (2009) Diagnosis and treatment of melanoma brain metastasis: a literature review. Cancer Control 16(3):248–255

91. Gampa G, Vaidhyanathan S, Resman BW, Parrish KE, Markovic SN, Sarkaria JN et al (2016) Challenges in the delivery of therapies to melanoma brain metastases. Curr Pharmacol Rep 2(6):309–325

92. Tawbi HA, Forsyth PA, Algazi A, Hamid O, Hodi FS, Moschos SJ et al (2018) Combined nivolumab and ipilimumab in melanoma metastatic to the brain. N Engl J Med 379(8):722–730

93. Long GV, Atkinson VG, Lo S, Sandhu SK, Brown M, Gonzalez M et al (2019) 1311O-Long-term outcomes from the randomized phase II study of nivolumab (nivo) or nivo+ipilimumab (ipi) in patients (pts) with melanoma brain metastases (mets): anti-PD1 brain collaboration (ABC). Ann Oncol 30:v534

94. Gonzalez M, Hong AM, Carlino MS, Atkinson V, Wang W, Lo S et al (2019) A phase II, open label, randomized controlled trial of nivolumab plus ipilimumab with stereotactic radiotherapy versus ipilimumab plus nivolumab alone in patients with melanoma brain metastases (ABC-X Trial). J Clin Oncol 37(15_suppl):TPS9600-TPS

95. Martin JH, Lewis LD (2020) Taking the brake off the immune system: Hypotheses, trials, tribulations, and the evolving discipline of clinical immunopharmacology. Br J Clin Pharmacol 86(9):1674–1677

96. Le Louedec F, Leenhardt F, Marin C, Chatelut E, Evrard A, Ciccolini J (2020) Cancer immunotherapy dosing: a pharmacokinetic/pharmacodynamic perspective. Vaccines (Basel) 8(4)

97. Chatelut E, Le Louedec F, Milano G (2020) Setting the dose of checkpoint inhibitors: the role of clinical pharmacology. Clin Pharmacokinet 59(3):287–296

98. Hurkmans DP, Basak EA, van Dijk T, Mercieca D, Schreurs MWJ, Wijkhuijs AJM et al (2019) A prospective cohort study on the pharmacokinetics of nivolumab in metastatic non-small cell lung cancer, melanoma, and renal cell cancer patients. J Immunother Cancer 7(1):192

99. Centanni M, Moes D, Troconiz IF, Ciccolini J, van Hasselt JGC (2019) Clinical pharmacokinetics and pharmacodynamics of immune checkpoint inhibitors. Clin Pharmacokinet 58(7):835–857

100. Keizer RJ, Huitema AD, Schellens JH, Beijnen JH (2010) Clinical pharmacokinetics of therapeutic monoclonal antibodies. Clin Pharmacokinet 49(8):493–507

101. Sheng J, Srivastava S, Sanghavi K, Lu Z, Schmidt BJ, Bello A et al (2017) Clinical pharmacology considerations for the development of immune checkpoint inhibitors. J Clin Pharmacol 57(Suppl 10):S26–S42

102. Kang SP, Gergich K, Lubiniecki GM, de Alwis DP, Chen C, Tice MAB et al (2017) Pembrolizumab KEYNOTE-001: an adaptive study leading to accelerated approval for two indications and a companion diagnostic. Ann Oncol 28(6):1388–1398

103. Leven C, Padelli M, Carre JL, Bellissant E, Misery L (2019) Immune checkpoint inhibitors in melanoma: a review of pharmacokinetics and exposure-response relationships. Clin Pharmacokinet

104. Ahamadi M, Freshwater T, Prohn M, Li CH, de Alwis DP, de Greef R et al (2017) Model-based characterization of the pharmacokinetics of pembrolizumab: a humanized anti-PD-1 monoclonal antibody in advanced solid tumors. CPT Pharmacometrics Syst Pharmacol 6(1):49–57

105. Jardim DL, de Melo GD, Giles FJ, Kurzrock R (2018) Analysis of drug development paradigms for immune checkpoint inhibitors. Clin Cancer Res 24(8):1785–1794

106. Elassaiss-Schaap J, Rossenu S, Lindauer A, Kang SP, de Greef R, Sachs JR et al (2017) Using model-based "Learn and Confirm" to reveal the pharmacokinetics-pharmacodynamics relationship of pembrolizumab in the KEYNOTE-001 Trial. CPT Pharmacometrics Syst Pharmacol 6(1):21–28

107. Merck. Keytruda (Pembrolizumab) [package insert]. US food and drug administration website

108. Freshwater T, Kondic A, Ahamadi M, Li CH, de Greef R, de Alwis D et al (2017) Evaluation of dosing strategy for pembrolizumab for oncology indications. J Immunother Cancer 5:43

109. Zhao X, Suryawanshi S, Hruska M, Feng Y, Wang X, Shen J et al (2017) Assessment of nivolumab benefit-risk profile of a 240-mg flat dose relative to a 3-mg/kg dosing regimen in patients with advanced tumors. Ann Oncol 28(8):2002–2008

110. Long GV, Tykodi SS, Schneider JG, Garbe C, Gravis G, Rashford M et al (2018) Assessment of nivolumab exposure and clinical safety of 480 mg every 4 weeks flat-dosing schedule in patients with cancer. Ann Oncol 29(11):2208–2213

111. Agrawal S, Feng Y, Roy A, Kollia G, Lestini B (2016) Nivolumab dose selection: challenges, opportunities, and lessons learned for cancer immunotherapy. J Immunother Cancer 4:72

112. Forde PM, Chaft JE, Smith KN, Anagnostou V, Cottrell TR, Hellmann MD et al (2018) Neoadjuvant PD-1 blockade in resectable lung cancer. N Engl J Med 378(21):1976–1986

113. Yoo SH, Keam B, Kim M, Kim SH, Kim YJ, Kim TM et al (2018) Low-dose nivolumab can be effective in non-small cell lung cancer: alternative option for financial toxicity. ESMO Open 3(5):e000332

114. Ascierto PA, Del Vecchio M, Robert C, Mackiewicz A, Chiarion-Sileni V, Arance A et al (2017) Ipilimumab 10 mg/kg versus ipilimumab 3 mg/kg in patients with unresectable or metastatic melanoma: a randomised, double-blind, multicentre, phase 3 trial. Lancet Oncol 18(5):611–622

115. Ascierto PA, Marabelle A (2018) How do immune checkpoint-targeted antibodies work? The need for improved pharmacokinetic evaluation in early phase studies. Ann Oncol 29(11):2157–2160

116. Lucas CJ, Martin JH (2017) Pharmacokinetic-guided dosing of new oral cancer agents. J Clin Pharmacol 57(Suppl 10):S78–S98

117. Bjarnason GA, Knox JJ, Kollmannsberger CK, Soulieres D, Ernst DS, Zalewski P et al (2019) The efficacy and safety of sunitinib given on an individualised schedule as first-line therapy for metastatic renal cell carcinoma: a phase 2 clinical trial. Eur J Cancer 108:69–77

118. Chatelut E, Hendrikx J, Martin J, Ciccolini J, Moes D (2021) Unraveling the complexity of therapeutic drug monitoring for monoclonal antibody therapies to individualize dose in oncology. Pharmacol Res Perspect 9(2):e00757

119. Oude Munnink TH, Henstra MJ, Segerink LI, Movig KL, Brummelhuis-Visser P (2016) Therapeutic drug monitoring of monoclonal antibodies in inflammatory and malignant disease: translating TNF-alpha experience to oncology. Clin Pharmacol Ther 99(4):419–431

120. Baverel PG, Dubois VFS, Jin CY, Zheng Y, Song X, Jin X et al (2018) Population pharmacokinetics of durvalumab in cancer patients and association with longitudinal biomarkers of disease status. Clin Pharmacol Ther 103(4):631–642

121. Liu C, Yu J, Li H, Liu J, Xu Y, Song P et al (2017) Association of time-varying clearance of nivolumab with disease dynamics and its implications on exposure response analysis. Clin Pharmacol Ther 101(5):657–666

122. Li H, Sun Y, Yu J, Liu C, Liu J, Wang Y (2019) Semimechanistically based modeling of pembrolizumab time-varying clearance using 4 longitudinal covariates in patients with non-small cell lung cancer. J Pharm Sci 108(1):692–700

123. Coss CC, Clinton SK, Phelps MA (2018) Cachectic cancer patients: immune to checkpoint inhibitor therapy? Clin Cancer Res 24(23):5787–5789

124. Navani V, Graves MC, Marchett GC, Mandaliya H, Bowden NA, van der Westhuizen A (2021) Overall survival in metastatic melanoma correlates with pembrolizumab exposure and T-cell exhaustion markers. Pharmacol Res Perspect 9(4):e00808

125. Turner DC, Kondic AG, Anderson KM, Robinson AG, Garon EB, Riess JW et al (2018) Pembrolizumab exposure-response assessments challenged by association of cancer cachexia and catabolic clearance. Clin Cancer Res 24(23):5841–5849

126. Wang X, Feng Y, Bajaj G, Gupta M, Agrawal S, Yang A et al (2017) Quantitative characterization of the exposure-response relationship for cancer immunotherapy: a case study of nivolumab in patients with advanced melanoma. CPT Pharmacometrics Syst Pharmacol 6(1):40–48

127. Bajaj G, Gupta M, Feng Y, Statkevich P, Roy A (2017) Exposure-response analysis of nivolumab in patients with previously treated or untreated advanced melanoma. J Clin Pharmacol 57(12):1527–1533

128. Chen DS, Mellman I (2017) Elements of cancer immunity and the cancer-immune set point. Nature 541(7637):321–330

129. Halse H, Colebatch AJ, Petrone P, Henderson MA, Mills JK, Snow H et al (2018) Multiplex immunohistochemistry accurately defines the immune context of metastatic melanoma. Sci Rep 8(1):11158

130. Murray HC, Maltby VE, Smith DW, Bowden NA (2015) Nucleotide excision repair deficiency in melanoma in response to UVA. Exp Hematol Oncol 5:6
131. Buttner R, Longshore JW, Lopez-Rios F, Merkelbach-Bruse S, Normanno N, Rouleau E et al (2019) Implementing TMB measurement in clinical practice: considerations on assay requirements. ESMO Open 4(1):e000442
132. Chabanon RM, Pedrero M, Lefebvre C, Marabelle A, Soria JC, Postel-Vinay S (2016) Mutational landscape and sensitivity to immune checkpoint blockers. Clin Cancer Res 22(17):4309–4321
133. Yadav M, Jhunjhunwala S, Phung QT, Lupardus P, Tanguay J, Bumbaca S et al (2014) Predicting immunogenic tumour mutations by combining mass spectrometry and exome sequencing. Nature 515(7528):572–576
134. Trujillo JA, Sweis RF, Bao R, Luke JJ (2018) T-cell-inflamed versus non-T-cell-inflamed tumors: a conceptual framework for cancer immunotherapy drug development and combination therapy selection. Cancer Immunol Res 6(9):990–1000
135. Herbst RS, Soria JC, Kowanetz M, Fine GD, Hamid O, Gordon MS et al (2014) Predictive correlates of response to the anti-PD-L1 antibody MPDL3280A in cancer patients. Nature 515(7528):563–567
136. Spranger S, Bao R, Gajewski TF (2015) Melanoma-intrinsic beta-catenin signalling prevents anti-tumour immunity. Nature 523(7559):231–235
137. Yarchoan M, Hopkins A, Jaffee EM (2017) Tumor mutational burden and response rate to PD-1 inhibition. N Engl J Med 377(25):2500–2501
138. FDA approves pembrolizumab for adults and children with TMB-H solid tumors [press release]. FDA, 16th June 2020
139. Strickler JH, Hanks BA, Khasraw M (2021) Tumor mutational burden as a predictor of immunotherapy response: is more always better? Clin Cancer Res 27(5):1236–1241
140. Chan TA, Yarchoan M, Jaffee E, Swanton C, Quezada SA, Stenzinger A et al (2019) Development of tumor mutation burden as an immunotherapy biomarker: utility for the oncology clinic. Ann Oncol 30(1):44–56
141. Sha D, Jin Z, Budczies J, Kluck K, Stenzinger A, Sinicrope FA (2020) Tumor Mutational burden as a predictive biomarker in solid tumors. Cancer Discov 10(12):1808–1825
142. Hugo W, Zaretsky JM, Sun L, Song C, Moreno BH, Hu-Lieskovan S et al (2016) Genomic and transcriptomic features of response to anti-PD-1 therapy in metastatic melanoma. Cell 165(1):35–44
143. Walker MS, Hughes TA (2008) Messenger RNA expression profiling using DNA microarray technology: diagnostic tool, scientific analysis or un-interpretable data? Int J Mol Med 21(1):13–17
144. Ayers M, Lunceford J, Nebozhyn M, Murphy E, Loboda A, Kaufman DR et al (2017) IFN-gamma-related mRNA profile predicts clinical response to PD-1 blockade. J Clin Invest 127(8):2930–2940
145. Cristescu R, Mogg R, Ayers M, Albright A, Murphy E, Yearley J et al (2018) Pan-tumor genomic biomarkers for PD-1 checkpoint blockade-based immunotherapy. Science 362(6411)
146. Anagnostou V, Bruhm DC, Niknafs N, White JR, Shao XM, Sidhom JW et al (2020) Integrative tumor and immune cell multi-omic analyses predict response to immune checkpoint blockade in melanoma. Cell Rep Med 1(8):100139
147. Cabrita R, Lauss M, Sanna A, Donia M, Skaarup Larsen M, Mitra S et al (2020) Tertiary iymphoid structures improve immunotherapy and survival in melanoma. Nature 577(7791):561–565
148. Schadendorf D, Ascierto PA, Haanen J, Espinosa E, Demidov L, Garbe C et al (2019) Safety and efficacy of nivolumab in challenging subgroups with advanced melanoma who progressed on or after ipilimumab treatment: a single-arm, open-label, phase II study (CheckMate 172). Eur J Cancer 121:144–153

149. Wong A, Williams M, Milne D, Morris K, Lau P, Spruyt O et al (2017) Clinical and palliative care outcomes for patients of poor performance status treated with antiprogrammed death-1 monoclonal antibodies for advanced melanoma. Asia Pac J Clin Oncol 13(6):385–390

150. Miller S, Senior PV, Prakash M, Apostolopoulos V, Sakkal S, Nurgali K (2016) Leukocyte populations and IL-6 in the tumor microenvironment of an orthotopic colorectal cancer model. Acta Biochim Biophys Sin (Shanghai) 48(4):334–341

151. Shibata M, Nezu T, Kanou H, Abe H, Takekawa M, Fukuzawa M (2002) Decreased production of IL-12 and type 2 immune responses are marked in cachectic patients with colorectal and gastric cancer. J Clin Gastroenterol 34(4):416–420

152. Kluger HM, Tawbi HA, Ascierto ML, Bowden M, Callahan MK, Cha E et al (2020) Defining tumor resistance to PD-1 pathway blockade: recommendations from the first meeting of the SITC immunotherapy resistance taskforce. J Immunother Cancer 8(1).

153. Sharma P, Hu-Lieskovan S, Wargo JA, Ribas A (2017) Primary, adaptive, and acquired resistance to cancer immunotherapy. Cell 168(4):707–723

154. Gubin MM, Zhang X, Schuster H, Caron E, Ward JP, Noguchi T et al (2014) Checkpoint blockade cancer immunotherapy targets tumour-specific mutant antigens. Nature 515(7528):577–581

155. Sucker A, Zhao F, Real B, Heeke C, Bielefeld N, Mabetaen S et al (2014) Genetic evolution of T-cell resistance in the course of melanoma progression. Clin Cancer Res 20(24):6593–6604

156. Peng W, Chen JQ, Liu C, Malu S, Creasy C, Tetzlaff MT et al (2016) Loss of PTEN promotes resistance to T-cell-mediated immunotherapy. Cancer Discov 6(2):202–216

157. Tumeh PC, Harview CL, Yearley JH, Shintaku IP, Taylor EJ, Robert L et al (2014) PD-1 blockade induces responses by inhibiting adaptive immune resistance. Nature 515(7528):568–571

158. Taube JM, Klein A, Brahmer JR, Xu H, Pan X, Kim JH et al (2014) Association of PD-1, PD-1 ligands, and other features of the tumor immune microenvironment with response to anti-PD-1 therapy. Clin Cancer Res 20(19):5064–5074

159. Shin DS, Zaretsky JM, Escuin-Ordinas H, Garcia-Diaz A, Hu-Lieskovan S, Kalbasi A et al (2017) Primary resistance to PD-1 blockade mediated by JAK1/2 mutations. Cancer Discov 7(2):188–201

160. Schachter J, Ribas A, Long GV, Arance A, Grob JJ, Mortier L et al (2017) Pembrolizumab versus ipilimumab for advanced melanoma: final overall survival results of a multicentre, randomised, open-label phase 3 study (KEYNOTE-006). Lancet 390(10105):1853–1862

161. Gide TN, Wilmott JS, Scolyer RA, Long GV (2018) Primary and acquired resistance to immune checkpoint inhibitors in metastatic melanoma. Clin Cancer Res 24(6):1260–1270

162. Lee JH, Shklovskaya E, Lim SY, Carlino MS, Menzies AM, Stewart A et al (2020) Transcriptional downregulation of MHC class I and melanoma de- differentiation in resistance to PD-1 inhibition. Nat Commun 11(1):1897

163. Huang AC, Postow MA, Orlowski RJ, Mick R, Bengsch B, Manne S et al (2017) T-cell invigoration to tumour burden ratio associated with anti-PD-1 response. Nature 545(7652):60–65

164. Pauken KE, Sammons MA, Odorizzi PM, Manne S, Godec J, Khan O et al (2016) Epigenetic stability of exhausted T-cells limits durability of reinvigoration by PD-1 blockade. Science 354(6316):1160–1165

165. Zaretsky JM, Garcia-Diaz A, Shin DS, Escuin-Ordinas H, Hugo W, Hu-Lieskovan S et al (2016) Mutations associated with acquired resistance to PD-1 blockade in melanoma. N Engl J Med 375(9):819–829

166. Sade-Feldman M, Jiao YJ, Chen JH, Rooney MS, Barzily-Rokni M, Eliane JP et al (2017) Resistance to checkpoint blockade therapy through inactivation of antigen presentation. Nat Commun 8(1):1136

167. Platanias LC (2005) Mechanisms of type-I- and type-II-interferon-mediated signalling. Nat Rev Immunol 5(5):375–386

168. Ribas A (2015) Adaptive immune resistance: how cancer protects from immune attack. Cancer Discov 5(9):915–919

169. Liu X, Hogg GD, DeNardo DG (2021) Rethinking immune checkpoint blockade: 'Beyond the T-cell'. J Immunother Cancer 9(1)
170. Linch SN, McNamara MJ, Redmond WL (2015) OX40 agonists and combination immunotherapy: putting the pedal to the metal. Front Oncol 5:34
171. Workman CJ, Wang Y, El Kasmi KC, Pardoll DM, Murray PJ, Drake CG et al (2009) LAG-3 regulates plasmacytoid dendritic cell homeostasis. J Immunol 182(4):1885–1891
172. Sarhan D, Cichocki F, Zhang B, Yingst A, Spellman SR, Cooley S et al (2016) Adaptive NK cells with Low TIGIT expression are inherently resistant to myeloid-derived suppressor cells. Cancer Res 76(19):5696–5706
173. Chiba S, Baghdadi M, Akiba H, Yoshiyama H, Kinoshita I, Dosaka-Akita H et al (2012) Tumor-infiltrating DCs suppress nucleic acid-mediated innate immune responses through interactions between the receptor TIM-3 and the alarmin HMGB1. Nat Immunol 13(9):832–842
174. Lipson EJ, Tawbi HA-H, Schadendorf D, Ascierto PA, Matamala L, Gutiérrez EC et al (2021) Relatlimab (RELA) plus nivolumab (NIVO) versus NIVO in first-line advanced melanoma: primary phase III results from RELATIVITY-047 (CA224–047). J Clin Oncol 39(15_suppl):9503
175. Kimura T, Kato Y, Ozawa Y, Kodama K, Ito J, Ichikawa K et al (2018) Immunomodulatory activity of lenvatinib contributes to antitumor activity in the Hepa1-6 hepatocellular carcinoma model. Cancer Sci 109(12):3993–4002
176. Arance AM, Cruz-Merino Ldl, Petrella TM, Jamal R, Ny L, Carneiro A et al (2021) Lenvatinib (len) plus pembrolizumab (pembro) for patients (pts) with advanced melanoma and confirmed progression on a PD-1 or PD-L1 inhibitor: updated findings of LEAP-004. J Clin Oncol 39(15_suppl):9504
177. Buchbinder EI, Dutcher JP, Daniels GA, Curti BD, Patel SP, Holtan SG et al (2019) Therapy with high-dose IL-2 (HD IL-2) in metastatic melanoma and renal cell carcinoma following PD1 or PDL1 inhibition. J Immunother Cancer 7(1):49
178. Kim ST, Smith SA, Mortimer P, Loembe AB, Cho H, Kim KM et al (2021) Phase I study of ceralasertib (AZD6738), a Novel DNA damage repair agent, in combination with weekly paclitaxel in refractory cancer. Clin Cancer Res 27(17):4700–4709
179. Rosenberg SA, Yang JC, Sherry RM, Kammula US, Hughes MS, Phan GQ et al (2011) Durable complete responses in heavily pretreated patients with metastatic melanoma using T-cell transfer immunotherapy. Clin Cancer Res 17(13):4550–4557
180. Sarnaik AA, Hamid O, Khushalani NI, Lewis KD, Medina T, Kluger HM et al (2021) Lifileucel, a tumor-infiltrating lymphocyte therapy metastatic melanoma. J Clin Oncol 39(24):2656–2666
181. Micevic G, Theodosakis N, Bosenberg M (2017) Aberrant DNA methylation in melanoma: biomarker and therapeutic opportunities. Clin Epigenetics 9:34
182. Kulis M, Esteller M (2010) 2-DNA methylation and cancer. In: Herceg Z, Ushijima T (eds) Advances in genetics, vol 70. Academic Press, pp 27–56
183. Chiappinelli KB, Zahnow CA, Ahuja N, Baylin SB (2016) Combining epigenetic and immunotherapy to combat cancer. Can Res 76(7):1683
184. Early phase II study of Azacitidine and Carboplatin priming for Avelumab in patients with advanced melanoma who are resistant to immunotherapy. [Available from: https://www.aus tralianclinicaltrials.gov.au/anzctr/trial/ACTRN12618000053224]

Engaging Pattern Recognition Receptors in Solid Tumors to Generate Systemic Antitumor Immunity

3

Michael Brown

3.1 Pattern Recognition Receptors (PRRs) Induce Innate Inflammation

Pattern recognition receptors (PRRs) recognize rigid features known as pathogen associated molecular patterns ('PAMPs') or damage associated molecular patterns ('DAMPs') either in the extracellular space, endosome, or cytoplasm to induce appropriate inflammation during pathogen infection and/or tissue damage. Canonical PRRs include Toll-like Receptors (TLRs), of which there are 10 (TLRs 1-10) in humans; RIG-I like receptors (RLRs) including MDA5 and RIG-I; cytosolic double stranded DNA sensors (e.g., cGAS-STING); the AIM2-like receptors; the NOD-like receptors; and C-type lectin receptors. For information on PRRs, their locations, and specificities, see Fig. 3.1. Upon recognition of PAMPs or DAMPs by PRRs, signaling to the Tank Binding Kinase 1 (TBK1) and IKK-α/β kinases primarily lead to IRF3 phosphorylation and NFκB activation, respectively [1, 2], to concertedly lead to the synthesis of pro-inflammatory cytokines (e.g., type I IFNs, TNF, IL-6) and, in DCs, induce co-stimulatory ligand expression (e.g. CD86, CD80).

3.2 PRR Signaling Dictates CD8$^+$ T Cell Priming, Recruitment, and Function During Viral Infection

Leveraging antitumor functions of CD8$^+$ T cells to eliminate malignant cells in an antigen-specific manner is the goal of most cancer immunotherapy strategies.

M. Brown (✉)
Department of Neurosurgery, Duke University, Durham, NC, USA
e-mail: mcb52@duke.edu

© The Author(s), under exclusive license to Springer Nature Switzerland AG 2022   91
P. Hays (ed.), *Cancer Immunotherapies*, Cancer Treatment and Research 183,
https://doi.org/10.1007/978-3-030-96376-7_3

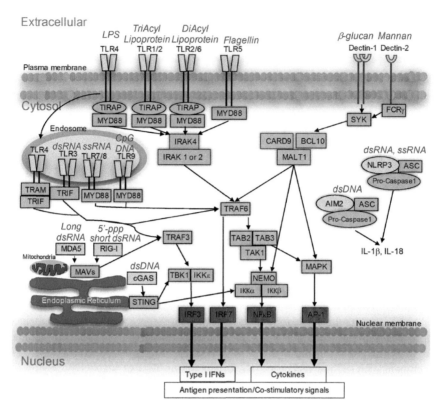

Fig. 3.1 PRR cellular locations, specificities, and downstream signaling. Yellow indicates PRRs, blue indicates signaling adapters/kinases, and red indicates transcription factors that mediate transcription of inflammatory genes. PAMPs that activate PRRs are depicted in italicized red text. Toll-like Receptors 1, 2 and 4–6 are located on the cell surface and recognize bacterial features such as lipids, proteins, and lipoproteins. TLRs 3 and 7–9 are localized to endosomes, and recognize viral nucleic acids. TLR4 is both extracellular and endosomal. The specificity and function of TLR10 (not shown) is currently obscure, but in contrast to other TLRs, may negatively regulate inflammation [3]. The RIG-I like receptors, RIG-I and MDA5, recognize cytosolic viral double stranded (ds) RNA and have recently been shown to become activated at endoplasmic reticulum (ER) derived microsomes [4]. Cytosolic DNA sensing by cGAS-STING is mediated at the ER. The AIM2 inflammasome recognizes dsDNA in the cytosol to initiate cleavage of caspase-1, followed by cleavage of pro-IL-1β and pro-IL-18 to their mature, secreted state. The NLRP3 inflammasome (a NOD like receptor) recognizes various DAMP and PAMP features, including viral dsRNA and single stranded RNA, and similarly leads to caspase-1 activation. Several variations of inflammasomes recognizing diverse features are not shown. The C-type lectin receptors Dectin-1 and 2 recognize bacterial and fungal features; several additional C-type lectins with various specificities are not shown. The transcription factors IRF3 and IRF7 largely drive transcription of type I interferons (IFNs) while NFkB and AP-1 induce other pro-inflammatory cyotkines. The activated transcription factors also induce co-stimulatory ligand expression, as well as anti-viral/anti-bacterial gene products

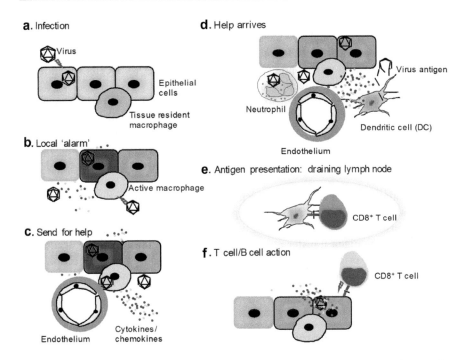

Fig. 3.2 Innate immunity engages CD8⁺ T cells during viral infection. See text for stepwise details

In a natural infectious setting, CD8⁺ T cells are enlisted to eliminate intracellular pathogens, e.g., viruses. Provided below, and depicted in Fig. 3.2, is an example of how a typical acute viral infection leads to priming of antiviral CD8⁺ T cells via the activation of PRRs. The concepts of antiviral CD8⁺ T cell priming and effector functions during an infectious process are analogous to events that must occur to enable priming and effector function of antitumor CD8⁺ T cells.

Step 1: Viral infection of epithelium occurs (Fig 3.2a).

Step 2: Local inflammation is induced after recognition of viral features (e.g. viral nucleic acids) by PRRs expressed on tissue resident macrophages, infected epithelium, or by other tissue resident innate immune populations (Fig 3.2b).

Step 3: PRR mediated inflammatory signals leads to surface expression of adhesion molecules on the local endothelium (Fig 3.2c).

Step 4: The induction of adhesion molecules on endothelium in concert with chemokines/cytokines facilitates recruitment of additional innate immune cells (Fig. 3.2d). Much of the innate inflammation at this stage functions to limit viral replication and spread, and may lead to killing of infected cells by innate immune populations including NK cells, neutrophils, and macrophages. In addition, during this process conventional DCs consume antigens in the infected site, while receiving activating signals from inflammatory cytokines from infected cells and/or direct PRR signaling (Fig 3.2d) that induces their 'activation'—which includes the induction of co-stimulatory signals.

Step 5: Antigen-bearing DCs then migrate to the draining lymph node or other lymphatic organ to prime and expand populations of CD8$^+$ T cells recognizing viral antigen from the infection site (Fig 3.2e).

Step 6: Activated CD8$^+$ T cells chemotract to recognize and kill remaining infected cells, and inflammation from the infection induces antigen presentation machinery and stress signals in infected cells that further enable T cell mediated killing. A pool of memory T cells persist after the infection is cleared for future pathogen recognition and elimination (Fig 3.2f).

Notably variations in the routes by which these steps occur are pathogen and tissue specific; alternate modes of CD8$^+$ T cell priming have been demonstrated, e.g. via antigen transfer between migratory vs lymph node resident DCs [5, 6]; and other mechanisms of antigen transfer at sites distant from the infection may occur [7]. These processes, originally defined in the context of natural viral infection [8], have been shown to largely apply to immune surveillance, that is, the recognition and elimination of malignant cells by the immune system.

3.3   The Cancer Immunity Cycle and PRR Signaling

The host immune system recognizes malignant cells on the basis of protein coding genetic mutations; abnormal post-translational modifications; aberrantly expressed proteins (e.g., cancer-testis antigens); and in some cancer types, oncogenic viral proteins. Recognition of such antigens is mediated by cell surface MHC class I for CD8$^+$ T cells, and surface MHC class II for CD4$^+$ T cells, with CD8$^+$ T cells typically being the primary antitumor effectors during immune surveillance. While co-evolution between malignant cells and the host immune system eliminates immunogenic malignant cells and results in outgrowth of 'immunoedited' tumors that are less immunogenic [9], and heterogeneity in the expression of tumor associated antigens is common [10], the success of immune checkpoint blockade in several tumor types implies that other potentially reversible regulatory nodes prevent immune recognition and destruction of solid tumors.

The cancer immunity cycle outlines established steps by which antitumor T cells can become activated endogenously to eliminate malignant cells [11] (Fig. 3.3). Dying cancer cells release antigens that are taken up by dendritic cells (DCs) to be loaded on to MHC-class I or II. Tumor antigen presenting DCs present antigen along with co-stimulatory signals to T cells, typically within the tumor draining lymph node. If the appropriate co-stimulatory signals along with cognate antigen are delivered to tumor antigen-specific T cells at this step, activated tumor-specific T cells may traffic to the tumor site, recognize tumor antigen-expressing malignant cells, and mediate killing of malignant cells through several mechanisms. The cytotoxic mechanisms of T cells include the release of perforin and

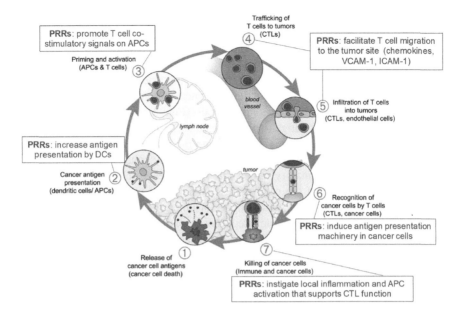

Fig. 3.3 PRR signaling supports the cancer immunity cycle through multiple mechanisms, adapted from Chen and Mellman *Immunity* 2013 [11] with permission. Red boxes denote mechanisms by which PRR signaling impacts cancer immune surveillance. (1) Cancer cells routinely die due to genotoxic stress, chemo/radiation, nutrient deprivation, hypoxia, and other reasons; leading to the release of tumor antigens. (2) Migrating antigen presenting cells, including migratory conventional dendritic cells capable of cross presenting antigens to CD8$^+$ T cells (see Fig. 3.4), take up antigen in the tumor microenvironment to be processed and presented on MHC-class I or II. PRRs potentiate this step by increasing antigen presentation on DCs, increasing antigen uptake (e.g. via calreticulin and HSP surface expression on apoptotic cells) and potentially recruiting additional migratory DCs into tumors. (3) Within the draining lymph node or other secondary lymphoid organ, tumor antigen loaded DCs present antigen to T cells, leading to their activation. PRR signaling induces the expression of co-stimulatory signals on DCs to potentiate T cell priming; see Fig. 3.4 for detailed explanation. (4) T cells traffic to the site of the tumor by surveying endothelial ligand (e.g. ICAM-1 and VCAM-1) expression and chemokine signals, and (5) infiltrate the tumor tissue. PRRs enhance endothelial cell T cell adhesion ligand expression and chemokine secretion from the tumor site to facilitate T cell infiltration. (6) T cells recognize cognate tumor antigen presented on tumor cells; PRRs facilitate this process by inducing inflammation that causes induction of antigen presentation machinery in cancer cells. (7) T cells kill cancer cells expressing their cognate antigen via granzymes and perforin, FAS ligand, and secretion of cytotoxic cytokines; PRRs induce inflammation that enhances antitumor T cell function and cytotoxicity

granzymes, apoptotic signals (FAS-ligand), cytokines that mediate apoptotic signaling, e.g., TNF, as well as cytokines that induce upregulation of MHC class I and antigen processing machinery in tumor cells, particularly IFN-γ.

However, several factors determine whether tumor-specific T cells will ultimately become tolerized and anergic after DC mediated antigen presentation, whether sufficient signals enable trafficking of tumor-reactive T cells to the tumor site, as well as whether antitumor T cells can function within the immune subversive tumor microenvironment (TME). These issues are dictated by that status of the innate immune system during the cancer immunity cycle, including that of DCs and tumor associated macrophages/myeloid derived suppressor cells (MDSCs). PRRs play a multifaceted role in determining the pace, and efficacy of the cancer immunity lifecycle. PRR signaling enhances antigen presentation and co-stimulatory signals on antigen presenting cells (Fig. 3.4), culminating in more effective priming of tumor antigen-specific T cells. In addition, intratumor activation of PRRs induces chemokines that facilitate the recruitment of newly primed antitumor T cells, and further supports their function by enhancing tumor antigen presentation and an inflammatory milieu that potentiates T cell effector functions. For detailed explanation of the role of PRR signaling at each step of the cancer immunity cycle, see Fig. 3.3. Accordingly, an emerging clinical strategy aimed at rectifying stalled cancer immune lifecycles in patients is that of targeting the activation of PRRs within the TME to provoke the expression of: co-stimulatory signals on DCs during antigen presentation, T cell recruiting chemokines within

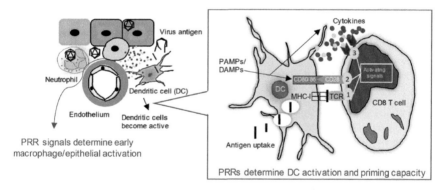

Fig. 3.4 Dendritic cell (DC) activation is dictated by PPR signaling, which enables CD8$^+$ T cell cross-priming. An example of cross presentation is shown, which occurs via loading of engulfed exogenous antigen onto MHC-class I for presentation to CD8$^+$ T cells, typically by cDC1s (CD103$^+$, BATF3$^+$ in mice; CD141$^+$ in humans). In the context of viral infection, PRRs expressed by viral antigen presenting DCs sense PAMPs and DAMPs at the site of inflammation/infection and receive signals from locally produced cytokines. These signals lead to upregulation of antigen processing and presentation machinery (signal 1), induction of co-stimulatory ligands including the CD28 ligands CD80 and CD86 (signal 2), and secretion of cytokines like IL-12 and type I IFNs that lead to further activation and differentiation of the cognate antigen-specific CD8$^+$ T cell (signal 3). Upon receiving these signals, CD8$^+$ T cells become activated and can traffic to the site of infection to eliminate virally infected cells. Similarly, during the cancer immunity cycle, migratory cDCs take up tumor associated antigens, traffic to the draining lymph node, and present exogenously acquired antigen on MHC-class I to T cells. As occurs during viral infection, PRR signaling bolsters the efficiency of tumor antigen uptake, processing, and presentation on MHC-class I on DCs (Signal 1); the induction of co-stimulatory ligands expressed on the DC surface (Signal 2); as well as the induction of pro-inflammatory cytokines required for optimal antitumor CD8$^+$ T cell priming (Signal 3)

the tumor site, and inflammation within the TME that supports antitumor CD8$^+$ T cell function.

3.4 Endogenous Activation of PRRs in Cancer

Endogenous signals from cells that are stressed or dying necrotically can induce DC activation in the absence of foreign pathogen associated features [12]; lending explanation for how T cell priming and activation in the contexts of spontaneous antitumor immunity, transplantation rejection, and/or autoimmunity occurs. Recent work has documented the importance of several PRRs in the context of tumor biology that lead to DC activation, and generalized inflammation within tumors. These include the following PAMPs and DAMPs, which represent only a subset of relevant documented endogenous PRR ligands:

1. *Double stranded DNA (dsDNA) from dying tumor cells* can be recognized by cGAS-STING, particularly within DCs that take up debris from dead tumor cells. STING signaling culminates in IFNβ/antiviral responses that mediates activation of systemic antitumor T cell immunity and tumor regression [13, 14].
2. *High-mobility-group box 1 (HMGB1)* is a nuclear protein that is released during cell death, including after chemotherapy/radiation, to engage TLR4 on DCs [15].
3. *Cell surface calreticulin* facilitates phagocytosis of apoptotic tumor cells by DCs and macrophages and determines the immunogenicity of phagocytosed cells [16–18].
4. *Heat Shock Proteins (HSPs)* released from dying cancer cells are widely reported to bind to TLRs 2 and 4 to induce inflammation [19–24].
5. *Endogenous retroviruses* have been shown to reactivate in some cancers presumably due to epigenetic dysregulation or loss of innate signaling in malignant cells [25], and can induce TLR and RLR signaling due to cytoplasmic presence of replicating retroviral RNA [26].
6. *Uric acid/monosodium urate crystals*, a byproduct of purine metabolism that causes gout, have been shown to induce activation of the NLRP3 inflammasome [27], TLR2, and TLR4 [28]. Uric acid mediates DC activation [29].
7. *The tumor microbiome* has recently been defined, demonstrating evidence of microbial presence (e.g., intracellular bacteria and viruses) in various tumor types at baseline [30, 31]. The presence of such microbes is likely to impact endogenous PRR signaling in tumors, though this remains to be determined.

Thus, PRRs in tumors are not inert, but rather may recognize features associated with cell death; tissue damage; and in some cases, endogenous pathogens. Collectively, their activity in cancer likely explains how spontaneous antitumor T cells are primed to eliminate malignant cells. In contrast, a lack of co-stimulation from DCs to antitumor T cells is frequent in cancer [32, 33], causing tolerance and suppression as opposed to activation. This contradiction may be due to insufficient

PRR signals from endogenous PAMPs/DAMPs, tolerance/desensitization of PRR signaling due to chronic PAMP exposure, or other mechanisms of tumor mediated immune suppression. Thus, given their roles in orchestrating co-stimulatory signal expression in antigen presenting cells, targeting PRRs to induce inflammation compatible with T cell priming and co-stimulation is a therapeutically viable strategy to engage antitumor CD8$^+$ T cell immunity.

3.5   Engaging PRRs for Cancer Immunotherapy

3.5.1   Inducing Innate Inflammation in Tumors: A Historical Perspective

The first widespread medical use of a PRR activator in cancer is that of Coley's toxin [34, 35] in the late 1800s, though the use of pathogens for cancer therapy, as well as anecdotal correlations of pathogen infection and spontaneous tumor regression, was documented much earlier [34, 36, 37]. Based upon clinical case review of a patient that experienced sarcoma tumor regression after bacterial infection at the tumor site, William Coley tested if bacterial infection of sarcomas may mediate tumor regression in patients. After initially using live bacterium, Coley switched to inactivated bacterium; which caused regression in some patients. This cocktail of inactivated bacteria became known as 'Coley's toxins', but the approach was ultimately overshadowed by advances in radiation therapy, and suffered from limitations in standardizing the treatment [34].

In the early 1900s *Mycobacterium bovis* was isolated from a cow with tuberculosis mastitis. Laboratory passaging in bovine bile (to prevent clumping) led to a loss of virulence, and the strain of *M Bovis* was named Bacillus of Calmette and Guerin (BCG) after the scientists that developed the strain [38]. Coincidentally, Tuberculosis infection was noted to be associated with a lower frequency of cancer [39], raising the possibility for using *M bovis* for cancer therapy. In 1969 the first report of BCG's use as a cancer therapy was reported by Mathe et al.in the treatment of lymphoblastoid leukemia where encouraging results were reported [40]. The first clinical trial of BCG for bladder cancer was published in 1976 [41] where a decrease in recurrence of superficial bladder cancer was observed. These observations were confirmed in 1980 [42], spurring widespread use of BCG as a intravesicular therapy for bladder cancer. BCG was FDA approved for the treatment of bladder cancer in 1990, and represents the first approved cancer immunotherapy. BCG mediates innate inflammation that engages CD4$^+$ and CD8$^+$ T cells with several TLRs being shown to mediate the initial innate response, including TLRs 2, 4, and 9 [43]. The success of BCG, along with discoveries on the role of PRR signaling in mediating immune surveillance, led to further studies applying intratumor PRR activators for cancer immunotherapy in several cancer types.

3.5.2 Non-Infectious Engagers of PRRs for Cancer Immunotherapy

PRR activating PAMPs have shown preclinical promise in engaging systemic anti-tumor immunity. In addition, therapies that evoke PRR signaling through indirect means are also being explored. Clinically tested approaches to engage immune surveillance through targeted PRR activation using non-replicating, non-infectious PAMPs are described below according to their PRR specificities.

3.5.2.1 TLR Agonists

The toll like receptors were first discovered in *Drosophila*, and later confirmed to induce innate inflammation in mammalian systems [8]. TLR signaling culminates in activation of TBK1 and IKKα/β to induce type I IFNs and NFkB dependent gene expression, respectively (Fig. 3.1).

TLR3

The double stranded RNA mimetics Poly I:C and Poly A:U, and derivatives thereof, have been widely tested as cancer immunotherapies in several solid tumor indications. Poly A:U was tested in the 1980–90 s wherein it was shown to extend relapse free survival after systemic delivery in breast cancer patients [44], but showed minimal efficacy in melanoma [45], and was associated with less favorable survival after systemic delivery in colorectal cancer patients [46].

The double stranded RNA mimetic, Poly I:C, effectively induces type I IFN in several tumor associated cell types and mediates generation of Th1 responses in mice [47, 48]. A poly-L-lysine stabilized version of Poly I:C in carboxymethyl-cellulose, Poly ICLC (Hiltonol), also engages MDA5 activation [49, 50] and has been tested in several trials. Poly ICLC has been delivered intramuscularly to boost systemic type I IFN responses [51, 52], as well as via intratumoral routes [53–55]. Poly ICLC was well tolerated in early trials, but limited efficacy as a monotherapy was reported overall. Ongoing work demonstrating potential clinical benefit focuses on combining Poly ICLC with other modalities including FLT3L, radiation, and PD-1 blockade [56]. Poly ICLC is also being used as a personalized cancer vaccine adjuvant, where sustained antitumor T cell responses were demonstrated [57].

TLR4

After the clinical use of Coley's toxin in sarcomas, it was proposed that TLR4 activation via bacterial polysaccharides mediated that anti-sarcoma effects of Coley's toxin in mice [58]. Lipopolysaccharide (LPS), also known as endotoxin, is the canonical bacterial polysaccharide used to activate TLR4 signaling in laboratory studies. Inducing TLR4 signaling leads to robust myeloid cell activation, particularly macrophages, and in preclinical models has been shown to induce robust antitumor effects [59–61]. The first clinical trial of LPS in cancer patients occurred via intravenous injection concomitant with ibuprofen to prevent inflammatory side effects, and showed induction of pro-inflammatory cytokines in the sera (TNF, IL-6 and MCSF) with moderate antitumor activity observed in 2 patients with colorectal cancer [62]. A follow-up trial of systemic LPS delivery showed only moderate antitumor efficacy [63]. Systemic toxicities associated with inflammation were a common issue for trials using LPS.

Usage of the lipid A subunit of LPS was later shown to induce antitumor activity with a more favorable toxicity profile [59, 64, 65]. Lipid A isolated from *Salmonella*, called monophosphoryl lipid A (MPL) was subsequently tested in cancer patients intravenously and was found to have minimal antitumor efficacy [66]. Several derivatives of MPL have been clinically tested in cancer patients with inconsistent or lacking indication of antitumor efficacy [67–69]. Despite limited efficacy of various MPL based strategies delivered intratumor, subcutaneous, or intravenously, MPL succeeded as an adjuvant for the HPV vaccine Cervarix® and was FDA approved for this use in 2009 [70]. Recent clinical efforts include the use of the TLR4 activating glycolipid (GSK1795091) in combination with an activating OX40 antibody, and a synthetic MPL mimetic (GLA-SE) in combination with radiation therapy [71, 72].

TLR7/8

Imiquimod is a small non-nucleoside TLR7/8 activator that originally demonstrated utility as an antiviral agent in preclinical models in the 1980s [73–75]. The antiviral effects observed were dependent upon mediating inflammation as opposed to direct action on viruses [74, 76–82]. Initial clinical trials delivering oral imiquimod in cancer patients failed to demonstrate efficacy beyond induction of inflammation [83, 84]. However, topical application of imiquimod cream for actinic keratosis [85–91] and basal cell carcinoma [92–100] was efficacious, and was FDA approved in 2004 for these indications. Several studies in various topical pre-cancerous and cancerous disease have since been conducted [67], with more recent testing occurring in breast cancer, melanoma, and other solid tumors alone or in combination with immune checkpoint blockade.

R848 (Resmiquimod) and motolimod (VTX-2337) are TLR7/8 agonists that are currently being clinically explored for cancer immunotherapy, in pre-cancerous actinic keratosis and head and neck squamous cell carcinoma, respectively [101]. Other agonists targeting TLR7 and TLR8 are currently being tested alone or in combination with immune checkpoint blockade in various solid tumors [102].

TLR9

TLR9 agonists induce potent type I IFN responses from plasmacytoid DCs (pDCs), and generally mimic unmethylated CpG DNA. Importantly, murine cell-type expression patterns of TLR9 is distinct from that of humans, with TLR9 largely being expressed in human pDCs and B cells, while murine expression of TLR9 is more ubiquitous in macrophage and DC populations [103]. Numerous clinical trials using TLR9 agonists have been conducted as monotherapy studies in solid tumors, exhibiting manageable safety profiles despite association with cytokine release syndrome related to IFN mediated inflammation [104]. Clinical efficacy signals in monotherapy trials have been limited, with more promising signals being observed when combined with other modalities [104].

Lefitolimod (MGN1703) was tested in two phase II trials in small cell lung cancer and metastatic colorectal cancer with subcutaneous delivery and did not meet survival endpoints [105, 106]. A phase III trial was conducted in a subgroup of metastatic colorectal cancer patients identified in the phase II trial, where negative results were posted.

Tilsotolimod (IMO-2125) has been tested in multiple solid tumors, most extensively in melanoma. A phase I/II trial in anti-PD-1 refractory melanoma showed evidence of efficacy of Tilsotolimod in combination with ipilimumab (anti-CTLA-4) or pembrolizumab (anti-PD-1) with an objective response rate (ORR) of 22% (News release, Idera Pharmaceuticals, April 21, 2020 press release). A follow-up phase III trial of Tilsotolimod in combination with ipilimumab in anti-PD-1 refractory melanoma was conducted but did not meet objective response rate endpoint (Idera pharmaceuticals, March 18, 2021 press release). A phase III trial in microsatellite stable colorectal cancer in combination with ipilumimab and nivolumab (anti-PD-1) is ongoing.

SD-101 has shown abscopal effects in indolent lymphoma patients after intratumor administration [107] and demonstrated an objective response rate (ORR) of 78% in treatment-naïve and 15% ORR in PD-1 refractory melanoma in combination with pembrolizumab [108]. Intratumor SD-101 in combination with pembrolizumab and paclitaxel in HER2-negative breast cancer showed non-significant improvement in pathological complete responses [109]. Several trials of SD-101 as an intratumor therapy are ongoing in combination with other modalities in melanoma, breast cancer, prostate cancer, and lymphoma.

CMP-001 is a virus like particle comprised of bacteriophage capsid with a CpG oligodeoxynucleotides. The drug is taken up by pDCs via FCgamma receptor anti-bacteriophage antibodies that bind the virus like particle leading to robust type-I IFN induction [110, 111]. In anti-PD-1 refractory melanoma patients, intratumor CMP-001 in combination with pembrolizumab achieved an ORR of 25% associated with abscopal effects noted [112, 113]. Ongoing clinical trials are testing CMP-001 in melanoma, head and neck squamous cell carcinoma, and lymphoma.

TLR9 agonists remain a very active area of clinical pursuit, particularly with newer routes of delivery, e.g., in the aforementioned case of bacteriophage-antibody mediated delivery via CMP-001; NZ-TLR, which uses a cold isostatic pressing to encapsulate a TLR9 agonist that permits extended release following intratumor injection; and AST-008, a spherical nucleic acid-based nanomaterial TLR9 agonist [104].

3.5.2.2 RLR Activation

The RLRs MDA5 and RIG-I are cytoplasmic sensors of viral RNA that have recently gained attention as potential targets for cancer immunotherapy.

MDA5

MDA5 recognizes long dsRNA in the cytosol, culminating a distinct type I IFN dominant activation of macrophages and DCs and cell death signaling in cancer cells [114–116]. Poly ICLC activates both TLR3 and MDA5 as mentioned in Sect. 3.5.2.1, with MDA5 activation being linked to the potent Th1 antitumor activity observed by Poly ICLC ([49, 50] (see TLR3 agonist description above for clinical status of Poly ICLC). An emerging route to target MDA5 activation is via synthetic RNA viruses and indirect reactivation of endogenous retroviruses (ERVs) using epigenetic modulators.

Synthetic positive-sense RNA viruses and replicons are commonly engineered from Semliki Forest virus, Sinbus virus, or Venezuelan equine encephalitis virus and delivers self-replicating RNA into the cytosol of cells, which can be recognized by both MDA5 and RIG-I [117]. Results from a phase I clinical trial testing a Simliki Forest virus-based HPV vaccine in HPV induced cancers efficiently induced HPV antigen-specific T cells and was well tolerated [118]. Synthetic coxsackievirus A21 RNA that engage MDA5 are currently in development for clinical use [119]. How well replicons/synthetic viral RNA engage MDA5 and other PRRs relative to oncolytic viruses/natural virus infection (see Sect. 3.5.3 below) remains unknown.

ERVs comprise up to 8% of the human genome [120] where they typically remain inactive, but have been shown to be reactivated in various cancer types [25]. Usage of demethylating agents was initially proposed to mediate antitumor effects by inducing expression of tumor suppressor genes [121]. However, several studies have shown that 5-aza-2-deoxycytidine, a DNA methylation inhibitor, causes re-expression of ERV gene products that induce dsRNA recognized by MDA5 in cancer cells, sensitizing tumor-bearing mice to anti-CTLA-4 therapy [122, 123]. Given that immunotherapy success is associated with ERV gene expression in tumors [124–126], it is possible that optimizing DNA demethylating agents for induction of ERV mediated MDA5 signaling will enhance immune checkpoint blockade therapy. DNA demethylating agents have been tested extensively in the clinic [127, 128], however it is unclear whether MDA5 engagement occurred and/or contributed to therapy effect.

RIG-I

In contrast to MDA5, RIG-I recognizes short dsRNA as well as 5'-ppp-RNA that lacks a 7-methylguanosine cap on the 5' end of RNA commonly added to endogenous mRNAs. Several RIG-I agonists have been developed and are in clinical testing.

MK-4621 is a 5'-ppp synthetic RNA oligonucleotide that was delivered intratumoral in various solid tumor types where an interim analysis showed a favorable safety profile and induction of serum chemokine levels [129]. A second study testing MK-4621 complexed with JetPEI™ and pembrolizumab is also ongoing [130]. CV8102 is a single stranded, uncapped RNA complexed with cationic peptides that activates RIG-I along with TLR7/8. This drug was tested by intratumor injection in solid tumors alone or in combination with PD-1 blockade wherein the drug was well tolerated and early responses were observed [131]. GEN0101 is a drug composed of inactivated Sendai virus particles that engage RIG-I and have shown immunological responses to treatment, declines in prostate-specific antigen, and potential disease stabilization after intratumoral and subcutaneous injection in castration-resistant prostate cancer patients [132, 133]. Several other studies of GEN0101 have been conducted in melanoma and mesothelioma, however results from these trials have not been reported [134].

3.5.2.3 STING Agonists

Due to the role of endogenous STING signaling in tumors leading to spontaneous antitumor immunity [13], STING agonists have gained considerable attention as intratumoral agonists for clinical cancer immunotherapy.

DMXAA was originally developed as an anti-vascular drug that was later found to activate TBK1-IRF3 signaling [135] via STING [136]. A large phase III study was conducted in non-small cell lung cancer patients, but was discontinued [137]. Other clinical efforts with this agent have failed to show compelling clinical responses. However, despite preclinical data indicating its ability to engage antitumor $CD8^+$ T cell immunity, it was later found to only induce mouse STING signaling, and not that of humans [138], possibly explaining its lack of clinical activity.

Other clinical trials of intratumor delivered STING agonists are ongoing and include GSK3745417, MK-2118, MK-1454, BMS-986301, IMSA-101, ADU-S100 and E7766; most of which are being combined with immune checkpoint blockade. At the time of writing, biological activity of STING agonists has been reported in patients [139, 140], but the clinical efficacy of these agents remains to be reported. Further development of modified versions of STING agonists in preclinical settings are ongoing that include the development of orally available STING agonists [141, 142], as well as higher potency STING agonists [143, 144].

3.5.3 Infectious Agents as Engagers of PRRs for Cancer Immunotherapy

In addition to BCG described above, attenuated, replication-competent viruses and bacteria being clinically developed for cancer immunotherapy also engage PRR signaling. A potential advantage of intratumoral therapy with infectious agents versus that of targeted PRR engagement with PAMPs is that infectious agents generally engage multiple PRRs within the spatiotemporal context of a natural infectious process, possibly recapitulating a more natural T cell priming scenario. However, infectious agents derived from natural animal viruses and bacterium also typically mediate some level of innate or adaptive immune interference (e.g. suppression of antiviral signaling or antigen presentation) that evolved to ensure the successful lifecycle of the pathogen [145–148]. It remains to be determined which approach (non-infectious agonists vs infectious agents) will be more effective in engaging immune surveillance and controlling tumor growth. Beyond the PRR-engaging attributes of these agents, it also must be noted that oncolytic viruses also mediate killing of cancer cells, adding an additional dimension of anticancer and immunogenic activity.

3.5.3.1 Oncolytic Viruses

While dubbed 'oncolytic' due to selective toxicity observed by various attenuated viruses in cancer cell lines [149], the antitumor potential of using viruses for cancer therapy may largely be due to their ability to elicit antitumor CD8$^+$ T cells through PRR activation [114, 150–152]. Diverse virus species have been developed for cancer immunotherapy, ranging from large DNA viruses to small RNA viruses, that have distinct tissue tropisms, viral replication strategies, and mechanisms of immune subversion. Thus, as with targeting distinct PRRs for cancer therapy, different virus contexts are likely to mediate antitumor efficacy through different routes, with differing efficiencies. Clinically tested viral cancer immunotherapies are described below, however, numerous virus contexts beyond these agents are being considered for future clinical testing.

Herpes Simplex Viruses (HSV). HSV is a dsDNA viruses that engage a number of PRRs including STING, TLR2, TLR3, and TLR9 [153]. Talimogene laherparepvec (T-VEC), an attenuated oncolytic HSV1 (oHSV) expressing GMCSF, is the only FDA approved oncolytic virus to date, which demonstrated a 16.3% durable response rate and 33% 5-year response rate in a randomized phase III clinical trial of melanoma [154]. An abscopal effect was noted, with regression of non-injected lesions occurring in some patients [155]. Other monotherapy clinical trials of oHSVs have shown evidence of efficacy similar to what has been observed for T-VEC in early stage clinical trials [156]. While initial observations in combination with immune checkpoint blockade suggested promise [157], a phase III clinical trial testing T-VEC combined with pembrolizumab was recently discontinued due to futility [158].

The next generation of herpesvirus-based immunotherapies have been developed with intentions of improving oHSV immunotherapy efficacy, particularly in regards to preventing oHSV mediated disruption of antigen presentation (G47Δ [159]); enhancing oHSV toxicity selectively in cancer cells (rQNestin34.5v2 [160], RP1 [161]); enhancing IFN resistance of oHSV (ONCR-177 [162]); and 'arming' oHSV with PD-1 or CTLA4 blocking antibodies and immunostimulatory cytokines, particularly IL-12 (M032 [163], ONCR-177 [164], RP2 [165], MVR-T3011 [166]). Several of these agents are moving into early stage clinical trials in various solid tumor types as intratumorally delivered therapies.

Adenovirus. Adenovirus has a dsDNA genome and is recognized by TLR9 on pDCs[167], STING [168, 169], NOD like receptors [170, 171], with evidence for roles of other TLRs in vivo [172]. Immunogenic cell death (e.g., release of HMGB1, calreticulin, ATP, and HSP70) has also been proposed as a key mechanism driving the immunogenicity of oncolytic adenoviruses [173–175]. In 2005 China approved the replicating adenovirus H101 (Oncorine) for the treatment of nasopharyngeal carcinoma [176]. DNX-2401 is a modified adenovirus that selectively replicates in cancer cells with defective Retinoblastoma (Rb) and has shown promising phase I results in recurrent glioblastoma wherein 20% of patients surviving > 3 years [177]. A follow-up phase II study of DNX-2401 delivered at the time of biopsy in recurrent glioblastoma patients was conducted in combination with pembrolizumab, wherein 5/42 patients receiving the full DNX-2401 dose had confirmed responses [178]. A randomized phase III study is in planning [178]. Several 'armed' adenoviruses are in clinical testing in various indications; armed with GMCSF (CG0070 [179, 180] and ONCOS-102 [181]), immunostimulatory ligands CD40L and 41BBL (LOAd-703 [182]); hyaluronidase (to facilitate viral spread and $CD8^+$ T cell recruitment within the tumor, VCN-01 [183]); IL-12 (AD5-yCD/mutTK$_{SR39}$rep-hIL-12, [184]); OX40L (DNX-2440 [185]); and CXCL9, CXCL10, and IFN-α (NG-641 [186]). In addition, combination strategies of modified oncolytic adenoviruses with CAR T cell therapy (CAdVec) and chemoradiation (Colo-AD1) are being pursued [187].

Poxviruses. Poxviruses are large dsDNA viruses (130–300 Kb) that have sophisticated replication strategies and mechanisms to evade viral elimination by the host immune system [188]; attenuated vaccina viruses, are the most extensively tested oncolytic poxviruses. Poxviruses are recognized by several PRRs, including TLR2, TLR6, MDA5, and the NALP3 inflammasome [189, 190]. Interestingly, UV and heat inactivated Vaccina virus was shown to mediate stronger innate inflammation through STING signaling compared to replicating Vaccina, possibly reflecting strategies by which Vaccina interferes with innate signaling [191]. Pexa-Vec (JX-594), a Vaccinia virus, was tested as an intratumor therapy in hepatocellular carcinoma in a phase II clinical trial where evidence of disease control was reported [192]. However, a follow-up study revealed lack of overall survival benefit in this patient population [193]. A trial of Pexa-Vec in colorectal cancer was pursued with immune checkpoint inhibitor combination, but failed to show a significant improval in response [194]. As with other DNA viruses used for virotherapy, a current emphasis on arming poxviruses is driving ongoing clinical efforts, including with GM-CSF, chemokines, IL-15 and PD-1 blocking antibodies [195]. GL-ONC1 and vvDD are vaccina viruses that were delivered intravenously [196, 197], other studies in solid tumors are ongoing and evaluation of antitumor effects have not yet been reported. Myxoma viruses are also being developed for virotherapy in preclinical settings [195].

PVSRIPO. PVSRIPO, a (+)stranded RNA picornavirus, is the live attenuated type I Sabin strain of Polio with exchange of the Sabin Internal Ribosomal Entry Site (IRES) with that of human rhinovirus type II [198, 199]. This substitution neuroattenuates the virus, but does not impair its ability to kill malignant cells [198]. PVSRIPO requires poliovirus receptor (PVR) expression for viral entry, which is highly expressed on both antigen presenting cells and malignant cells [114, 198, 200]. PVSRIPO infection activates MDA5, leading to a sustained type I/III IFN dominant IFN signature in tumor-associated macrophages and dendritic cells that culminates in antitumor CD8$^+$ T cell immunity [114, 115, 201]. Importantly, in preclinical models the antitumor efficacy of PVS-RIPO was primarily dependent upon viral infection of TME constituents as opposed to malignant cells, indicating that PVSRIPO may function primarily as an engager of MDA5 within the TME [114]. A phase I clinical trial in recurrent GBM demonstrated a 21% survival rate at 36 months, relative to 4% survival in an eligibility criteria-matched historical control cohort of patients [202]. A small phase I trial in anti-PD-1 refractory melanoma demonstrated antitumor responses in both injected and non-injected lesions in 4/12 patients, with 6/12 patients resuming immune checkpoint blockade after PVSRIPO having durable disease control at 18 months of follow-up [203]. Ongoing clinical studies are focused on combining PVSRIPO with PD-1 blockade in melanoma, GBM, and other solid tumors [203–205].

Reovirus. Reoviruses are segmented dsRNA viruses with a long history of preclinical investigation backing its utility as an immunovirotherapy agent [206, 207]. Reovirus is recognized primarily by RIG-I and MDA5 [208]. The antitumor efficacy of Reovirus is independent of viral replication in preclinical models [209], implying that PRR recognition occurs upon viral entry, leading to antitumor CD8$^+$ T cell priming [210]. Reolysin (aka pelareorep) has been delivered both intravenously and intratumorally in clinical trials. An initial phase I study observed local tumor responses in 7/19 patients, with one complete response in advanced solid tumors [211]. Phase II trials in combination with chemotherapy for malignant melanoma [212], breast cancer [213], non-small cell lung cancer [214], head and neck cancer [215], metastatic pancreatic cancer [216] have been conducted with some indication of efficacy in subsets of patients. Recent work has demonstrated that intravenously delivered Reovirus reaches brain tumors in patients and induces PD-1/PD-L1, possibly indicating its potential use as a systemic agent in combination with immune checkpoint blockade [217]. As with other oncolytic viruses, the ongoing focus of current Reovirus virotherapy is focused on combining with other immunomodulatory agents [218].

Coxsackievirus A21 is a (+)stranded RNA picornavirus primarily sensed by MDA5 [219], that has been clinically tested in both intravenous and intratumoral contexts as V937 (aka CAVATAK). Indications tested include non-muscle-invasive bladder cancer [220], in which tumor associated inflammation was observed, and melanoma [221], wherein 43.2% of patients had progression free survival at 1 year post treatment. Ongoing studies combining V937 and immune checkpoint blockade are being conducted in unresectable melanoma, and early results indicate the combination of V937 and pembrolizumab has a 47% ORR [222].

Vesicular Stomatitis Virus (VSV) is a (−) stranded RNA virus that was developed as an oncolytic virus due to the lack of type I IFN mediated suppression of attenuated VSV replication in human cancer cells [223]. VSV is recognized by RIG-I [224] and TLR7 [225]. However, VSV vectors are capable of causing neurological disease in non-human primates [226]; thus an interferon-β expressing VSV (VSV-IFNβ) was developed to restrict VSV replication beyond normal cells and was found to not cause neurotoxicity in non-human primates [227–229]. VSV-IFNβ was further modified with expression of a sodium iodide symporter (NIS) to enable imaging. Ongoing clinical studies in various indications include IV infusion in multiple myeloma, T cell lymphoma, and acute myeloid leukemia [230].

Measles Virus is a (−) stranded RNA virus that was originally proposed as an oncolytic virus candidate due to case reports of measles infection being linked to tumor regression [231]. Measles is recognized by MDA5 and RIG-I [232], however it is reported to intercept RLR recognition in antigen presenting cells [233]. Indeed natural (wildtype) measles infection also suppresses adaptive immunity [234]. The live attenuated vaccine strain of Measles (Edmonston-Zagreb strain) has been developed for cancer immunotherapy and tested in early stage clinical trials of T cell lymphoma [235], ovarian cancer [236], glioblastoma, breast cancer, head and neck squamous cell carcinoma, malignant peripheral nerve sheath tumors, bladder cancer, and multiple myeloma [237]. A NIS-expressing version of measles was also generated and tested in patients after intravenous administration [238].

Newcastle's Disease Virus (NDV) is a (−) stranded RNA virus recognized primarily by RIG-I [239, 240] that naturally infects chickens. NDV has been shown to mediate both oncolysis and type I IFN-dependent priming of anti-tumor T cells in preclinical models [150, 241]. Several clinical trials used NDV-treated oncolysate, or lysed cancer cells, for vaccination in cancer patients [242], most of which were in melanoma where improved overall survival was demonstrated relative to historical controls. A phase III clinical trial demonstrated longer survival after NDV-pulsed autologous vaccine compared to surgical resection alone in colorectal cancer patients [243]. The MTH-68 strain of NDV was tested in various advanced cancer types in small cohorts of patients, including glioblastoma, with potential evidence of efficacy after intravenous administration [244, 245]. The PV701 strain was likewise tested intravenously in small cohorts patients with advanced cancers, documenting some objective responses [246, 247]. Extension of these studies have been complicated by changing regulatory guidelines restricting the use of NDV strains [241]. Ongoing clinical efforts to test NDV include a GMCSF expressing NDV variant (MEDI5395) being tested in various advanced cancers in combination with durvalumab (anti-PD-L1 antibody) [248].

3.5.3.2 Intracellular Bacterium

Bacille Calmette-Guerin (BCG) is standard of care therapy for non-muscle inva-sive bladder cancer, and was the first FDA approved cancer immunotherapy. See Sect. 3.5.1 for a description of the use of BCG in cancer therapy.

Listeria monocytogenes is a gram positive, intracellular bacterium that causes listeriosis, a foodborne illness. *Listeria* is recognized by TLR2, TLR5, NOD-like receptors, and STING [249]. Strains of *Listeria* have been developed for use as cancer vaccine vectors, with its intracellular lifecycle being an asset to deliver tumor associated antigens and engage antitumor T cell responses [250]. *Listeria* vaccine clinical trials have been conducted via intravenous delivery in pancre-atic cancer, against mesothelin (CRS-207 [251]); in cervical cancer against HPV antigens [252, 253], and in mesothelioma against mesothelin (CRS-207 [254]). Encouraging objective responses have been observed in early stage clinical tri-als of mesothelioma and cervical cancer; however a phase III trial of *Listeria* E7 vaccine (AIM2CERV) in cervical cancer was closed by the sponsor [250], and CRS-207 development was recently discontinued after a failed lacking activity in combination with pembrolizumab [255]. Several *Listeria*-based approaches are in development with ongoing clinical trials.

3.6   The Role of Type I IFN in Mediating the Antitumor Efficacy of PRR Agonists

Type I IFNs are critical toward engaging DC priming of antitumor T cells [256]. Indeed, the efficacy of several PRR activators (both non-infectious agonists and infectious agents) has been shown to be dependent upon eliciting type I IFN signaling in tumors, including: Poly IC/Poly IC-LC [49, 114, 257, 258], RLR agonists [50, 259], TLR7/8 agonists [260], TLR9 agonists [261], STING agonists [262–264], PVSRIPO [114], and NDV [150]. IFNAR signaling both primes DC differentiation and expression of costimulatory ligands [47], while also boosting cytolytic function of antitumor CD8$^+$ T cells locally [114, 265, 266]. However, it is critical to note that out-of-context type I IFN does not recapitulate the anti-tumor efficacy of broader signals delivered by type I IFN during PRR signaling [114, 267], which encompasses a myriad of other pro-inflammatory signals coin-ciding with type I IFN (Fig. 3.1). Indeed, while exogenous type I IFN treatment in cancer patients has shown some activity in the clinic, the efficacy of type I IFN treatment of tumors/cancer patients was limited [268]. PRR agonists, either infectious or non-infectious, offer potential to contextualize type I IFN signaling, and its T cell engaging capacity, within an inflammatory milieu supporting the production of chemokines, other DC/T cell modulating cytokines, and induction of pro-inflammatory signals within the TME that support CD8$^+$ T cell effector functions. Whether this potential is fully realized clinically remains to be determined.

Yet, IFN signaling also mediates cancer cell chemoradiation resistance and induction of immune checkpoints that prevent antitumor T cell function [269–271]. Moreover, type I IFN contributes to T cell exhaustion and dysfunction during chronic viral infection [272], and type I IFN signaling in CD4$^+$ T cells has been shown to negatively associate with immunotherapy response [273]. The context of IFN signaling may well determine whether it promotes or desensitizes antitumor immunity [274]: tumors with active IFN signaling at baseline may be resistant or non-responsive to PRR agonist therapy; and due to the role of IFNs in inducing immune checkpoint ligands [275], combination strategies to mitigate such negative feedback may be necessary to empower the antitumor effects of PRR engaging therapies. Indeed, most clinically pursued PRR engagers have been shown to induce PD-L1 and other immune checkpoint ligands, and are potentiated by immune checkpoint blockade in preclinical models [49, 114, 115, 150, 157, 267, 276].

3.7 Comparison of PRR Activators to Other Immunotherapies and Their Utility in Combination

PRR engaging immunotherapies intended to mediate in situ vaccination differ mechanistically from other anticancer modalities in several complementary ways. Complementary and distinct attributes of PRR activating immunotherapies compared to other established immunotherapy approaches are presented below:

Immune checkpoint blockade (ICB): Blockade of PD-L1, PD-1, CTLA4, and other immune inhibitory receptors function to resuscitate antitumor T cell function, and has achieved unprecedented clinical responses in immunogenic tumors with high mutation loads and/or oncogenic viral gene expression [277]. Generally, these modalities rely upon the presence of pre-existing antitumor T cells and are more efficacious in tumors that have higher baseline inflammation [278]. Inclusive of resistance mechanisms to ICB is that of innate immunosuppression, which limits co-stimulation, infiltration, and effector function of antitumor T cells [279]. In contrast, intratumoral therapy with PRR activators mediates innate inflammation within tumors, that enhances expression of co-stimulatory signals on antigen presenting cells, causes chemokine induction that enables trafficking of T cells to the site of the tumor, and directly bolsters the function of antitumor T cells, e.g., via type I IFNs [114, 280]. Numerous pre-clinical studies have shown that PRR activation within tumors leads to priming of antitumor T cells, which may broaden the potential of ICB therapy by bolstering antitumor T cell populations, while supporting their recruitment and function within the tumor microenvironment. Indeed, several studies have demonstrated synergy between PRR activators and ICB [281].

Cancer vaccines: Various cancer vaccine modalities have been developed including peptide vaccines, autologous dendritic cell vaccines, and mRNA vaccines. Indicating their clinical potential, prophylactic vaccination against HPV antigens has been remarkably successful in preventing cervical cancer [282]. Traditionally, cancer vaccines have been restricted to 'shared' tumor associated antigens common across numerous patients, e.g., HER2, EGFRviii, and MART1. With more recent feasibility of whole exome sequencing of biopsy tissue, personalized vaccines based upon patient-specific neoantigens are in development [57]. However, as with immune checkpoint blockade, cancer vaccines require that primed and expanded antitumor T cell populations induced by the vaccine traffic and function within the tumor. Moreover, these strategies require knowledge and accurate prediction of effectively presented, homogenously expressed, and targetable neoantigens. Notably, PRR activating adjuvants are used in cancer vaccines to enable priming and expansion of antitumor T cells in the periphery. In contrast, intratumoral delivery of PRR engaging therapies function to mediate vaccination using the tumor site in an antigen agnostic manner, by activating innate immunity and antigen presentation to prime T cells against antigens present within the tumor. Moreover, PRR agonists induce inflammation that enable trafficking and potentiation of antitumor T cell function. Intratumoral PRR agonist therapy is anticipated to complement cancer vaccines by enabling the recruitment, further tumor/tumor draining lymph node localized expansion of tumor antigen-specific T cells, and by providing inflammation in the tumor that supports antitumor T cell function. *Adoptive T cell transfer/CAR T cells:* A direct route to bolster antitumor T cell populations in cancer patients is to deliver either expanded autologous antitumor T cells (or tumor infiltrating T cells) or autologous chimeric antigen receptor (CAR) T cells against specific tumor antigens. These approaches have shown promising antitumor efficacy in some cancer types [283, 284]. Distinct from a T cell-based approach to induce antitumor immunity in patients, intratumoral PRR activation leads to priming of T cells in the tumor bed and tumor draining lymph node while providing a supportive innate inflammatory framework for antitumor T cells to function. Intratumoral PRR activation has been shown to potentiate adoptive T cell therapy and CAR T cell therapy in pre-clinical models, primarily by enhancing recruitment of the ex-vivo expanded or engineered autologous T cells to the site of the tumor [285, 286].

3.8 The Future of PRR-Targeted Cancer Immunotherapies: Hurdles and Limitations

Beyond logistical regulatory and manufacturing issues, several hurdles remain for the success of PRR engaging immunotherapy to be realized. First, the optimal dosing of PRR engagers remains unclear, and is likely to be specific to each agent. For example, administration of higher doses of a STING agonist in mice impaired systemic antitumor immunity [287], implying that exaggerated activation of intratumor STING signaling may mediate a deleterious effect on antitumor T cell function. Whether this is true for other PRR engagers remains unknown. It also remains unclear as to which tumor types may benefit most from PRR engaging therapy: should immunologically quiescent tumors be targeted to enhanced intratumor inflammation and engage T cells? Or are immunologically active tumors more responsive to PRR-induced inflammation? As presented in this chapter, PRR agonists have been tested in both notoriously immunosuppressed tumors (e.g., glioblastoma) as well as immunogenic tumors (e.g., melanoma).

In addition, PRR-induced inflammation plays both anti- and pro-tumor roles [288]. For example, TLR3 signaling in the tumor microenvironment has been shown to enhance cancer metastases [288]; VEGF, matrix metalloproteinases, and other inflammatory features induced by PRR signaling may facilitate tumor vascularization; interferon responses induce APOBEC which can add to the evolutionary potential of cancer cells by increasing mutation rates [289]; PRR signaling promotes NFkB signaling, which can enable cancer cell survival and resistance to T cell mediated killing [290, 291]; and PRR signaling may exacerbate T cell exhaustion and dysfunction. In some respects, combination therapies like immune checkpoint blockade, anti-VEGF therapies, and other mechanisms may complement PRR engaging therapies to mitigate these effects. Overcoming and defining these limitations will be critical to optimize PRR activation for future cancer therapy.

Competing Interests M.C.B. is an inventor on intellectual property licensed to, and a paid advisor of, Istari Oncology, which is developing a recombinant poliovirus, PVSRIPO, for cancer immunotherapy.

References

1. Fitzgerald KA, McWhirter SM, Faia KL, Rowe DC, Latz E, Golenbock DT et al (2003) IKKepsilon and TBK1 are essential components of the IRF3 signaling pathway. Nat Immunol 4(5):491–496. https://doi.org/10.1038/ni921
2. Kawai T, Akira S (2007) Signaling to NF-kappaB by toll-like receptors. Trends Mol Med 13(11):460–469. https://doi.org/10.1016/j.molmed.2007.09.002
3. Fore F, Indriputri C, Mamutse J, Nugraha J (2020) TLR10 and Its unique anti-inflammatory properties and potential use as a target in therapeutics. Immune Netw. 20(3):e21. https://doi.org/10.4110/in.2020.20.e21
4. Esser-Nobis K, Hatfield LD, Gale M Jr (2020) Spatiotemporal dynamics of innate immune signaling via RIG-I-like receptors. Proc Natl Acad Sci USA 117(27):15778–15788. https://doi.org/10.1073/pnas.1921861117
5. Allan RS, Waithman J, Bedoui S, Jones CM, Villadangos JA, Zhan Y et al (2006) Migratory dendritic cells transfer antigen to a lymph node-resident dendritic cell population for efficient CTL priming. Immunity 25(1):153–162. https://doi.org/10.1016/j.immuni.2006.04.017
6. Yewdall AW, Drutman SB, Jinwala F, Bahjat KS, Bhardwaj N (2010) CD8+ T cell priming by dendritic cell vaccines requires antigen transfer to endogenous antigen presenting cells. PLoS ONE 5(6):e11144. https://doi.org/10.1371/journal.pone.0011144
7. Grabowska J, Lopez-Venegas MA, Affandi AJ, den Haan JMM (2018) CD169(+) macrophages capture and dendritic cells instruct: the interplay of the gatekeeper and

the general of the immune system. Front Immunol 9:2472. https://doi.org/10.3389/fimmu.
2018.02472

8. Janeway CA Jr TP, Walport M, Shlomchik MJ (2001) Immunobiology. Garland Science, New York

9. Dunn GP, Old LJ, Schreiber RD (2004) The three Es of cancer immunoediting. Annu Rev Immunol 22:329–360. https://doi.org/10.1146/annurev.immunol.22.012703.104803

10. Dagogo-Jack I, Shaw AT (2018) Tumour heterogeneity and resistance to cancer therapies. Nat Rev Clin Oncol 15(2):81–94. https://doi.org/10.1038/nrclinonc.2017.166

11. Chen DS, Mellman I (2013) Oncology meets immunology: the cancer-immunity cycle. Immunity 39(1):1–10. https://doi.org/10.1016/j.immuni.2013.07.012

12. Gallucci S, Lolkema M, Matzinger P (1999) Natural adjuvants: endogenous activators of dendritic cells. Nat Med 5(11):1249–1255. https://doi.org/10.1038/15200

13. Corrales L, Glickman LH, McWhirter SM, Kanne DB, Sivick KE, Katibah GE et al (2015) Direct activation of STING in the tumor microenvironment leads to potent and systemic tumor regression and immunity. Cell Rep 11(7):1018–1030. https://doi.org/10.1016/j.celrep.2015.04.031

14. Corrales L, McWhirter SM, Dubensky TW Jr, Gajewski TF (2016) The host STING pathway at the interface of cancer and immunity. J Clin Invest 126(7):2404–2411. https://doi.org/10.1172/JCI86892

15. Apetoh L, Ghiringhelli F, Tesniere A, Obeid M, Ortiz C, Criollo A et al (2007) Toll-like receptor 4-dependent contribution of the immune system to anticancer chemotherapy and radiotherapy. Nat Med 13(9):1050–1059. https://doi.org/10.1038/nm1622

16. Gardai SJ, McPhillips KA, Frasch SC, Janssen WJ, Starefeldt A, Murphy-Ullrich JE et al (2005) Cell-surface calreticulin initiates clearance of viable or apoptotic cells through trans-activation of LRP on the phagocyte. Cell 123(2):321–334. https://doi.org/10.1016/j.cell.2005.08.032

17. Obeid M, Tesniere A, Ghiringhelli F, Fimia GM, Apetoh L, Perfettini JL et al (2007) Cal-reticulin exposure dictates the immunogenicity of cancer cell death. Nat Med 13(1):54–61. https://doi.org/10.1038/nm1523

18. Chao MP, Jaiswal S, Weissman-Tsukamoto R, Alizadeh AA, Gentles AJ, Volkmer J et al (2010) Calreticulin is the dominant pro-phagocytic signal on multiple human cancers and is counterbalanced by CD47. Sci Transl Med 2(63):63ra94. https://doi.org/10.1126/scitranslmed.3001375

19. Ohashi K, Burkart V, Flohe S, Kolb H (2000) Cutting edge: heat shock protein 60 is a putative endogenous ligand of the toll-like receptor-4 complex. J Immunol 164(2):558–561. https://doi.org/10.4049/jimmunol.164.2.558

20. Vabulas RM, Ahmad-Nejad P, da Costa C, Miethke T, Kirschning CJ, Hacker H et al (2001) Endocytosed HSP60s use toll-like receptor 2 (TLR2) and TLR4 to activate the toll/IL-1 receptor signaling pathway in innate immune cells. J Biol Chem 276(33):31332–31339. https://doi.org/10.1074/jbc.M103217200

21. Vabulas RM, Ahmad-Nejad P, Ghose S, Kirschning CJ, Issels RD, Wagner H (2002) HSP70 as endogenous stimulus of the toll/IL-1 receptor signal pathway. J Biol Chem 277(17):15107–15112. https://doi.org/10.1074/jbc.M111204200

22. Asea A, Rehli M, Kabingu E, Boch JA, Bare O, Auron PE et al (2002) Novel signal transduction pathway utilized by extracellular HSP70: role of toll-like receptor (TLR) 2 and TLR4. J Biol Chem 277(17):15028–15034. https://doi.org/10.1074/jbc.M200497200

23. Dybdahl B, Wahba A, Lien E, Flo TH, Waage A, Qureshi N et al (2002) Inflammatory response after open heart surgery: release of heat-shock protein 70 and signaling through toll-like receptor-4. Circulation 105(6):685–690. https://doi.org/10.1161/hc0602.103617

24. Roelofs MF, Boelens WC, Joosten LA, Abdollahi-Roodsaz S, Geurts J, Wunderink LU et al (2006) Identification of small heat shock protein B8 (HSP22) as a novel TLR4 ligand and potential involvement in the pathogenesis of rheumatoid arthritis. J Immunol 176(11):7021–7027. https://doi.org/10.4049/jimmunol.176.11.7021

25. Rooney MS, Shukla SA, Wu CJ, Getz G, Hacohen N (2015) Molecular and genetic properties of tumors associated with local immune cytolytic activity. Cell 160(1–2):48–61. https://doi.org/10.1016/j.cell.2014.12.033

26. Smith CC, Beckermann KE, Bortone DS, De Cubas AA, Bixby LM, Lee SJ et al (2018) Endogenous retroviral signatures predict immunotherapy response in clear cell renal cell carcinoma. J Clin Invest 128(11):4804–4820. https://doi.org/10.1172/JCI121476

27. Gasse P, Riteau N, Charron S, Girre S, Fick L, Petrilli V et al (2009) Uric acid is a danger signal activating NALP3 inflammasome in lung injury inflammation and fibrosis. Am J Respir Crit Care Med 179(10):903–913. https://doi.org/10.1164/rccm.200808-1274OC

28. Liu-Bryan R, Scott P, Sydlaske A, Rose DM, Terkeltaub R (2005) Innate immunity conferred by Toll-like receptors 2 and 4 and myeloid differentiation factor 88 expression is pivotal to monosodium urate monohydrate crystal-induced inflammation. Arthritis Rheum 52(9):2936–2946. https://doi.org/10.1002/art.21238

29. Kool M, Soullie T, van Nimwegen M, Willart MA, Muskens F, Jung S et al (2008) Alum adjuvant boosts adaptive immunity by inducing uric acid and activating inflammatory dendritic cells. J Exp Med 205(4):869–882. https://doi.org/10.1084/jem.20071087

30. Nejman D, Livyatan I, Fuks G, Gavert N, Zwang Y, Geller LT et al (2020) The human tumor microbiome is composed of tumor type-specific intracellular bacteria. Science 368(6494):973–980. https://doi.org/10.1126/science.aay9189

31. Dohlman AB, Arguijo Mendoza D, Ding S, Gao M, Dressman H, Iliev ID, et al (2021) The cancer microbiome atlas: a pan-cancer comparative analysis to distinguish tissue-resident microbiota from contaminants. Cell Host Microbe 29(2):281–298 e5. https://doi.org/10.1016/j.chom.2020.12.001

32. Zang X, Allison JP (2007) The B7 family and cancer therapy: costimulation and coinhibition. Clin Cancer Res 13(18 Pt 1):5271–5279. https://doi.org/10.1158/1078-0432.CCR-07-1030

33. Driessens G, Kline J, Gajewski TF (2009) Costimulatory and coinhibitory receptors in antitumor immunity. Immunol Rev 229(1):126–144. https://doi.org/10.1111/j.1600-065X.2009.00771.x

34. Hoption Cann SA, van Netten JP, van Netten C (2003) Dr William Coley and tumour regression: a place in history or in the future. Postgrad Med J 79(938):672–680

35. Coley WB (1910) The Treatment of Inoperable Sarcoma by Bacterial Toxins (the Mixed Toxins of the Streptococcus erysipelas and the Bacillus prodigiosus). Proc R Soc Med 3(Surg Sect):1–48

36. Jackson R (1974) Saint Peregrine, O.S.M.—the patron saint of cancer patients. Can Med Assoc J. 111(8):824

37. Hoption Cann SA, van Netten JP, van Netten C, Glover DW (2002) Spontaneous regression: a hidden treasure buried in time. Med Hypotheses 58(2):115–119. https://doi.org/10.1054/mehy.2001.1469

38. Meyer JP, Persad R, Gillatt DA (2002) Use of bacille Calmette-Guerin in superficial bladder cancer. Postgrad Med J 78(922):449–454. https://doi.org/10.1136/pmj.78.922.449

39. Peral R (1929) Cancer and tuberculosis. Am J Hyg 9:97–159

40. Mathe G, Amiel JL, Schwarzenberg L, Schneider M, Cattan A, Schlumberger JR et al (1969) Active immunotherapy for acute lymphoblastic leukaemia. Lancet 1(7597):697–699. https://doi.org/10.1016/s0140-6736(69)92648-8

41. Morales A, Eidinger D, Bruce AW (1976) Intracavitary Bacillus Calmette-Guerin in the treatment of superficial bladder tumors. J Urol 116(2):180–183. https://doi.org/10.1016/s0022-5347(17)58737-6

42. Lamm DL, Thor DE, Harris SC, Reyna JA, Stogdill VD, Radwin HM (1980) Bacillus Calmette-Guerin immunotherapy of superficial bladder cancer. J Urol 124(1):38–40. https://doi.org/10.1016/s0022-5347(17)55282-9

43. Redelman-Sidi G, Glickman MS, Bochner BH (2014) The mechanism of action of BCG therapy for bladder cancer–a current perspective. Nat Rev Urol 11(3):153–162. https://doi.org/10.1038/nrurol.2014.15

44. Lacour J, Lacour F, Spira A, Michelson M, Petit JY, Delage G et al (1980) Adjuvant treatment with polyadenylic-polyuridylic acid (Polya.Polyu) in operable breast cancer. Lancet 2(8187):161–4. https://doi.org/10.1016/s0140-6736(80)90057-4

45. Pawlicki M, Jonca M, Krzemieniecki K, Zuchowska-Vogelgesang B (1993) Results of adjuvant therapy with the preparation Poly-a Poly-u in patients with malignant melanoma during a 10-year observation. Wiad Lek 46(23–24):912–914

46. Lacour J, Laplanche A, Malafosse M, Gallot D, Julien M, Rotman N et al (1992) Polyadenylic-polyuridylic acid as an adjuvant in resectable colorectal carcinoma: a 6 1/2 year follow-up analysis of a multicentric double blind randomized trial. Eur J Surg Oncol 18(6):599–604

47. Longhi MP, Trumpfheller C, Idoyaga J, Caskey M, Matos I, Kluger C et al (2009) Dendritic cells require a systemic type I interferon response to mature and induce CD4+ Th1 immunity with poly IC as adjuvant. J Exp Med 206(7):1589–1602. https://doi.org/10.1084/jem.200 90247

48. Trumpfheller C, Caskey M, Nchinda G, Longhi MP, Mizenina O, Huang Y et al (2008) The microbial mimic poly IC induces durable and protective CD4+ T cell immunity together with a dendritic cell targeted vaccine. Proc Natl Acad Sci U S A 105(7):2574–2579. https://doi.org/10.1073/pnas.0711976105

49. Sultan H, Wu J, Fesenkova VI, Fan AE, Addis D, Salazar AM, et al (2020) Poly-IC enhances the effectiveness of cancer immunotherapy by promoting T cell tumor infiltration. J Immunother Cancer 8(2). https://doi.org/10.1136/jitc-2020-001224

50. Sultan H, Wu J, Kumai T, Salazar AM, Celis E (2018) Role of MDA5 and interferon-I in dendritic cells for T cell expansion by anti-tumor peptide vaccines in mice. Cancer Immunol Immunother 67(7):1091–1103. https://doi.org/10.1007/s00262-018-2164-6

51. Salazar AM, Levy HB, Ondra S, Kende M, Scherokman B, Brown D, et al (1996) Long-term treatment of malignant gliomas with intramuscularly administered polyinosinic-polycytidylic acid stabilized with polylysine and carboxymethylcellulose: an open pilot study. Neurosurgery 38(6):1096–1103; discussion 103–104

52. Okada H, Butterfield LH, Hamilton RL, Hoji A, Sakaki M, Ahn BJ et al (2015) Induction of robust type-I CD8+ T-cell responses in WHO grade 2 low-grade glioma patients receiving peptide-based vaccines in combination with poly-ICLC. Clin Cancer Res 21(2):286–294. https://doi.org/10.1158/1078-0432.CCR-14-1790

53. Salazar AM, Erlich RB, Mark A, Bhardwaj N, Herberman RB (2014) Therapeutic in situ autovaccination against solid cancers with intratumoral poly-ICLC: case report, hypothesis, and clinical trial. Cancer Immunol Res 2(8):720–724. https://doi.org/10.1158/2326-6066.CIR-14-0024

54. de la Torre AN, Contractor S, Castaneda I, Cathcart CS, Razdan D, Klyde D et al (2017) A Phase I trial using local regional treatment, nonlethal irradiation, intratumoral and systemic polyinosinic-polycytidylic acid polylysine carboxymethylcellulose to treat liver cancer: in search of the abscopal effect. J Hepatocell Carcinoma 4:111–121. https://doi.org/10.2147/JHC.S136652

55. Kyi C, Roudko V, Sabado R, Saenger Y, Loging W, Mandeli J et al (2018) Therapeutic immune modulation against solid cancers with intratumoral poly-ICLC: a pilot trial. Clin Cancer Res 24(20):4937–4948. https://doi.org/10.1158/1078-0432.CCR-17-1866

56. Hammerich L, Marron TU, Upadhyay R, Svensson-Arvelund J, Dhainaut M, Hussein S et al (2019) Systemic clinical tumor regressions and potentiation of PD-1 blockade with in situ vaccination. Nat Med 25(5):814–824. https://doi.org/10.1038/s41591-019-0410-x

57. Hu Z, Leet DE, Allesoe RL, Oliveira G, Li S, Luoma AM et al (2021) Personal neoantigen vaccines induce persistent memory T cell responses and epitope spreading in patients with melanoma. Nat Med 27(3):515–525. https://doi.org/10.1038/s41591-020-01206-4

58. Shear M, Perrault A (1944) Chemical treatment of tumors. IX. Reactions of mice with primary subcutaneous tumors to injection of a hemorrhage-producing bacterial polysaccharide. J National Cancer Inst 4(5):461–476

59. Won EK, Zahner MC, Grant EA, Gore P, Chicoine MR (2003) Analysis of the antitumoral mechanisms of lipopolysaccharide against glioblastoma multiforme. Anticancer Drugs 14(6):457–466. https://doi.org/10.1097/00001813-200307000-00012
60. Chicoine MR, Won EK, Zahner MC (2001) Intratumoral injection of lipopolysaccharide causes regression of subcutaneously implanted mouse glioblastoma multiforme. Neurosurgery 48(3):607–614; discussion 14–15. https://doi.org/10.1097/00006123-200103000-00032
61. Berendt MJ, North RJ, Kirstein DP (1978) The immunological basis of endotoxin-induced tumor regression. Requirement for a pre-existing state of concomitant anti-tumor immunity. J Exp Med 148(6):1560–1569. https://doi.org/10.1084/jem.148.6.1560
62. Engelhardt R, Mackensen A, Galanos C (1991) Phase I trial of intravenously administered endotoxin (Salmonella abortus equi) in cancer patients. Cancer Res 51(10):2524–2530
63. Otto F, Schmid P, Mackensen A, Wehr U, Seiz A, Braun M et al (1996) Phase II trial of intravenous endotoxin in patients with colorectal and non-small cell lung cancer. Eur J Cancer 32A(10):1712–1718. https://doi.org/10.1016/0959-8049(96)00186-4
64. Ha DK, Leung SW, Fung KP, Choy YM, Lee CY (1985) Role of lipid A of endotoxin in the production of tumour necrosis factor. Mol Immunol 22(3):291–294. https://doi.org/10.1016/0161-5890(85)90164-6
65. Nowotny A, Golub S, Key B (1971) Fate and effect of endotoxin derivtives in tumor-bearing mice. Proc Soc Exp Biol Med 136(1):66–69. https://doi.org/10.3181/00379727-136-35194
66. Vosika GJ, Barr C, Gilbertson D (1984) Phase-I study of intravenous modified lipid A. Cancer Immunol Immunother 18(2):107–112. https://doi.org/10.1007/BF00205743
67. Vacchelli E, Galluzzi L, Eggermont A, Fridman WH, Galon J, Sautes-Fridman C et al (2012) Trial watch: FDA-approved Toll-like receptor agonists for cancer therapy. Oncoimmunology. 1(6):894–907. https://doi.org/10.4161/onci.20931
68. de Bono JS, Dalgleish AG, Carmichael J, Diffley J, Lofts FJ, Fyffe D et al (2000) Phase I study of ONO-4007, a synthetic analogue of the lipid A moiety of bacterial lipopolysaccharide. Clin Cancer Res 6(2):397–405
69. Isambert N, Fumoleau P, Paul C, Ferrand C, Zanetta S, Bauer J et al (2013) Phase I study of OM-174, a lipid A analogue, with assessment of immunological response, in patients with refractory solid tumors. BMC Cancer 13:172. https://doi.org/10.1186/1471-2407-13-172
70. Schiffman M, Wacholder S (2012) Success of HPV vaccination is now a matter of coverage. Lancet Oncol 13(1):10–12. https://doi.org/10.1016/S1470-2045(11)70324-2
71. Seo YD, Zhou J, Morse K, Patino J, Mackay S, Kim EY et al (2018) Effect of intratumoral (IT) injection of the toll-like receptor 4 (TLR4) agonist G100 on a clinical response and CD4 T-cell response locally and systemically. J Clin Oncol 36(5):71
72. Shetab Boushehri MA, Lamprecht A (2018) TLR4-based immunotherapeutics in cancer: a review of the achievements and shortcomings. Mol Pharm 15(11):4777–4800. https://doi.org/10.1021/acs.molpharmaceut.8b00691
73. Chen M, Griffith BP, Lucia HL, Hsiung GD (1988) Efficacy of S26308 against guinea pig cytomegalovirus infection. Antimicrob Agents Chemother 32(5):678–683. https://doi.org/10.1128/AAC.32.5.678
74. Harrison CJ, Jenski L, Voychehovski T, Bernstein DI (1988) Modification of immunological responses and clinical disease during topical R-837 treatment of genital HSV-2 infection. Antiviral Res 10(4–5):209–223. https://doi.org/10.1016/0166-3542(88)90032-0
75. Bernstein DI, Harrison CJ (1989) Effects of the immunomodulating agent R837 on acute and latent herpes simplex virus type 2 infections. Antimicrob Agents Chemother 33(9):1511–1515. https://doi.org/10.1128/AAC.33.9.1511
76. Harrison CJ, Stanberry LR, Bernstein DI (1991) Effects of cytokines and R-837, a cytokine inducer, on UV-irradiation augmented recurrent genital herpes in guinea pigs. Antiviral Res 15(4):315–322. https://doi.org/10.1016/0166-3542(91)90012-g
77. Bernstein DI, Miller RL, Harrison CJ (1993) Adjuvant effects of imiquimod on a herpes simplex virus type 2 glycoprotein vaccine in guinea pigs. J Infect Dis 167(3):731–735. https://doi.org/10.1093/infdis/167.3.731

78. Bernstein DI, Miller RL, Harrison CJ (1993) Effects of therapy with an immunomodulator (imiquimod, R-837) alone and with acyclovir on genital HSV-2 infection in guinea-pigs when begun after lesion development. Antiviral Res 20(1):45–55. https://doi.org/10.1016/0166-354 2(93)90058-q

79. Reiter MJ, Testerman TL, Miller RL, Weeks CE, Tomai MA (1994) Cytokine induction in mice by the immunomodulator imiquimod. J Leukoc Biol 55(2):234–240. https://doi.org/10. 1002/jlb.55.2.234

80. Gibson SJ, Imbertson LM, Wagner TL, Testerman TL, Reiter MJ, Miller RL et al (1995) Cellular requirements for cytokine production in response to the immunomodulators imiquimod and S-27609. J Interferon Cytokine Res 15(6):537–545. https://doi.org/10.1089/jir.1995. 15.537

81. Megyeri K, Au WC, Rosztoczy I, Raj NB, Miller RL, Tomai MA et al (1995) Stimulation of interferon and cytokine gene expression by imiquimod and stimulation by Sendai virus utilize similar signal transduction pathways. Mol Cell Biol 15(4):2207–2218. https://doi.org/ 10.1128/MCB.15.4.2207

82. Testerman TL, Gerster JF, Imbertson LM, Reiter MJ, Miller RL, Gibson SJ et al (1995) Cytokine induction by the immunomodulators imiquimod and S-27609. J Leukoc Biol 58(3):365–372. https://doi.org/10.1002/jlb.58.3.365

83. Witt PL, Ritch PS, Reding D, McAuliffe TL, Westrick L, Grossberg SE et al (1993) Phase I trial of an oral immunomodulator and interferon inducer in cancer patients. Cancer Res 53(21):5176–5180

84. Savage P, Horton V, Moore J, Owens M, Witt P, Gore ME (1996) A phase I clinical trial of imiquimod, an oral interferon inducer, administered daily. Br J Cancer 74(9):1482–1486. https://doi.org/10.1038/bjc.1996.569

85. Persaud AN, Shamuelova E, Sherer D, Lou W, Singer G, Cervera C et al (2002) Clinical effect of imiquimod 5% cream in the treatment of actinic keratosis. J Am Acad Dermatol 47(4):553–556. https://doi.org/10.1067/mjd.2002.123492

86. Salasche SJ, Levine N, Morrison L (2002) Cycle therapy of actinic keratoses of the face and scalp with 5% topical imiquimod cream: an open-label trial. J Am Acad Dermatol 47(4):571–577. https://doi.org/10.1067/mjd.2002.126257

87. Harrison LI, Skinner SL, Marbury TC, Owens ML, Kurup S, McKane S et al (2004) Pharmacokinetics and safety of imiquimod 5% cream in the treatment of actinic keratoses of the face, scalp, or hands and arms. Arch Dermatol Res 296(1):6–11. https://doi.org/10.1007/s00 403-004-0465-4

88. Lebwohl M, Dinehart S, Whiting D, Lee PK, Tawfik N, Jorizzo J et al (2004) Imiquimod 5% cream for the treatment of actinic keratosis: results from two phase III, randomized, double-blind, parallel group, vehicle-controlled trials. J Am Acad Dermatol 50(5):714–721. https:// doi.org/10.1016/j.jaad.2003.12.010

89. Stockfleth E, Christophers E, Benninghoff B, Sterry W (2004) Low incidence of new actinic keratoses after topical 5% imiquimod cream treatment: a long-term follow-up study. Arch Dermatol 140(12):1542. https://doi.org/10.1001/archderm.140.12.1542-a

90. Szeimies RM, Gerritsen MJ, Gupta G, Ortonne JP, Serresi S, Bichel J et al (2004) Imiquimod 5% cream for the treatment of actinic keratosis: results from a phase III, randomized, double-blind, vehicle-controlled, clinical trial with histology. J Am Acad Dermatol 51(4):547–555. https://doi.org/10.1016/j.jaad.2004.02.022

91. Korman N, Moy R, Ling M, Matheson R, Smith S, McKane S et al (2005) Dosing with 5% imiquimod cream 3 times per week for the treatment of actinic keratosis: results of two phase 3, randomized, double-blind, parallel-group, vehicle-controlled trials. Arch Dermatol 141(4):467–473. https://doi.org/10.1001/archderm.141.4.467

92. Beutner KR, Geisse JK, Helman D, Fox TL, Ginkel A, Owens ML (1999) Therapeutic response of basal cell carcinoma to the immune response modifier imiquimod 5% cream. J Am Acad Dermatol 41(6):1002–1007. https://doi.org/10.1016/s0190-9622(99)70261-6

93. Marks R, Gebauer K, Shumack S, Amies M, Bryden J, Fox TL et al (2001) Imiquimod 5% cream in the treatment of superficial basal cell carcinoma: results of a multicenter 6-week

dose-response trial. J Am Acad Dermatol 44(5):807–813. https://doi.org/10.1067/mjd.2001. 113689

94. Shumack S, Robinson J, Kossard S, Golitz L, Greenway H, Schroeter A et al (2002) Efficacy of topical 5% imiquimod cream for the treatment of nodular basal cell carcinoma: comparison of dosing regimens. Arch Dermatol 138(9):1165–1171. https://doi.org/10.1001/archderm.138.9.1165

95. Sterry W, Ruzicka T, Herrera E, Takwale A, Bichel J, Andres K et al (2002) Imiquimod 5% cream for the treatment of superficial and nodular basal cell carcinoma: randomized studies comparing low-frequency dosing with and without occlusion. Br J Dermatol 147(6):1227–1236. https://doi.org/10.1046/j.1365-2133.2002.05069.x

96. Geisse J, Caro I, Lindholm J, Golitz L, Stampone P, Owens M (2004) Imiquimod 5% cream for the treatment of superficial basal cell carcinoma: results from two phase III, randomized, vehicle-controlled studies. J Am Acad Dermatol 50(5):722–733. https://doi.org/10.1016/j.jaad.2003.11.066

97. Huber A, Huber JD, Skinner RB Jr, Kuwahara RT, Haque R, Amonette RA (2004) Topical imiquimod treatment for nodular basal cell carcinomas: an open-label series. Dermatol Surg 30(3):429–430. https://doi.org/10.1111/j.1524-4725.2004.30116.x

98. Marks R, Owens M, Walters SA (2004) Australian multi-centre trial G. Efficacy and safety of 5% imiquimod cream in treating patients with multiple superficial basal cell carcinomas. Arch Dermatol 140(10):1284–1285. https://doi.org/10.1001/archderm.140.10.1284-b

99. Vidal D, Alomar A (2004) Efficacy of imiquimod 5% cream for basal cell carcinoma in transplant patients. Clin Exp Dermatol 29(3):237–239. https://doi.org/10.1111/j.1365-2230.2004.01456.x

100. Vidal D, Matias-Guiu X, Alomar A (2004) Open study of the efficacy and mechanism of action of topical imiquimod in basal cell carcinoma. Clin Exp Dermatol 29(5):518–525. https://doi.org/10.1111/j.1365-2230.2004.01601.x

101. Ferris RL, Saba NF, Gitlitz BJ, Haddad R, Sukari A, Neupane P et al (2018) Effect of adding motolimod to standard combination chemotherapy and cetuximab treatment of patients with squamous cell carcinoma of the head and neck: the active8 randomized clinical trial. JAMA Oncol 4(11):1583–1588. https://doi.org/10.1001/jamaoncol.2018.1888

102. Frega G, Wu Q, Le Naour J, Vacchelli E, Galluzzi L, Kroemer G et al (2020) Trial watch: experimental TLR7/TLR8 agonists for oncological indications. Oncoimmunology. 9(1):1796002. https://doi.org/10.1080/2162402X.2020.1796002

103. Rehli M (2002) Of mice and men: species variations of Toll-like receptor expression. Trends Immunol 23(8):375–378. https://doi.org/10.1016/s1471-4906(02)02259-7

104. Karapetyan L, Luke JJ, Davar D (2020) Toll-like receptor 9 agonists in cancer. Onco Targets Ther 13:10039–10060. https://doi.org/10.2147/OTT.S247050

105. Thomas M, Ponce-Aix S, Navarro A, Riera-Knorrenschild J, Schmidt M, Wiegert E et al (2018) Immunotherapeutic maintenance treatment with toll-like receptor 9 agonist lefitolimod in patients with extensive-stage small-cell lung cancer: results from the exploratory, controlled, randomized, international phase II IMPULSE study. Ann Oncol 29(10):2076–2084. https://doi.org/10.1093/annonc/mdy326

106. Schmoll HJ, Wittig B, Arnold D, Riera-Knorrenschild J, Nitsche D, Kroening H et al (2014) Maintenance treatment with the immunomodulator MGN1703, a Toll-like receptor 9 (TLR9) agonist, in patients with metastatic colorectal carcinoma and disease control after chemotherapy: a randomised, double-blind, placebo-controlled trial. J Cancer Res Clin Oncol 140(9):1615–1624. https://doi.org/10.1007/s00432-014-1682-7

107. Frank MJ, Reagan PM, Bartlett NL, Gordon LI, Friedberg JW, Czerwinski DK et al (2018) In Situ vaccination with a TLR9 agonist and local low-dose radiation induces systemic responses in untreated indolent lymphoma. Cancer Discov 8(10):1258–1269. https://doi.org/10.1158/2159-8290.CD-18-0743

108. Ribas A, Medina T, Kummar S, Amin A, Kalbasi A, Drabick JJ et al (2018) SD-101 in combination with pembrolizumab in advanced melanoma: results of a phase Ib. Multicenter Study. Cancer Discov. 8(10):1250–1257. https://doi.org/10.1158/2159-8290.CD-18-0280

109. Chien AJ, Soliman HH, Ewing CA, Boughey JC, Campbell MJ, Rugo HS et al (2021) Evaluation of intra-tumoral (IT) SD-101 and pembrolizumab (Pb) in combination with paclitaxel (P) followed by AC in high-risk HER2-negative (HER2-) stage II/III breast cancer: Results from the I-SPY 2 trial. J Clin Oncol 39(15):508. https://doi.org/10.1200/JCO.2021.39.15_suppl.508

110. Lemke-Miltner CD, Blackwell SE, Yin C, Krug AE, Morris AJ, Krieg AM et al (2020) Antibody opsonization of a TLR9 agonist-containing virus-like particle enhances In Situ immunization. J Immunol 204(5):1386–1394. https://doi.org/10.4049/jimmunol.1900742

111. Sabree SA, Voigt AP, Blackwell SE, Vishwakarma A, Chimenti MS, Salem AK, et al (2021) Direct and indirect immune effects of CMP-001, a virus-like particle containing a TLR9 agonist. J Immunother Cancer 9(6). https://doi.org/10.1136/jitc-2021-002484

112. Milhem M, Zakharia Y, Davar D, Buchbinder E, Medina T, Daud A et al (2020) O85 durable responses in anti-PD-1 refractory melanoma following intratumoral injection of a toll-like receptor 9 (TLR9) agonist, CMP-001, in combination with pembrolizumab. J Immunother Cancer 8(Suppl 1):A2. https://doi.org/10.1136/LBA2019.4

113. Ribas A, Medina T, Kirkwood JM, Zakharia Y, Gonzalez R, Davar D et al (2021) Overcoming PD-1 blockade resistance with CpG-A Toll-like receptor 9 agonist vidutolimod in patients with metastatic melanoma. Cancer Discov. https://doi.org/10.1158/2159-8290.CD-21-0425

114. Brown MC, Mosaheb MM, Mohme M, McKay ZP, Holl EK, Kastan JP et al (2021) Viral infection of cells within the tumor microenvironment mediates antitumor immunotherapy via selective TBK1-IRF3 signaling. Nat Commun 12(1):1858. https://doi.org/10.1038/s41467-021-22088-1

115. Brown MC, Holl EK, Boczkowski D, Dobrikova E, Mosaheb M, Chandramohan V, et al (2017) Cancer immunotherapy with recombinant poliovirus induces IFN-dominant activation of dendritic cells and tumor antigen-specific CTLs. Sci Transl Med 9(408). https://doi.org/10.1126/scitranslmed.aan4220

116. Yu X, Wang H, Li X, Guo C, Yuan F, Fisher PB et al (2016) Activation of the MDA-5-IPS-1 viral sensing pathway induces cancer cell death and Type I IFN-dependent antitumor immunity. Cancer Res 76(8):2166–2176. https://doi.org/10.1158/0008-5472.CAN-15-2142

117. Akhrymuk I, Frolov I, Frolova EI (2016) Both RIG-I and MDA5 detect alphavirus replication in concentration-dependent mode. Virology 487:230–241. https://doi.org/10.1016/j.virol.2015.09.023

118. Komdeur FL, Singh A, van de Wall S, Meulenberg JJM, Boerma A, Hoogeboom BN et al (2021) First-in-human Phase I clinical trial of an SFV-based RNA replicon cancer vaccine against HPV-induced cancers. Mol Ther 29(2):611–625. https://doi.org/10.1016/j.ymthe.2020.11.002

119. Edward K, Agnieszka D, Jacqueline H, Lingxin K, De Ana A, Jeffrey B et al (2021). Nature Portfolio. https://doi.org/10.21203/rs.3.rs-523458/v1

120. Feschotte C, Gilbert C (2012) Endogenous viruses: insights into viral evolution and impact on host biology. Nat Rev Genet 13(4):283–296. https://doi.org/10.1038/nrg3199

121. Navada SC, Steinmann J, Lubbert M, Silverman LR (2014) Clinical development of demethylating agents in hematology. J Clin Invest 124(1):40–46. https://doi.org/10.1172/JCI69739

122. Roulois D, Loo Yau H, Singhania R, Wang Y, Danesh A, Shen SY et al (2015) DNA-demethylating agents target colorectal cancer cells by inducing viral mimicry by endogenous transcripts. Cell 162(5):961–973. https://doi.org/10.1016/j.cell.2015.07.056

123. Chiappinelli KB, Strissel PL, Desrichard A, Li H, Henke C, Akman B et al (2015) Inhibiting DNA methylation causes an interferon response in cancer via dsRNA including endogenous retroviruses. Cell 162(5):974–986. https://doi.org/10.1016/j.cell.2015.07.011

124. Rycaj K, Plummer JB, Yin B, Li M, Garza J, Radvanyi L et al (2015) Cytotoxicity of human endogenous retrovirus K-specific T cells toward autologous ovarian cancer cells. Clin Cancer Res 21(2):471–483. https://doi.org/10.1158/1078-0432.CCR-14-0388

125. Wang-Johanning F, Rycaj K, Plummer JB, Li M, Yin B, Frerich K et al (2012) Immunotherapeutic potential of anti-human endogenous retrovirus-K envelope protein antibodies in targeting breast tumors. J Natl Cancer Inst 104(3):189–210. https://doi.org/10.1093/jnci/djr540

126. Krishnamurthy J, Rabinovich BA, Mi T, Switzer KC, Olivares S, Maiti SN et al (2015) Genetic engineering of T cells to target HERV-K, an ancient retrovirus on melanoma. Clin Cancer Res 21(14):3241–3251. https://doi.org/10.1158/1078-0432.CCR-14-3197

127. Wrangle J, Wang W, Koch A, Easwaran H, Mohammad HP, Vendetti F et al (2013) Alterations of immune response of non-small cell lung cancer with azacytidine. Oncotarget 4(11):2067–2079. https://doi.org/10.18632/oncotarget.1542

128. Linnekamp JF, Butter R, Spijker R, Medema JP, van Laarhoven HWM (2017) Clinical and biological effects of demethylating agents on solid tumours—a systematic review. Cancer Treat Rev 54:10–23. https://doi.org/10.1016/j.ctrv.2017.01.004

129. Middleton MR, Wermke M, Calvo E, Chartash E, Zhou H, Zhao X et al (2018) LBA16—Phase I/II, multicenter, open-label study of intratumoral/intralesional administration of the retinoic acid–inducible gene I (RIG-I) activator MK-4621 in patients with advanced or recurrent tumors. Ann Oncol 29:viii712. https://doi.org/10.1093/annonc/mdy424.016

130. Elion DL, Cook RS (2019) Activation of RIG-I signaling to increase the pro-inflammatory phenotype of a tumor. Oncotarget 10(24):2338–2339. https://doi.org/10.18632/oncotarget. 26729

131. Eigentler T, Bauernfeind FG, Becker JC, Brossart P, Fluck M, Heinzerling L et al (2020) A phase I dose-escalation and expansion study of intratumoral CV8102 as single-agent or in combination with anti-PD-1 antibodies in patients with advanced solid tumors. J Clin Oncol 38(15_suppl):3096. https://doi.org/10.1200/JCO.2020.38.15_suppl.3096

132. Fujita K, Nakai Y, Kawashima A, Ujike T, Nagahara A, Nakajima T et al (2017) Phase I/II clinical trial to assess safety and efficacy of intratumoral and subcutaneous injection of HVJ-E in castration-resistant prostate cancer patients. Cancer Gene Ther 24(7):277–281. https://doi.org/10.1038/cgt.2017.15

133. Fujita K, Kato T, Hatano K, Kawashima A, Ujike T, Uemura M et al (2020) Intratumoral and s.c. injection of inactivated hemagglutinating virus of Japan envelope (GEN0101) in metastatic castration-resistant prostate cancer. Cancer Sci 111(5):1692–1698. https://doi.org/10.1111/cas.14366

134. Iurescia S, Fioretti D, Rinaldi M (2020) The innate immune signalling pathways: turning RIG-I sensor activation against cancer. Cancers (Basel) 12(11):3158. https://doi.org/10.3390/cancers12113158

135. Roberts ZJ, Goutagny N, Perera PY, Kato H, Kumar H, Kawai T et al (2007) The chemotherapeutic agent DMXAA potently and specifically activates the TBK1-IRF-3 signaling axis. J Exp Med 204(7):1559–1569. https://doi.org/10.1084/jem.20061845

136. Prantner D, Perkins DJ, Lai W, Williams MS, Sharma S, Fitzgerald KA et al (2012) 5,6-Dimethylxanthenone-4-acetic acid (DMXAA) activates stimulator of interferon gene (STING)-dependent innate immune pathways and is regulated by mitochondrial membrane potential. J Biol Chem 287(47):39776–39788. https://doi.org/10.1074/jbc.M112.382986

137. Le Naour J, Zitvogel L, Galluzzi L, Vacchelli E, Kroemer G (2020) Trial watch: STING agonists in cancer therapy. Oncoimmunology 9(1):1777624. https://doi.org/10.1080/2162402X. 2020.1777624

138. Conlon J, Burdette DL, Sharma S, Bhat N, Thompson M, Jiang Z et al (2013) Mouse, but not human STING, binds and signals in response to the vascular disrupting agent 5,6-dimethylxanthenone-4-acetic acid. J Immunol 190(10):5216–5225. https://doi.org/10.4049/jimmunol.1300097

139. Harrington KJ, Brody J, Ingham M, Strauss J, Cemerski S, Wang M et al (2018) LBA15—preliminary results of the first-in-human (FIH) study of MK-1454, an agonist of stimulator of interferon genes (STING), as monotherapy or in combination with pembrolizumab (pembro) in patients with advanced solid tumors or lymphomas. Ann Oncol 29:viii712. https://doi.org/10.1093/annonc/mdy424.015

140. Meric-Bernstam F, Sandhu SK, Hamid O, Spreafico A, Kasper S, Dummer R et al (2019) Phase Ib study of MIW815 (ADU-S100) in combination with spartalizumab (PDR001) in patients (pts) with advanced/metastatic solid tumors or lymphomas. J Clin Oncol 37(15_suppl):2507. https://doi.org/10.1200/JCO.2019.37.15_suppl.2507

141. Chin EN, Yu C, Vartabedian VF, Jia Y, Kumar M, Gamo AM et al (2020) Antitumor activity of a systemic STING-activating non-nucleotide cGAMP mimetic. Science 369(6506):993–999. https://doi.org/10.1126/science.abb4255

142. Pan BS, Perera SA, Piesvaux JA, Presland JP, Schroeder GK, Cumming JN et al (2020) An orally available non-nucleotide STING agonist with antitumor activity. Science 369(6506):aba6098. https://doi.org/10.1126/science.aba6098

143. Ager CR, Boda A, Rajapakshe K, Lea ST, Di Francesco ME, Jayaprakash P, et al (2021) High potency STING agonists engage unique myeloid pathways to reverse pancreatic cancer immune privilege. J Immunother Cancer 9(8). https://doi.org/10.1136/jitc-2021-003246

144. Ager CR, Reilley MJ, Nicholas C, Bartkowiak T, Jaiswal AR, Curran MA (2017) Intratumoral STING activation with T-cell checkpoint modulation generates systemic antitumor immunity. Cancer Immunol Res 5(8):676–684. https://doi.org/10.1158/2326-6066.CIR-17-0049

145. Mosaheb MM, Brown MC, Dobrikova EY, Dobrikov MI, Gromeier M (2020) Harnessing virus tropism for dendritic cells for vaccine design. Curr Opin Virol 44:73–80. https://doi.org/10.1016/j.coviro.2020.07.012

146. Vossen MT, Westerhout EM, Soderberg-Naucler C, Wiertz EJ (2002) Viral immune evasion: a masterpiece of evolution. Immunogenetics 54(8):527–542. https://doi.org/10.1007/s00251-002-0493-1

147. Yewdell JW, Bennink JR (1999) Mechanisms of viral interference with MHC class I antigen processing and presentation. Annu Rev Cell Dev Biol 15:579–606. https://doi.org/10.1146/annurev.cellbio.15.1.579

148. Bowie AG, Unterholzner L (2008) Viral evasion and subversion of pattern-recognition receptor signalling. Nat Rev Immunol 8(12):911–922. https://doi.org/10.1038/nri2436

149. Parato KA, Senger D, Forsyth PA, Bell JC (2005) Recent progress in the battle between oncolytic viruses and tumours. Nat Rev Cancer 5(12):965–976. https://doi.org/10.1038/nrc1750

150. Zamarin D, Holmgaard RB, Subudhi SK, Park JS, Mansour M, Palese P et al (2014) Localized oncolytic virotherapy overcomes systemic tumor resistance to immune checkpoint blockade immunotherapy. Sci Transl Med 6(226):226ra32. https://doi.org/10.1126/scitranslmed.3008095

151. Li X, Wang P, Li H, Du X, Liu M, Huang Q et al (2017) The efficacy of oncolytic adenovirus is mediated by t-cell responses against virus and tumor in Syrian Hamster model. Clin Cancer Res 23(1):239–249. https://doi.org/10.1158/1078-0432.CCR-16-0477

152. Kaufman HL, Kohlhapp FJ, Zloza A (2015) Oncolytic viruses: a new class of immunotherapy drugs. Nat Rev Drug Discov 14(9):642–662. https://doi.org/10.1038/nrd4663

153. Paludan SR, Bowie AG, Horan KA, Fitzgerald KA (2011) Recognition of herpesviruses by the innate immune system. Nat Rev Immunol 11(2):143–154. https://doi.org/10.1038/nri2937

154. Andtbacka RH, Kaufman HL, Collichio F, Amatruda T, Senzer N, Chesney J et al (2015) Talimogene laherparepvec improves durable response rate in patients with advanced melanoma. J Clin Oncol 33(25):2780–2788. https://doi.org/10.1200/JCO.2014.58.3377

155. Andtbacka RH, Ross M, Puzanov I, Milhem M, Collichio F, Delman KA et al (2016) Patterns of clinical response with talimogene laherparepvec (T-VEC) in patients with melanoma treated in the OPTiM Phase III clinical trial. Ann Surg Oncol 23(13):4169–4177. https://doi.org/10.1245/s10434-016-5286-0

156. Koch MS, Lawler SE, Chiocca EA (2020) HSV-1 oncolytic viruses from bench to bedside: an overview of current clinical trials. Cancers (Basel) 12(12):3514. https://doi.org/10.3390/cancers12123514

157. Ribas A, Dummer R, Puzanov I, VanderWalde A, Andtbacka RHI, Michielin O, et al (2017) Oncolytic virotherapy promotes intratumoral T cell infiltration and improves anti-PD-1 immunotherapy. Cell 170(6):1109–1119 e10. https://doi.org/10.1016/j.cell.2017.08.027

158. Najjar YG (2021) Search for effective treatments in patients with advanced refractory melanoma continues: can novel intratumoral therapies deliver? J Immunother Cancer 9(7). https://doi.org/10.1136/jitc-2021-002820

159. Todo T (2019) ATIM-14. Results of phase II clinical trial of oncolytic herpes virus G47Δ in patients with glioblastoma. Neuro-Oncology 21(Supplement_6):vi4-vi. https://doi.org/10.1093/neuonc/noz175.014

160. Chiocca EA, Nakashima H, Kasai K, Fernandez SA, Oglesbee M (2020) Preclinical Toxicology of rQNestin34.5v.2: An Oncolytic Herpes Virus with Transcriptional Regulation of the ICP34.5 Neurovirulence Gene. Mol Ther Meth Clin Dev 17:871–893. https://doi.org/10.1016/j.omtm.2020.03.028

161. Thomas S, Kuncheria L, Roulstone V, Kyula JN, Mansfield D, Bommareddy PK et al (2019) Development of a new fusion-enhanced oncolytic immunotherapy platform based on herpes simplex virus type 1. J Immunother Cancer 7(1):214. https://doi.org/10.1186/s40425-019-0682-1

162. Kennedy EM, Farkaly T, Grzesik P, Lee J, Denslow A, Hewett J et al (2020) Design of an interferon-resistant oncolytic HSV-1 incorporating redundant safety modalities for improved tolerability. Mol Ther Oncolytics 18:476–490. https://doi.org/10.1016/j.omto.2020.08.004

163. Patel DM, Foreman PM, Nabors LB, Riley KO, Gillespie GY, Markert JM (2016) Design of a Phase I clinical trial to evaluate M032, a genetically engineered HSV-1 expressing IL-12, in patients with recurrent/progressive glioblastoma multiforme, anaplastic astrocytoma, or gliosarcoma. Hum Gene Ther Clin Dev 27(2):69–78. https://doi.org/10.1089/humc.2016.031

164. Haines BB, Denslow A, Grzesik P, Lee JS, Farkaly T, Hewett J et al (2021) ONCR-177, an oncolytic HSV-1 designed to potently activate systemic antitumor immunity. Cancer Immunol Res 9(3):291–308. https://doi.org/10.1158/2326-6066.CIR-20-0609

165. Harrington KJ, Aroldi F, Sacco JJ, Milhem MM, Curti BD, Vanderwalde AM et al (2021) Abstract LB180: Clinical biomarker studies with two fusion-enhanced versions of oncolytic HSV (RP1 and RP2) alone and in combination with nivolumab in cancer patients indicate potent immune activation. Cancer Res 81(13 Supplement):LB180. https://doi.org/10.1158/1538-7445.AM2021-LB180

166. Zheng Y, Yan R, Tang Y, Zhan B, Huang Y, Ni D et al (2021) Abstract 2597: non-clinical studies of systemic delivery of oncolytic virus arms with IL-12 and anti-PD-1 antibody. Can Res 81(13 Supplement):2597. https://doi.org/10.1158/1538-7445.AM2021-2597

167. Zhu J, Huang X, Yang Y (2007) Innate immune response to adenoviral vectors is mediated by both Toll-like receptor-dependent and -independent pathways. J Virol 81(7):3170–3180. https://doi.org/10.1128/JVI.02192-06

168. Anghelina D, Lam E, Falck-Pedersen E (2016) Diminished innate antiviral response to adenovirus vectors in cGAS/STING-deficient mice minimally impacts adaptive immunity. J Virol 90(13):5915–5927. https://doi.org/10.1128/JVI.00500-16

169. Wang F, Alain T, Szretter KJ, Stephenson K, Pol JG, Atherton MJ et al (2016) S6K-STING interaction regulates cytosolic DNA-mediated activation of the transcription factor IRF3. Nat Immunol 17(5):514–522. https://doi.org/10.1038/ni.3433

170. Suzuki M, Cela R, Bertin TK, Sule G, Cerullo V, Rodgers JR et al (2011) NOD2 signaling contributes to the innate immune response against helper-dependent adenovirus vectors independently of MyD88 in vivo. Hum Gene Ther 22(9):1071–1082. https://doi.org/10.1089/hum.2011.002

171. Muruve DA, Petrilli V, Zaiss AK, White LR, Clark SA, Ross PJ et al (2008) The inflammasome recognizes cytosolic microbial and host DNA and triggers an innate immune response. Nature 452(7183):103–107. https://doi.org/10.1038/nature06664

172. Atasheva S, Yao J, Shayakhmetov DM (2019) Innate immunity to adenovirus: lessons from mice. FEBS Lett 593(24):3461–3483. https://doi.org/10.1002/1873-3468.13696

173. Di Somma S, Iannuzzi CA, Passaro C, Forte IM, Iannone R, Gigantino V et al (2019) The oncolytic virus dl922-947 triggers immunogenic cell death in mesothelioma and reduces xenograft growth. Front Oncol 9:564. https://doi.org/10.3389/fonc.2019.00564

174. Liikanen I, Ahtiainen L, Hirvinen ML, Bramante S, Cerullo V, Nokisalmi P et al (2013) Oncolytic adenovirus with temozolomide induces autophagy and antitumor immune responses in cancer patients. Mol Ther 21(6):1212–1223. https://doi.org/10.1038/mt.2013.51

175. Ma J, Ramachandran M, Jin C, Quijano-Rubio C, Martikainen M, Yu D et al (2020) Characterization of virus-mediated immunogenic cancer cell death and the consequences for oncolytic virus-based immunotherapy of cancer. Cell Death Dis 11(1):48. https://doi.org/10.1038/s41419-020-2236-3

176. Garber K (2006) China approves world's first oncolytic virus therapy for cancer treatment. J Natl Cancer Inst 98(5):298–300. https://doi.org/10.1093/jnci/djj111

177. Lang FF, Conrad C, Gomez-Manzano C, Yung WKA, Sawaya R, Weinberg JS et al (2018) Phase I study of DNX-2401 (Delta-24-RGD) oncolytic adenovirus: replication and immunotherapeutic effects in recurrent malignant glioma. J Clin Oncol 36(14):1419–1427. https://doi.org/10.1200/JCO.2017.75.8219

178. Zadeh G, Daras M, Cloughesy TF, Colman H, Kumthekar PU, Chen CC, et al (2020) LTBK-04. phase 2 multicenter study of the oncolytic adenovirus DNX-2401 (TASADE-NOTUREV) in combination with pembrolizumab for recurrent glioblastoma; captive study (KEYNOTE-192). Neuro-Oncology. 22(Supplement_2):ii237-ii. https://doi.org/10.1093/neuonc/noaa215.989

179. Friedlander TW, Weinberg VK, Yeung A, Burke J, Lamm DL, McKiernan JM et al (2012) Activity of intravesical CG0070 in Rb-inactive superficial bladder cancer after BCG failure: updated results of a phase I/II trial. J Clin Oncol 30(15_suppl):4593. https://doi.org/10.1200/jco.2012.30.15_suppl.4593

180. Packiam Vignesh T, Barocas Daniel A, Chamie K, Davis Ronald L, Karim Kader A, Lamm Donald L et al (2019) MP43–02 CG0070, an oncolytic adenovirus, for bcg-unresponsive non-muscle-invasive bladder cancer (NMIBC): 18 month follow-up from a multicenter phase II trial. J Urol 201(4):617. https://doi.org/10.1097/01.JU.0000556225.57786.36

181. Ranki T, Pesonen S, Hemminki A, Partanen K, Kairemo K, Alanko T et al (2016) Phase I study with ONCOS-102 for the treatment of solid tumors—an evaluation of clinical response and exploratory analyses of immune markers. J Immunother Cancer 4:17. https://doi.org/10.1186/s40425-016-0121-5

182. Musher BL, Smaglo BG, Abidi W, Othman M, Patel K, Jing J et al (2020) A phase I/II study combining a TMZ-CD40L/4–1BBL-armed oncolytic adenovirus and nab-paclitaxel/gemcitabine chemotherapy in advanced pancreatic cancer: an interim report. J Clin Oncol 38(4_suppl):716. https://doi.org/10.1200/JCO.2020.38.4_suppl.716

183. Garcia-Carbonero R, Gil Martín M, Alvarez Gallego R, Macarulla Mercade T, Riesco Martinez MC, Guillen-Ponce C et al (2019) Systemic administration of the hyaluronidase-expressing oncolytic adenovirus VCN-01 in patients with advanced or metastatic pancreatic cancer: first-in-human clinical trial. Ann Oncol 30:v271–v272. https://doi.org/10.1093/annonc/mdz247.037

184. Barton KN, Siddiqui F, Pompa R, Freytag SO, Khan G, Dobrosotskaya I et al (2021) Phase I trial of oncolytic adenovirus-mediated cytotoxic and IL-12 gene therapy for the treatment of metastatic pancreatic cancer. Mol Ther Oncolytics 20:94–104. https://doi.org/10.1016/j.omto.2020.11.006

185. Jiang H, Rivera-Molina Y, Gomez-Manzano C, Clise-Dwyer K, Bover L, Vence LM et al (2017) Oncolytic adenovirus and tumor-targeting immune modulatory therapy improve autologous cancer vaccination. Cancer Res 77(14):3894–3907. https://doi.org/10.1158/0008-5472.CAN-17-0468

186. Champion BR, Besneux M, Patsalidou M, Silva A, Zonca M, Marino N et al (2019) Abstract 5013: NG-641: an oncolytic T-SIGn virus targeting cancer-associated fibroblasts in the stromal microenvironment of human carcinomas. Can Res 79(13 Supplement):5013. https://doi.org/10.1158/1538-7445.AM2019-5013

187. Peter M, Kuhnel F (2020) Oncolytic adenovirus in cancer immunotherapy. Cancers (Basel) 12(11):3354. https://doi.org/10.3390/cancers12113354

188. Seet BT, Johnston JB, Brunetti CR, Barrett JW, Everett H, Cameron C et al (2003) Poxviruses and immune evasion. Annu Rev Immunol 21:377–423. https://doi.org/10.1146/annurev.immunol.21.120601.141049

189. Delaloye J, Roger T, Steiner-Tardivel QG, Le Roy D, Knaup Reymond M, Akira S et al (2009) Innate immune sensing of modified vaccinia virus Ankara (MVA) is mediated by TLR2-TLR6, MDA-5 and the NALP3 inflammasome. PLoS Pathog 5(6):e1000480. https://doi.org/10.1371/journal.ppat.1000480

190. Yang Y, Huang CT, Huang X, Pardoll DM (2004) Persistent Toll-like receptor signals are required for reversal of regulatory T cell-mediated CD8 tolerance. Nat Immunol 5(5):508–515. https://doi.org/10.1038/ni1059

191. Dai P, Wang W, Yang N, Serna-Tamayo C, Ricca JM, Zamarin D et al (2017) Intratumoral delivery of inactivated modified vaccinia virus Ankara (iMVA) induces systemic antitumor immunity via STING and Batf3-dependent dendritic cells. Sci Immunol. 2(11):1713. https://doi.org/10.1126/sciimmunol.aal1713

192. Heo J, Reid T, Ruo L, Breitbach CJ, Rose S, Bloomston M et al (2013) Randomized dose-finding clinical trial of oncolytic immunotherapeutic vaccinia JX-594 in liver cancer. Nat Med 19(3):329–336. https://doi.org/10.1038/nm.3089

193. Moehler M, Heo J, Lee HC, Tak WY, Chao Y, Paik SW et al (2019) Vaccinia-based oncolytic immunotherapy Pexastimogene Devacirepvec in patients with advanced hepatocellular carcinoma after sorafenib failure: a randomized multicenter Phase IIb trial (TRAVERSE). Oncoimmunology. 8(8):1615817. https://doi.org/10.1080/2162402X.2019.1615817

194. Monge C, Xie C, Brar G, Akoth E, Webb S, Mabry D et al (2020) A phase I/II study of JX-594 oncolytic virus in combination with immune checkpoint inhibition in refractory colorectal cancer. Eur J Cancer 138:S57–S58. https://doi.org/10.1016/S0959-8049(20)31231-4

195. Torres-Dominguez LE, McFadden G (2019) Poxvirus oncolytic virotherapy. Expert Opin Biol Ther 19(6):561–573. https://doi.org/10.1080/14712598.2019.1600669

196. Mell LK, Brumund KT, Daniels GA, Advani SJ, Zakeri K, Wright ME et al (2017) Phase I trial of intravenous oncolytic vaccinia virus (GL-ONC1) with cisplatin and radiotherapy in patients with locoregionally advanced head and neck carcinoma. Clin Cancer Res 23(19):5696–5702. https://doi.org/10.1158/1078-0432.CCR-16-3232

197. Downs-Canner S, Guo ZS, Ravindranathan R, Breitbach CJ, O'Malley ME, Jones HL et al (2016) Phase 1 study of intravenous oncolytic poxvirus (vvDD) in patients with advanced solid cancers. Mol Ther 24(8):1492–1501. https://doi.org/10.1038/mt.2016.101

198. Gromeier M, Lachmann S, Rosenfeld MR, Gutin PH, Wimmer E (2000) Intergeneric poliovirus recombinants for the treatment of malignant glioma. Proc Natl Acad Sci U S A 97(12):6803–6808

199. Brown MC, Dobrikova EY, Dobrikov MI, Walton RW, Gemberling SL, Nair SK et al (2014) Oncolytic polio virotherapy of cancer. Cancer 120(21):3277–3286. https://doi.org/10.1002/cncr.28862

200. McKay ZP, Brown MC, Gromeier M (2021) Aryl hydrocarbon receptor signaling controls CD155 expression on macrophages and mediates tumor immunosuppression. J Immunol 206(6):1385–1394. https://doi.org/10.4049/jimmunol.2000792

201. Mosaheb MM, Dobrikova EY, Brown MC, Yang Y, Cable J, Okada H et al (2020) Genetically stable poliovirus vectors activate dendritic cells and prime antitumor CD8 T cell immunity. Nat Commun 11(1):524. https://doi.org/10.1038/s41467-019-13939-z

202. Desjardins A, Gromeier M, Herndon JE 2nd, Beaubier N, Bolognesi DP, Friedman AH et al (2018) Recurrent glioblastoma treated with recombinant poliovirus. N Engl J Med 379(2):150–161. https://doi.org/10.1056/NEJMoa1716435

203. Beasley GM, Nair SK, Farrow NE, Landa K, Selim MA, Wiggs CA, et al (2021) Phase I trial of intratumoral PVSRIPO in patients with unresectable, treatment-refractory melanoma. J Immunother Cancer. 9(4). https://doi.org/10.1136/jitc-2020-002203

204. Inman BA, Balar AV, Milowsky MI, Pruthi RS, Polasek MJ, Morris SR et al (2021) Abstract CT242: LUMINOS-103: A basket trial evaluating the safety and efficacy of PVSRIPO in patients with advanced solid tumors. Cancer Res 81(13 Supplement):CT242. https://doi.org/10.1158/1538-7445.AM2021-CT242

205. Neighbours L, McKay ZP, Gromeier M, Nichols G, Kelly AT, Corum D et al (2021) Safety and efficacy of murine PVSRIPO plus anti-PD-1 immune checkpoint inhibitor (ICI) in a

melanoma tumor model. J Clin Oncol 39(15):2560. https://doi.org/10.1200/JCO.2021.39.15_suppl.2560

206. Norman KL, Hirasawa K, Yang AD, Shields MA, Lee PW (2004) Reovirus oncolysis: the Ras/RalGEF/p38 pathway dictates host cell permissiveness to reovirus infection. Proc Natl Acad Sci U S A 101(30):11099–11104. https://doi.org/10.1073/pnas.0404310101

207. Norman KL, Lee PW (2000) Reovirus as a novel oncolytic agent. J Clin Invest 105(8):1035–1038. https://doi.org/10.1172/JCI9871

208. Abad AT, Danthi P (2020) Recognition of reovirus RNAs by the innate immune system. Viruses 12(6):667. https://doi.org/10.3390/v12060667

209. Prestwich RJ, Ilett EJ, Errington F, Diaz RM, Steele LP, Kottke T et al (2009) Immune-mediated antitumor activity of reovirus is required for therapy and is independent of direct viral oncolysis and replication. Clin Cancer Res 15(13):4374–4381. https://doi.org/10.1158/1078-0432.CCR-09-0334

210. Prestwich RJ, Errington F, Ilett EJ, Morgan RS, Scott KJ, Kottke T et al (2008) Tumor infection by oncolytic reovirus primes adaptive antitumor immunity. Clin Cancer Res 14(22):7358–7366. https://doi.org/10.1158/1078-0432.CCR-08-0831

211. Morris DG, Feng X, DiFrancesco LM, Fonseca K, Forsyth PA, Paterson AH et al (2013) REO-001: a phase I trial of percutaneous intralesional administration of reovirus type 3 dearing (Reolysin(R)) in patients with advanced solid tumors. Invest New Drugs 31(3):696–706. https://doi.org/10.1007/s10637-012-9865-z

212. Mahalingam D, Fountzilas C, Moseley J, Noronha N, Tran H, Chakrabarty R et al (2017) A phase II study of REOLYSIN((R)) (pelareorep) in combination with carboplatin and paclitaxel for patients with advanced malignant melanoma. Cancer Chemother Pharmacol 79(4):697–703. https://doi.org/10.1007/s00280-017-3260-6

213. Bernstein V, Ellard SL, Dent SF, Tu D, Mates M, Dhesy-Thind SK et al (2018) A randomized phase II study of weekly paclitaxel with or without pelareorep in patients with metastatic breast cancer: final analysis of Canadian Cancer Trials Group IND.213. Breast Cancer Res Treat 167(2):485–93. https://doi.org/10.1007/s10549-017-4538-4

214. Morris D, Tu D, Tehfe MA, Nicholas GA, Goffin JR, Gregg RW et al (2016) A randomized phase II study of Reolysin in patients with previously treated advanced or metatstatic non small cell lung cancer (NSCLC) receiving standard salvage chemotherapy—canadian cancer trials group IND 211. J Clin Oncol 34(15):e20512. https://doi.org/10.1200/JCO.2016.34.15_suppl.e20512

215. Karnad AB, Haigentz M, Miley T, Coffey M, Gill G, Mita M (2011) Abstract C22: a phase II study of intravenous wild-type reovirus (Reolysin®) in combination with paclitaxel plus carboplatin in patients with platinum refractory metastatic and/or recurrent squamous cell carcinoma of the head and neck. Mol Cancer Ther 10(11 Supplement):C22. https://doi.org/10.1158/1535-7163.TARG-11-C22

216. Noonan AM, Farren MR, Geyer SM, Huang Y, Tahiri S, Ahn D et al (2016) Randomized phase 2 trial of the oncolytic virus pelareorep (Reolysin) in upfront treatment of metastatic pancreatic adenocarcinoma. Mol Ther 24(6):1150–1158. https://doi.org/10.1038/mt.2016.66

217. Samson A, Scott KJ, Taggart D, West EJ, Wilson E, Nuovo GJ et al (2018) Intravenous delivery of oncolytic reovirus to brain tumor patients immunologically primes for subsequent checkpoint blockade. Sci Transl Med. 10(422):aam7577. https://doi.org/10.1126/scitranslmed.aam7577

218. Muller L, Berkeley R, Barr T, Ilett E, Errington-Mais F (2020) Past, present and future of oncolytic reovirus. Cancers (Basel) 12(11):3219. https://doi.org/10.3390/cancers12113219

219. Feng Q, Hato SV, Langereis MA, Zoll J, Virgen-Slane R, Peisley A et al (2012) MDA5 detects the double-stranded RNA replicative form in picornavirus-infected cells. Cell Rep 2(5):1187–1196. https://doi.org/10.1016/j.celrep.2012.10.005

220. Annels NE, Mansfield D, Arif M, Ballesteros-Merino C, Simpson GR, Denyer M et al (2019) Phase I trial of an ICAM-1-targeted immunotherapeutic-coxsackievirus A21 (CVA21) as an oncolytic agent against non muscle-invasive bladder cancer. Clin Cancer Res 25(19):5818. https://doi.org/10.1158/1078-0432.CCR-18-4022

221. Andtbacka RHI, Curti B, Daniels GA, Hallmeyer S, Whitman ED, Lutzky J, et al (2021) Clinical responses of oncolytic coxsackievirus A21 (V937) in patients with unresectable melanoma. J Clin Oncol. JCO2003246. https://doi.org/10.1200/JCO.20.03246

222. Silk AW, Day SJ, Kaufman HL, Bryan J, Norrell JT, Imbergamo C et al (2021) Abstract CT139: intratumoral oncolytic virus V937 in combination with pembrolizumab (pembro) in patients (pts) with advanced melanoma: Updated results from the phase 1b CAPRA study. Cancer Res 81(13 Supplement):CT139. https://doi.org/10.1158/1538-7445.AM2021-CT139

223. Stojdl DF, Lichty BD, tenOever BR, Paterson JM, Power AT, Knowles S et al (2003) VSV strains with defects in their ability to shutdown innate immunity are potent systemic anticancer agents. Cancer Cell 4(4):263–275. https://doi.org/10.1016/s1535-6108(03)00241-1

224. Yoneyama M, Kikuchi M, Natsukawa T, Shinobu N, Imaizumi T, Miyagishi M et al (2004) The RNA helicase RIG-I has an essential function in double-stranded RNA-induced innate antiviral responses. Nat Immunol 5(7):730–737. https://doi.org/10.1038/ni1087

225. Lund JM, Alexopoulou L, Sato A, Karow M, Adams NC, Gale NW et al (2004) Recognition of single-stranded RNA viruses by Toll-like receptor 7. Proc Natl Acad Sci U S A 101(15):5598–5603. https://doi.org/10.1073/pnas.0400937101

226. Johnson JE, Nasar F, Coleman JW, Price RE, Javadian A, Draper K et al (2007) Neurovirulence properties of recombinant vesicular stomatitis virus vectors in non-human primates. Virology 360(1):36–49. https://doi.org/10.1016/j.virol.2006.10.026

227. Jenks N, Myers R, Greiner SM, Thompson J, Mader EK, Greenslade A et al (2010) Safety studies on intrahepatic or intratumoral injection of oncolytic vesicular stomatitis virus expressing interferon-beta in rodents and nonhuman primates. Hum Gene Ther 21(4):451–462. https://doi.org/10.1089/hum.2009.111

228. Obuchi M, Fernandez M, Barber GN (2003) Development of recombinant vesicular stomatitis viruses that exploit defects in host defense to augment specific oncolytic activity. J Virol 77(16):8843–8856. https://doi.org/10.1128/jvi.77.16.8843-8856.2003

229. Patel MR, Jacobson BA, Ji Y, Drees J, Tang S, Xiong K et al (2015) Vesicular stomatitis virus expressing interferon-beta is oncolytic and promotes antitumor immune responses in a syngeneic murine model of non-small cell lung cancer. Oncotarget 6(32):33165–33177. https://doi.org/10.18632/oncotarget.5320

230. Cook J, Peng KW, Geyer SM, Ginos BF, Dueck AC, Packiriswamy N et al (2021) Clinical activity of systemic VSV-IFNβ-NIS oncolytic virotherapy in patients with relapsed refractory T-cell lymphoma. J Clin Oncol 39(15_suppl):2500. https://doi.org/10.1200/JCO.2021.39.15_suppl.2500

231. Bluming AZ, Ziegler JL (1971) Regression of Burkitt's lymphoma in association with measles infection. Lancet 2(7715):105–106. https://doi.org/10.1016/s0140-6736(71)92086-1

232. Runge S, Sparrer KM, Lassig C, Hembach K, Baum A, Garcia-Sastre A et al (2014) In vivo ligands of MDA5 and RIG-I in measles virus-infected cells. PLoS Pathog 10(4):e1004081. https://doi.org/10.1371/journal.ppat.1004081

233. Mesman AW, Zijlstra-Willems EM, Kaptein TM, de Swart RL, Davis ME, Ludlow M et al (2014) Measles virus suppresses RIG-I-like receptor activation in dendritic cells via DC-SIGN-mediated inhibition of PP1 phosphatases. Cell Host Microbe 16(1):31–42. https://doi.org/10.1016/j.chom.2014.06.008

234. Mina MJ, Kula T, Leng Y, Li M, de Vries RD, Knip M et al (2019) Measles virus infection diminishes preexisting antibodies that offer protection from other pathogens. Science 366(6465):599–606. https://doi.org/10.1126/science.aay6485

235. Heinzerling L, Kunzi V, Oberholzer PA, Kundig T, Naim H, Dummer R (2005) Oncolytic measles virus in cutaneous T-cell lymphomas mounts antitumor immune responses in vivo and targets interferon-resistant tumor cells. Blood 106(7):2287–2294. https://doi.org/10.1182/blood-2004-11-4558

236. Galanis E, Hartmann LC, Cliby WA, Long HJ, Peethambaram PP, Barrette BA et al (2010) Phase I trial of intraperitoneal administration of an oncolytic measles virus strain engineered to express carcinoembryonic antigen for recurrent ovarian cancer. Cancer Res 70(3):875–882. https://doi.org/10.1158/0008-5472.CAN-09-2762

237. Engeland CE, Ungerechts G (2021) Measles virus as an oncolytic immunotherapy. Cancers (Basel). 13(3):544. https://doi.org/10.3390/cancers13030544

238. Galanis E, Atherton PJ, Maurer MJ, Knutson KL, Dowdy SC, Cliby WA et al (2015) Oncolytic measles virus expressing the sodium iodide symporter to treat drug-resistant ovarian cancer. Cancer Res 75(1):22–30. https://doi.org/10.1158/0008-5472.CAN-14-2533

239. Rehwinkel J, Tan CP, Goubau D, Schulz O, Pichlmair A, Bier K et al (2010) RIG-I detects viral genomic RNA during negative-strand RNA virus infection. Cell 140(3):397–408. https://doi.org/10.1016/j.cell.2010.01.020

240. Kato H, Sato S, Yoneyama M, Yamamoto M, Uematsu S, Matsui K et al (2005) Cell type-specific involvement of RIG-I in antiviral response. Immunity 23(1):19–28. https://doi.org/10.1016/j.immuni.2005.04.010

241. Burman B, Pesci G, Zamarin D (2020) Newcastle disease virus at the forefront of cancer immunotherapy. Cancers (Basel) 12(12):3552. https://doi.org/10.3390/cancers12123552

242. Schlag P, Manasterski M, Gerneth T, Hohenberger P, Dueck M, Herfarth C et al (1992) Active specific immunotherapy with Newcastle-disease-virus-modified autologous tumor cells following resection of liver metastases in colorectal cancer. First evaluation of clinical response of a phase II-trial. Cancer Immunol Immunother 35(5):325–330. https://doi.org/10.1007/BF01741145

243. Liang W, Wang H, Sun TM, Yao WQ, Chen LL, Jin Y et al (2003) Application of autologous tumor cell vaccine and NDV vaccine in treatment of tumors of digestive tract. World J Gastroenterol 9(3):495–498. https://doi.org/10.3748/wjg.v9.i3.495

244. Csatary LK, Gosztonyi G, Szeberenyi J, Fabian Z, Liszka V, Bodey B et al (2004) MTH-68/H oncolytic viral treatment in human high-grade gliomas. J Neurooncol 67(1–2):83–93. https://doi.org/10.1023/b:neon.0000021735.85511.05

245. Csatary LK, Moss RW, Beuth J, Torocsik B, Szeberenyi J, Bakacs T (1999) Beneficial treatment of patients with advanced cancer using a newcastle disease virus vaccine (MTH-68/H). Anticancer Res 19(1B):635–638

246. Pecora AL, Rizvi N, Cohen GI, Meropol NJ, Sterman D, Marshall JL et al (2002) Phase I trial of intravenous administration of PV701, an oncolytic virus, in patients with advanced solid cancers. J Clin Oncol 20(9):2251–2266. https://doi.org/10.1200/JCO.2002.08.042

247. Laurie SA, Bell JC, Atkins HL, Roach J, Bamat MK, O'Neil JD et al (2006) A phase 1 clinical study of intravenous administration of PV701, an oncolytic virus, using two-step desensitization. Clin Cancer Res 12(8):2555–2562. https://doi.org/10.1158/1078-0432.CCR-05-2038

248. Dy GK, Davar D, Galanis E, Townsley D, Karanovic D, Schwaederle M et al (2020) Abstract CT244: a phase 1 study of IV MEDI5395, an oncolytic virus, in combination with durvalumab in patients with advanced solid tumors. Cancer Res 80(16 Supplement):CT244. https://doi.org/10.1158/1538-7445.AM2020-CT244

249. Eitel J, Suttorp N, Opitz B (2010) Innate immune recognition and inflammasome activation in listeria monocytogenes infection. Front Microbiol 1:149. https://doi.org/10.3389/fmicb.2010.00149

250. Oladejo M, Paterson Y, Wood LM (2021) Clinical experience and recent advances in the development of listeria-based tumor immunotherapies. Front Immunol 12:642316. https://doi.org/10.3389/fimmu.2021.642316

251. Tsujikawa T, Crocenzi T, Durham JN, Sugar EA, Wu AA, Onners B et al (2020) Evaluation of cyclophosphamide/GVAX pancreas followed by listeria-mesothelin (CRS-207) with or without nivolumab in patients with pancreatic cancer. Clin Cancer Res 26(14):3578–3588. https://doi.org/10.1158/1078-0432.CCR-19-3978

252. Basu P, Mehta A, Jain M, Gupta S, Nagarkar RV, John S et al (2018) A randomized phase 2 study of ADXS11-001 listeria monocytogenes-listeriolysin O immunotherapy with or without cisplatin in treatment of advanced cervical cancer. Int J Gynecol Cancer 28(4):764–772. https://doi.org/10.1097/IGC.0000000000001235

253. Huh WK, Brady WE, Fracasso PM, Dizon DS, Powell MA, Monk BJ et al (2020) Phase II study of axalimogene filolisbac (ADXS-HPV) for platinum-refractory cervical carcinoma: an

NRG oncology/gynecologic oncology group study. Gynecol Oncol 158(3):562–569. https:// doi.org/10.1016/j.ygyno.2020.06.493

254. Hassan R, Alley E, Kindler H, Antonia S, Jahan T, Honarmand S et al (2019) Clinical response of live-attenuated, listeria monocytogenes expressing mesothelin (CRS-207) with chemotherapy in patients with malignant pleural mesothelioma. Clin Cancer Res 25(19):5787–5798. https://doi.org/10.1158/1078-0432.CCR-19-0070

255. Alley EW, Tanvetyanon T, Jahan TM, Gandhi L, Peikert T, Stevenson J et al (2019) A phase II single-arm study of CRS-207 with pembrolizumab (pembro) in previously treated malignant pleural mesothelioma (MPM). J Clin Oncol 37(8):29. https://doi.org/10.1200/JCO.2019.37. 8_suppl.29

256. Diamond MS, Kinder M, Matsushita H, Mashayekhi M, Dunn GP, Archambault JM et al (2011) Type I interferon is selectively required by dendritic cells for immune rejection of tumors. J Exp Med 208(10):1989–2003. https://doi.org/10.1084/jem.20101158

257. Di S, Zhou M, Pan Z, Sun R, Chen M, Jiang H et al (2019) Combined adjuvant of poly I: C improves antitumor effects of CAR-T cells. Front Oncol 9:241. https://doi.org/10.3389/fonc. 2019.00241

258. Ngoi SM, Tovey MG, Vella AT (2008) Targeting poly(I:C) to the TLR3-independent pathway boosts effector CD8 T cell differentiation through IFN-alpha/beta. J Immunol 181(11):7670–7680. https://doi.org/10.4049/jimmunol.181.11.7670

259. Poeck H, Besch R, Maihoefer C, Renn M, Tormo D, Morskaya SS et al (2008) 5'-Triphosphate-siRNA: turning gene silencing and Rig-I activation against melanoma. Nat Med 14(11):1256–1263. https://doi.org/10.1038/nm.1887

260. Inglefield JR, Dumitru CD, Alkan SS, Gibson SJ, Lipson KE, Tomai MA et al (2008) TLR7 agonist 852A inhibition of tumor cell proliferation is dependent on plasmacytoid dendritic cells and type I IFN. J Interferon Cytokine Res 28(4):253–263. https://doi.org/10.1089/jir. 2007.0097

261. Mai J, Li Z, Xia X, Zhang J, Li J, Liu H et al (2021) Synergistic activation of antitumor immunity by a particulate therapeutic vaccine. Adv Sci (Weinh). 8(12):2100166. https://doi.org/10. 1002/advs.202100166

262. Deng L, Liang H, Xu M, Yang X, Burnette B, Arina A et al (2014) STING-dependent cytosolic DNA sensing promotes radiation-induced type i interferon-dependent antitumor immunity in immunogenic tumors. Immunity 41(5):843–852. https://doi.org/10.1016/j.immuni.2014. 10.019

263. Klarquist J, Hennies CM, Lehn MA, Reboulet RA, Feau S, Janssen EM (2014) STING-mediated DNA sensing promotes antitumor and autoimmune responses to dying cells. J Immunol 193(12):6124–6134. https://doi.org/10.4049/jimmunol.1401869

264. Sivick KE, Desbien AL, Glickman LH, Reiner GL, Corrales L, Surh NH et al (2019) Magnitude of therapeutic STING activation determines CD8(+) T cell-mediated anti-tumor immunity. Cell Rep 29(3):785–789. https://doi.org/10.1016/j.celrep.2019.09.089

265. Kolumam GA, Thomas S, Thompson LJ, Sprent J, Murali-Krishna K (2005) Type I interferons act directly on CD8 T cells to allow clonal expansion and memory formation in response to viral infection. J Exp Med 202(5):637–650. https://doi.org/10.1084/jem.20050821

266. Le Bon A, Durand V, Kamphuis E, Thompson C, Bulfone-Paus S, Rossmann C et al (2006) Direct stimulation of T cells by type I IFN enhances the CD8+ T cell response during cross-priming. J Immunol 176(8):4682–4689. https://doi.org/10.4049/jimmunol.176.8.4682

267. Zamarin D, Ricca JM, Sadekova S, Oseledchyk A, Yu Y, Blumenschein WM et al (2018) PD-L1 in tumor microenvironment mediates resistance to oncolytic immunotherapy. J Clin Invest 128(4):1413–1428. https://doi.org/10.1172/JCI98047

268. Kirkwood J (2002) Cancer immunotherapy: the interferon-alpha experience. Semin Oncol 29(3 Suppl 7):18–26. https://doi.org/10.1053/sonc.2002.33078

269. Benci JL, Johnson LR, Choa R, Xu Y, Qiu J, Zhou Z et al (2019) Opposing functions of interferon coordinate adaptive and innate immune responses to cancer immune checkpoint blockade. Cell 178(4):933–948. https://doi.org/10.1016/j.cell.2019.07.019

270. Benci JL, Xu B, Qiu Y, Wu TJ, Dada H, Twyman-Saint Victor C et al (2016) Tumor interferon signaling regulates a multigenic resistance program to immune checkpoint blockade. Cell 167(6):1540–1554. https://doi.org/10.1016/j.cell.2016.11.022

271. Boukhaled GM, Harding S, Brooks DG (2021) Opposing roles of type I interferons in cancer immunity. Annu Rev Pathol 16:167–198. https://doi.org/10.1146/annurev-pathol-031920-093932

272. Snell LM, McGaha TL, Brooks DG (2017) Type I interferon in chronic virus infection and cancer. Trends Immunol 38(8):542–557. https://doi.org/10.1016/j.it.2017.05.005

273. Magen A, Nie J, Ciucci T, Tamoutounour S, Zhao Y, Mehta M et al (2019) Single-Cell Profiling Defines Transcriptomic Signatures Specific to Tumor-Reactive versus Virus-Responsive CD4(+) T Cells. Cell Rep 29(10):3019–3032. https://doi.org/10.1016/j.celrep.2019.10.131

274. Trinchieri G (2010) Type I interferon: friend or foe? J Exp Med 207(10):2053–2063. https://doi.org/10.1084/jem.20101664

275. Terawaki S, Chikuma S, Shibayama S, Hayashi T, Yoshida T, Okazaki T et al (2011) IFN-alpha directly promotes programmed cell death-1 transcription and limits the duration of T cell-mediated immunity. J Immunol 186(5):2772–2779. https://doi.org/10.4049/jimmunol.1003208

276. Reilley MJ, Morrow B, Ager CR, Liu A, Hong DS, Curran MA (2019) TLR9 activation cooperates with T cell checkpoint blockade to regress poorly immunogenic melanoma. J Immunother Cancer 7(1):323. https://doi.org/10.1186/s40425-019-0811-x

277. Yarchoan M, Hopkins A, Jaffee EM (2017) Tumor mutational burden and response rate to PD-1 Inhibition. N Engl J Med 377(25):2500–2501. https://doi.org/10.1056/NEJMc1713444

278. Sharma P, Allison JP (2015) The future of immune checkpoint therapy. Science 348(6230):56–61. https://doi.org/10.1126/science.aaa8172

279. Kalbasi A, Ribas A (2020) Tumour-intrinsic resistance to immune checkpoint blockade. Nat Rev Immunol 20(1):25–39. https://doi.org/10.1038/s41577-019-0218-4

280. Le Bon A, Etchart N, Rossmann C, Ashton M, Hou S, Gewert D et al (2003) Cross-priming of CD8+ T cells stimulated by virus-induced type I interferon. Nat Immunol 4(10):1009–1015. https://doi.org/10.1038/ni978

281. Shekarian T, Valsesia-Wittmann S, Brody J, Michallet MC, Depil S, Caux C et al (2017) Pattern recognition receptors: immune targets to enhance cancer immunotherapy. Ann Oncol 28(8):1756–1766. https://doi.org/10.1093/annonc/mdx179

282. Lei J, Ploner A, Elfstrom KM, Wang J, Roth A, Fang F et al (2020) HPV vaccination and the risk of invasive cervical cancer. N Engl J Med 383(14):1340–1348. https://doi.org/10.1056/NEJMoa1917338

283. Yee C, Thompson JA, Byrd D, Riddell SR, Roche P, Celis E et al (2002) Adoptive T cell therapy using antigen-specific CD8+ T cell clones for the treatment of patients with metastatic melanoma: in vivo persistence, migration, and antitumor effect of transferred T cells. Proc Natl Acad Sci U S A 99(25):16168–16173. https://doi.org/10.1073/pnas.242600099

284. Mullard A (2021) FDA approves fourth CAR-T cell therapy. Nat Rev Drug Discov 20(3):166. https://doi.org/10.1038/d41573-021-00031-9

285. Guedan S, Alemany R (2018) CAR-T cells and oncolytic viruses: joining forces to overcome the solid tumor challenge. Front Immunol 9:2460. https://doi.org/10.3389/fimmu.2018.02460

286. Rosewell Shaw A, Suzuki M (2018) Oncolytic viruses partner with T-Cell therapy for solid tumor treatment. Front Immunol 9:2103. https://doi.org/10.3389/fimmu.2018.02103

287. Sivick KE, Desbien AL, Glickman LH, Reiner GL, Corrales L, Surh NH et al (2018) Magnitude of therapeutic sting activation determines CD8(+) T cell-mediated anti-tumor immunity. Cell Rep. 25(11):3074–3085. https://doi.org/10.1016/j.celrep.2018.11.047

288. Huang B, Zhao J, Unkeless JC, Feng ZH, Xiong H (2008) TLR signaling by tumor and immune cells: a double-edged sword. Oncogene 27(2):218–224. https://doi.org/10.1038/sj.onc.1210904

289. Huff AL, Wongthida P, Kottke T, Thompson JM, Driscoll CB, Schuelke M et al (2018) APOBEC3 ssserapy. Mol Ther Oncolytics. 11:1–13. https://doi.org/10.1016/j.omto.2018.08.003

290. Verzella D, Pescatore A, Capece D, Vecchiotti D, Ursini MV, Franzoso G et al (2020) Life, death, and autophagy in cancer: NF-kappaB turns up everywhere. Cell Death Dis 11(3):210. https://doi.org/10.1038/s41419-020-2399-y

291. Luo JL, Kamata H, Karin M (2005) IKK/NF-kappaB signaling: balancing life and death–a new approach to cancer therapy. J Clin Invest 115(10):2625–2632. https://doi.org/10.1172/JCI26322

Allogeneic Tumor Antigen-Specific T Cells for Broadly Applicable Adoptive Cell Therapy of Cancer

4

Zaki Molvi and Richard J. O'Reilly

4.1 Introduction

Tumor-associated antigens (TAA) are peptides recognized by T cells when presented by alleles of the major histocompatibility complex that are derived from proteins selectively or differentially expressed by malignant cells. Ideal tumor antigens are restricted in their expression to malignant tissue with little or no expression on healthy tissue, potentially excepting immune privileged sites such as the central nervous system, eyes, testes, and fetal tissue. Preexisting immune responses to such tumor antigens were thought to be detectable only in patients with established antigen-expressing tumors. However, as we will discuss, functional T cells of varying frequency specific for several differentially expressed antigens exist in healthy individuals who have never had cancer. It has been hypothesized that T cells recognizing such antigens contribute to tumor surveillance in an immunocompetent host, and further, that certain of these tumor-reactive T cells can be harnessed for allogeneic cell therapy. In this chapter, we review evidence supporting this hypothesis, the array of differentially expressed self-antigens and tumor-specific neoantigens that are known to induce T cells with antitumor activity from healthy allogeneic donors, and recent technological advances that can streamline discovery of immunogenic tumor antigens and their cognate T cells. We also summarize the clinical experience with such T cells when used for adoptive immunotherapy.

Z. Molvi (✉) · R. J. O'Reilly
Immunology Program, Memorial Sloan Kettering Cancer Center, New York, NY, USA
e-mail: molviz@mskcc.org

R. J. O'Reilly
e-mail: oreillyr@mskcc.org

The first evidence that T cells from healthy allogeneic donors could exert an antitumor effect in man was derived from early clinical trials of marrow transplants applied to the treatment of acute lymphoblastic leukemia which demonstrated a markedly lower incidence of leukemic relapse among recipients of transplants from HLA-matched siblings than that observed following transplant from genotypically identical twins administered after the same pre-transplant conditioning [1]. This graft versus leukemia (GvL) effect was initially ascribed largely or exclusively to the response of donor T cells in the marrow graft against alloantigens unique to the host, a response clinically manifested as graft versus host disease (GvH).

The idea that the GvL effect depended on a GvH response was initially supported by clinical findings in recipients of unmodified marrow grafts [2], and by the observation in patients transplanted for CML, that T cell-depleted marrow transplants, while markedly reducing the incidence of GvHD, were associated with a doubling of the incidence of relapse post-transplant [3]. Thereafter, the therapeutic potential of HLA-matched lymphocytes as antileukemic effectors was clearly illustrated by the initial demonstration [4] that infusions of unrelated lymphocytes from the transplant donor alone could induce durable remissions of CML relapsing after allogeneic transplant, and, further that limited doses of donor leukocyte infusions (DLI) could induce such remissions without causing GvHD [5]. Strikingly, however, infusions of such unselected lymphocytes from the donor, while capable of inducing GvHD, have not been effective in the treatment of acute lymphoblastic leukemia (ALL) relapsing post-HCT, and have induced usually short-term remissions of AML in only a minority of cases [6]. Conversely, recent randomized trials have shown that T cell-depleted grafts are not associated with an increase in the incidence of relapse when applied to AML. These latter findings have spawned extensive research into the nature of the antigens on CML, AML, and ALL cells that can be recognized and targeted by HCT donor T cells and their expression on leukemic blasts or their progenitors.

4.2   Tumor Antigen-Reactive T Cells Emerging Post-DLI

Initially, clinical responses to DLI were shown to be correlated with the frequency of peripheral lymphocyte precursors capable of inhibiting clonogenic CML progenitor cells in an HLA class I- or II-restricted manner [7]. While the identities of the T cells and their specificities in DLI infusions that exert antileukemic activity are still only partly elucidated, there is significant evidence for a handful of well-studied antigens. These can be grouped into three categories: (1) alloantigens selectively expressed by hematopoietic cells of a transplant recipient; (2) non-mutated self-peptides that are differentially highly expressed by tumor cells but either not expressed or minimally expressed by normal tissues; and (3) tumor-exclusive antigens, termed neoantigens, which result from mutations or other alterations unique to the tumor cell.

HCT-donor derived T cell responses to specific well-characterized allelic polymorphism-derived minor alloantigens selectively expressed on hematopoietic cells of the transplant recipient have been repeatedly detected following DLI and have been closely correlated with inductions of complete or partial remissions of CML, AML, or myeloma. The minor alloantigens targeted by these T cells include HA-1H and HA-2V presented by HLA-A0201 [8], and LRH-1 (P2X5) presented by HLA-B0702 [9]. Notably, these T cells, isolated from the responding patient's blood post-DLI, have been shown to selectively inhibit the in vitro growth of CD34+hematopoietic cells and leukemic cells of host type [8]. Recent studies suggest that other leukemia-reactive T cells specific for as yet incompletely characterized minor alloantigens differentially expressed by leukemic cells also contribute to the hematologic responses observed. For example, Bergen et al. analyzed CD8 T cells in 11 DLI-responders treated for CML, AML, or MDS clinically presenting with either selective GvL, i.e. GvL without concurrent GvHD, or GvL + GvHD [10]. Analyzing the post-DLI peripheral blood, they found an increase in CD8 T cells expressing HLA-DR in post-DLI samples relative to pre-DLI, and further, that such HLA-DR+CD8 T cells were less frequently alloreactive in patients with selective GvL compared to those with GvL with concurrent GvHD as measured by reactivity to the respective patient's transformed cells (0.6% vs. 30% median alloreactivity for selective GvL and GvL + GvHD, respectively). In those with only GvL, whole genome data was used to infer minor histocompatibility antigens (MiHA) arising from single nucleotide polymorphisms (SNPs) present in each patient's, but not their respective donor's genome, yielding 19 unique peptides coded by SNPs preferentially recognized by donor-derived HLA-DR+CD8 T cell clones compared to their parental wildtype peptides. Thus, these MiHA recognized in selective GvL cases may also be tumor-selective alloantigens since their recognition does not drive GvHD.

The PR1 antigen, an HLA-A2-presented peptide derived from proteinase 3 that is an enzyme differentially expressed at high levels in myeloid leukemias, was one of the first tumor antigens found to be recognized by T cells in DLI responders with antileukemic activity [11]. Interestingly, PR1 peptide sensitization was first reported to elicit T cells in healthy HLA-A2+individuals that were capable of inhibiting CFU-GM formation in marrow from CML patients but not normal donors in an HLA-A2-restricted manner [12]. This was felt to reflect the expression of the parental protein of PR1, proteinase 3, which is high in CML blasts but low in myeloid progenitors. Notably, PR1-specific T cells are also markedly elevated post-HCT in CML patients. Indeed, PR1-specific T cells have been reported to make up 0.11–12.8% of total peripheral CD8 T cells in 75% of allogeneic HCT recipients with CML responding to DLI. In contrast, T cells that exhibit antileukemic activity are undetectable directly ex vivo in healthy HCT donors without repeated sensitization [11, 13]. T cells specific for peptides derived from the Wilms' tumor protein 1 (WT1) have also been found by our group and others in healthy individuals [14–16] as well as in DLI responders treated for myeloma or leukemia [17, 18]. Similar to PR1 responses in CML, the presence of circulating WT1 responses in patients with multiple myeloma was found to be associated with lower disease burden [17]. Peripheral T cell responses to the cancer-testis antigen CT7 were also found by our group to be associated with clinical response to post-HCT DLI in multiple myeloma patients [19]. Notably, tetramer staining revealed that CT7-specific T cells were detectable 3 years post-T cell-depleted HCT in the periphery and bone marrow of a multiple myeloma patient, with an effector memory phenotype in the periphery and central memory phenotype in the marrow [19].

T cells specific for several tumor neoantigens have also been detected in DLI recipients. Antileukemic T cells specific for peptides derived from the BCR-ABL oncogene can be induced in healthy individuals and are detectable in CML patients [20–22]. In CML patients, an inverse relationship has been reported between BCR-ABL-specific T cells and BCR-ABL gene copies [23]. Another tumor-specific protein, CML66, has been shown to be overexpressed in leukemias and is critical to cell survival [24, 25]. CML66-specific T cells have been found in patient peripheral blood post-DLI, but were not present in the donor graft or patient peripheral blood pre-DLI [26]. CML66-specific T cells restricted by HLA-A24 have been shown to lyse myeloid leukemia cell lines and patient-derived AML blasts but not normal cells [27].

Another tumor antigen, PRAME, is overexpressed in melanomas and leukemias [28], with myeloid leukemias displaying progressively increased expression with increased disease severity [29, 30]. PRAME-specific T cells capable of killing tumors can be induced from the blood of healthy individuals [10, 31]. NYESO1 is a tumor antigen presented by a variety of solid and liquid tumors. Though its recognition by normal donor CD8 T cells appears limited [32], NYESO1 epitopes promiscuously bind multiple HLA class II alleles and induce CD4 T cell responses in normal donors capable of specifically producing cytokines and lysing tumor cells [33–36]. The aforementioned antigens encompass the spectrum of well-characterized antigens targeted by allogeneic HCT donor-derived T cells thus far detected in patients responding to DLI [37].

4.3 Development of Tumor Antigen-Targeted Immunotherapies

Given that T cells recognizing certain tumor antigens in healthy donors emerge in post-HCT patients, considerable efforts have been made to develop allogeneic T cells targeting such antigens for adoptive cell therapy of cancer, many of which are now in clinical trials (Table 4.1). For the purposes of our discussion, we again divide the classes of tumor antigens that can be targeted by donor T cells into minor alloantigens, antigens differentially highly expressed by tumors but either not expressed or minimally expressed by normal tissues, and neoantigens exclusive to tumor cells resulting from mutations or other alterations unique to the tumor cell.

Table 4.1 Summary of clinical trials utilizing tumor antigen-specific T cells from an allogeneic source for cancer immunotherapy

Reference	Study center	Tumor antigen	HLA	Indications	Format of allogeneic T cells	Results
NCT03326921; EudraCT 2010–024625-20; Krakow et al. (2020); van Balen et al. [38]	Fred Hutchinson Cancer Research Center	HA-1	A0201	Multiple leukemias	HCT donor memory T cells transduced with HA-1 TCR	
NCT03091933	CIUSSS d l'Est-de-l'Île-de-Montréal	MiHa	Multiple	Multiple hematologic	Donor T cells primed against MiHa identified through genomic SNP analysis	
NCT00107354; Warren et al. [39]	Fred Hutchinson Cancer Research Center	MiHa: P2RX7, DPH1, DDX3Y	ALL, MDS		Donor CD8 T cells expanded with recipient EBV-BLCL	5/7 patients achieved complete transient remissions. 3/7 experienced pulmonary toxicity
Meij et al. [40]	Leiden University Medical Center	HA-1	A0201	AML, CML, ALL	Donor T cells sensitized with HA-1 peptide pulsed DC	
Franssen et al. (2017)	University Medical Center Utrecht; VU University Medical Center	MiHa: LRH-1, UTA2-1, HA-1	A2, A24, B7, B44	MM	Donor T cells sensitized with autologous peptide-pulsed DC	5/9 patients achieved SD
NCT01640301; Chapuis et al. [41]	Fred Hutchinson Cancer Research Center	WT1	A0201	AML	WT1 TCR-transduced donor EBV-specific T cells	12/12 patients relapse-free at follow up

(continued)

Table 4.1 (continued)

Reference	Study center	Tumor antigen	HLA	Indications	Format of allogeneic T cells	Results
Comoli [42]	Policlinico San Matteo; Policlinico di Modena	BCR-ABL	Multiple	Ph+ ALL	HCT donor T cells sensitized in vitro with BCR-ABL peptide-pulsed autologous DC and subsequently restimulated with autologous PBMC	2/2 patients treated with allogeneic cells achieved CR
NCT03027102; Lee et al. (2018)	Princess Margaret Cancer Centre	N.D	N.D	AML	Healthy donor double negative T cells capable of lysing AML cell lines	
NCT04180059	CHU de Nantes	HLA-DPB1*0401	DPB1*0401	Multiple hematologic post HCT	HLA-DPB1*0401-negative HCT donor T cells transduced with suicide gene	
NCT04511130	Multiple	N.D	N.D	AML	Zelenoleucel / MT-401: multi tumor-associated antigen-specific T cells	
NCT05001451	City of Hope	N.D	N/A	AML	GDX012: Allogeneic Vδ1 + γδ T cells	
Chapuis et al. [43]	Fred Hutchinson Cancer Research Center	WT1	Multiple	AML, ALL		5/11 CR. No evidence of new-onset GvHD. CR associated with >230 day T cell persistence
NCT02895412	Westmead Hospital	WT1	Multiple	AML, MDS	Donor T cells sensitized with autologous peptide-pulsed APC	4 of 7 patients remain in remission without GvHD

(continued)

Table 4.1 (continued)

Reference	Study center	Tumor antigen	HLA	Indications	Format of allogeneic T cells	Results
NCT05015426	Moffitt Cancer Center		N/A	AML	Donor γδ T cells expanded with artificial APC	
NCT01758328	Memorial Sloan Kettering Cancer Center	WT1	Multiple	MM	WT1-specific donor T cells	
NCT00620633	Memorial Sloan Kettering Cancer Center	WT1	Multiple	Leukemias, MDS	Donor T cells sensitized with autologous WT1 peptide-pulsed autologous EBV-BLCL	
Prockop et al. [44]	Memorial Sloan Kettering Cancer Center	WT1	DRB1*0801	AML	HCT Donor T cells stimulated with WT1 peptide-pulsed autologous EBV-BLCL	Leukemia cutis patient treated with multiple cycles of WT1-specific T cells which infiltrated skin lesions. Patient in continuous remission > 10 years without GvHD
NCT04284228; Malki et al. (2021)	Multiple	WT1, PRAME, Cyclin A1	A0201	AML, MDS	NEXI-001: HCT donor T cells stimulated with artificial APC	Blast decrease in 4/5 patients; no evidence of GvHD
NCT04679194	Multiple	WT1, PRAME, Survivin	Multiple	AML, MDS	Mana 312: HCT donor T cells sensitized with peptide pulsed DC	

(continued)

Table 4.1 (continued)

Reference	Study center	Tumor antigen	HLA	Indications	Format of allogeneic T cells	Results
NCT02494167; Lulla et al. [45]	Houston Methodist Hospital; Texas Children's Hospital	WT1, NYESO1, PRAME, Survivin	Multiple	AML, MDS	HCT donor T cells sensitized with autologous peptide pulsed DC	25 patients treated without Grade > 2 GvHD. 11/17 treated in CR alive at 1.9 years. 2/8 with active disease experience objective responses
NCT00460629; Bornhaeuser et al. (2011)	University Hospital Carl Gustav Carus	BCR-ABL, WT1, PR1	Multiple	CML	Donor T cells sensitized with peptide pulsed DC	7/14 patients in molecular remission at follow-up
NCT02475707	Houston Methodist Hospital; Texas Children's Hospital	WT1, PRAME, Survivin	Multiple	ALL	Donor T cells sensitized with peptide pulsed DC	

Abbreviations: N.D.—not disclosed, AML—acute myeloid leukemia, ALL—acute lymphoblastic leukemia, MM—multiple myeloma, MDS—myelodisplastic syndrome, CML—chronic myeloid leukemia, MiHa—minor histocompatibility antigen, CR—complete remission, SD—stable disease, PR—partial remission, GvHD—graft versus host disease

4.3.1 Minor Histocompatibility Antigens

Minor histocompatibility antigens (MiHa) refer to peptides produced by nonsynonymous mutations present in an HCT recipient's genome, but not the donor's, thus giving rise to HLA-binding peptides recognized as non-self by the HCT donor's T cells. T cells specific for several MiHa derived from proteins with expression restricted to hematopoietic cells have not caused GvHD and these have been considered to be suitably specific antigens for targeting hematologic cancers. Recently identified MiHa with hematopoietic tissue-restricted expression include mutant peptides produced by SNPs in HMHA1 and ARHGDIB presented by HLA-A2 and –B7 respectively [46]. A comprehensive review of other MiHa is discussed elsewhere [47]. However, a recent in silico analysis of 100 virtually paired genomes found that 0.5% of self-peptides presented by HLA alleles of a given recipient would be unique to the recipient and would constitute immunogenic MiHa for an HLA-matched donor [48]. Allelic variants of MiHa selectively expressed by hematopoietic cells are even more limited, making selection of HLA-matched donors disparate only for such MiHas difficult at this time.

Given their in vivo association with post-DLI leukemia control and in vitro leukemia-specificity [8, 49], allogeneic HA-1-specific T cells are being studied for their efficacy in adoptive T cell therapy. Following generation in vitro from HCT donors by repeated HA-1 peptide-pulsed dendritic cell sensitization, HA-1-specific T cells have been safely administered to relapsed leukemic patients without evidence of GvHD [40]. However, none of the first 3 patients in this phase I trial responded. More recently, an HA-1-specific A2-restricted TCR was transduced into virus-specific donor T cells and administered to 5 patients at high risk of leukemic relapse as post-HCT prophylaxis [38]. Again, no toxicity or Grade 2–4 GvHD was observed in any patient. In 3 of 5 patients, HA-1 TCR+T cells were found circulating at peak frequencies between $3 \times 10^{-7}\%$ and 1.7% of total peripheral blood mononuclear cells, indicating persistence without significant expansion. This prophylaxis study could not assess the therapeutic benefit of adoptively transferred HA-1-specific donor T cells. However, ongoing clinical trials are expected to provide more clarity (NCT03326921). Warren et al. generated MiHa-specific CD8 T cell clones by stimulating HLA-matched donor T cells with recipient EBV-transformed B cells (EBV-BLCL) and subsequently treated 7 post-HCT relapsed leukemic patients [39]. 5 of 7 patients achieved complete remissions, but 3 of 7 developed pulmonary toxicity attributed to administration. In the most severe case, the P2RX7-specific T cells administered were detected in bronchoalveolar lavage fluid. The P2RX7 protein targeted was subsequently shown to be highly expressed in healthy pulmonary epithelial cells, thus providing a non-leukemic target for the T cells infused.

4.3.2   Self-antigens Differentially Highly Expressed by Tumors

4.3.2.1 PR1

The PR1 peptide derived from redundant sequences in the serine proteases proteinase 3 and elastase was one of the first leukemia antigens recognized by T cells in DLI responders and healthy individuals [11, 50]. In the healthy state, proteinase 3 is expressed in polymorphonuclear neutrophils localized within granules where it degrades extracellular matrix proteins [51]. AML cells express high levels of proteinase 3 and inhibit autologous T cell proliferation in a proteinase 3-dependent manner [52]. These data have provided a strong indication for targeting PR1 in the leukemic host. To this end, PR1 peptide vaccines have been tested in AML, MDS, and CML patients. In this trial, 12 of 53 patients (24%) had an objective response [53]. Clinical responders more frequently mounted post-vaccine PR1 T cell responses without increased severity of GvHD [53, 54]. While trials of PR1-specific T cells alone have not been conducted, in vitro expanded PR1-specific donor T cells have been used together with T cells specific for WT1 and BCR-ABL for prevention of CML relapse after T cell-depleted HCT. In this trial, all 4 patients who received T cells displaying cytotoxic activity in vitro and expansion of T cells specific for one or more of these antigens were in remissions 31–63 months post-HCT and -T cell infusions [55].

4.3.2.2 WT1

The Wilms tumor protein (WT1) is a zinc finger transcription factor essential to the embryonal development of the urogenital system [56–58]. Postnatally, WT1 is expressed at low levels only in the gonads, kidney podocytes, and mesothelial linings of the pleura and peritoneum. WT1 is also expressed by CD34+ hematopoietic progenitors at low levels, and has been reported to induce quiescence of CD34+Lin- cells while promoting differentiation of precursors later in their development. On the other hand, in cancer cells, WT1 has been hypothesized to act as an oncogene important to their survival and growth. Indeed, antisense oligonucleotides used in siRNA mediated suppression of WT1 expression induce apoptosis of clonogenic leukemic progenitors [59, 60]. Furthermore, treatment of AML cell lines and primary AML cultures with T cells specific for WT1 peptides induces apoptosis of the clonogenic leukemic cells in vitro and full regressions of leukemic xenografts in NOD/SCID mice [61–63]. Thus, the role of WT1 as a tumor suppressor versus an oncogene remains controversial. However, it is also recognized that WT1 is subject to extensive splicing, resulting in 64 different isoforms, the balance of which appears to determine its function as a regulator of transcription. Further analysis of the function of these isoforms and their contribution to the sum of WT1 activities in normal and malignant cells may clarify this conundrum.

An HLA-A24 peptide derived from $WT1_{235-243}$ was first characterized as a T cell epitope by Ohminami et al. [22]. A T cell clone, TAK-1, was established from a healthy donor via repeated sensitization with synthetic WT1 peptide-pulsed autologous dendritic cells and was shown to preferentially kill in a WT1-

and HLA-A24-dependent manner with sub-micromolar avidity. Its antileukemic potential was determined by its ability to lyse all HLA-A24+WT1+leukemia cell lines and primary samples tested but not normal HLA-A24+fibroblasts or PBMC. Shortly thereafter, WT1 T cell epitopes with leukemia specificity were described in association with other common alleles, namely HLA-A0201 [23–25], –A0101 [26], and –DRB1*0405 [27]. Our group generated WT1 T cell lines from 56 healthy donors, mapping their HLA restriction and peptide specificity to 36 class I and 5 class II alleles [28]. The majority of these lines displayed effector activity including WT1-specific IFNg production and in vitro and in vivo leukemia-specific cytotoxicity without alloreactivity [14].

Despite the controversial role of WT1 in leukemogenesis, allogeneic WT1-specific T cells are being evaluated either for treatment or prophylaxis of WT1+acute leukemias or myelomas. Chapuis et al. treated 11 patients for WT1+AML or ALL who had either relapsed post-HCT ($N = 6$) or were at high risk of relapse post-HCT ($N = 5$). The T cells were well tolerated without toxicities. One patient developed chronic GvHD deemed unlikely to have been caused by the T cells. 7 patients (5 with disease) received cloned WT1-specific T cells up to 10^{10} cells/m^2 not cultured with IL-21. Of those, 6 had progression of disease or relapsed early in the course of. One patient treated in CR was disease-free 4.9 years post-infusion. In contrast, of 4 whose WT1-specific T cells were cultured with IL-21 to prevent terminal differentiation and apoptosis, all 4, including one who had residual disease prior to infusion, survived disease free >18 to >30 months post infusion. Strikingly, the T cells were also detected $\geq$430 days post-infusion in 3 of these patients. In our own trials, polyclonal WT1-specific T cells sensitized with a pool of overlapping peptides spanning the sequence of WT1 have been administered at much lower doses, 10^6–10^8 cells/kg, to patients who have relapsed post-HCT without other therapy. Of 16 treated, none developed toxicities or GvHD. Two patients achieved CR. One of these responses was brief (2 months). However, the other patient, treated for AML presenting as leukemia cutis, achieved a CR which has lasted >5 years (47, unpublished data). Chapuis et al. have used donor-derived T cells with either native specificity for WT1 or transduced with a high affinity WT1 TCR to treat high-risk leukemic patients [41, 43]. In the most recent report, 12 AML patients were treated with allogeneic HCT donor-derived EBV-specific T cells transduced with a WT1 TCR as post-HCT prophylaxis. These patients did not develop GvHD or on-target toxicity and experienced an impressive relapse-free survival of 100% at follow-up despite the cytogenetically diverse cohort [41]. Notably, EBV-specific T cells bearing the WT1 TCR, which had also been cultured with IL-21, were detected in the blood at frequencies >10% of T cells in 4/12 patients >1 year post-infusion. A comparable concurrent cohort exhibited 54% relapse-free survival. A WT1 heteroclitic peptide vaccine was also found to induce WT1-specific T cells in AML, multiple myeloma, mesothelioma, and ovarian cancer [64–66], with AML patient survival positively associated with induction of CD4 or CD8 T cell responses to WT1 [64].

4.3.2.3 PRAME

The cancer testis antigen PRAME was first identified as a tumor antigen when T cell clones recognizing PRAME in a melanoma patient were able to lyse autologous melanoma tissue in an HLA-A24-restricted manner [67]. Shortly thereafter, Kessler et al. identified A0201-presented T cell epitopes of PRAME via a reverse immunology approach [31]. 128 HLA-A2-binding peptides were inferred from the PRAME amino acid sequence, ranked by in vitro binding affinity for HLA-A2, screened for proteasomal cleavage sites, and used to prime T cells in healthy A0201 donors. The peptide with the highest affinity for HLA-A2 was a 10-mer peptide ALYVDSLFFL derived from amino acids 300–309 of PRAME, but was not as efficiently cleaved by the proteasome as other PRAME peptides, namely SLYSFPEPEA (aa 142–151) and SLLQHLIGL (aa 425–433). Despite these differences, T cell clones sensitized to each individual peptide displayed peptide-specific, HLA-restricted lytic capacity with half-maximal lysis observed with targets loaded with peptide at 12 nM or lower. They also exhibited HLA- and PRAME-dependent cytotoxicity toward melanoma cell lines in vitro. PRAME peptides presented were also further found to differ between those cleaved in the constitutive proteasome versus the immunoproteasome [68]. In particular, presentation of the A2-presented ALYVDSLFFL peptide was found to be induced by the immunoproteasome and attenuated by the constitutive proteasome [69].

Thus, despite the overexpression of PRAME protein in solid and hematologic malignancies, alterations in the immunoproteasome may dictate tumor susceptibility to PRAME-specific T cells, a feature already recognized in melanoma to be prognostic of response to checkpoint blockade and a determinant of tumor responses [70, 71]. Stanojevic et al. recently expanded the known landscape of antigenic PRAME peptides, generating PRAME-specific T cell lines from 12 healthy donors via sensitization with overlapping PRAME peptides spanning the entire protein sequence, mapping their specificity to 11 class I and 16 class II peptides [72]. These multi-PRAME epitope-specific T cells were capable of killing PRAME+ Wilms' tumor cell lines while sparing autologous healthy mononuclear cells.

4.3.2.4 Phosphopeptides

An emerging group of differentially expressed tumor antigens, phosphopeptides, may also be promising immunotherapy targets [73]. Compelling evidence points to phosphopeptides as ideal tumor targets by virtue of their presentation on multiple tumor types, including melanoma, AML, liver cancer, colorectal cancer, and ovarian cancer, but not normal hematopoietic cells [73–75]. Furthermore, phosphopeptide-specific T cells kill malignant cells in vitro and are present in most healthy donors, but depleted or absent in leukemic patients until after HCT [74]. In addition, phosphopeptide vaccines have been found to induce specific T cell responses in melanoma patients without causing significant toxicity [76]. A characteristic feature of phosphopeptides is their post-translational modification which alters or enhances their binding to MHC molecules for a number of HLA alleles [77]. As a result, these phosphopeptides are thought to present a fundamentally different surface to cognate TCRs, thus enhancing their antigenicity. Given that aberrant phosphorylation is a hallmark of cancer, phosphopeptides differentially expressed by tumor cells are expected to be an important immunotherapy target and warrant further investigation for use in allogeneic T cell therapy.

4.3.3 Tumor-Exclusive Neoantigens

Tumor-exclusive antigens are those peptides derived from nonsynonymous mutations in tumor tissue, but not in healthy tissues, that give rise to an immunogenic peptide presented by the tumor's MHC molecules. Also referred to as "neoantigens", this class of antigens can uniquely distinguish a tumor from normal tissue thereby rendering it the ideal target for cancer immunotherapy. A major obstacle in the development of broadly applicable cell therapies for tumor neoantigens is the variability in their presentation between patients due to differing mutational burdens, HLA genotypings, and antigen escape mechanisms, even between patients presenting with the same tumor type. Still, shared neoantigens have been defined due to the recurrence of frequently mutated genes across tumor types, giving rise to peptides capable of being presented by common HLA alleles.

4.3.3.1 BCR-ABL

In Philadelphia chromosome-positive CML, the BCR-ABL junction and its splice variants produce an oncoprotein that gives rise to neoantigens capable of being presented by the common HLA molecules A2, A3, A11, B8, and Cw4 [20, 21, 78]. These neoantigens elicit antileukemic T cell responses in healthy individuals and CML patients [20, 21, 23, 79, 80]. BCR-ABL-specific T cells in CML patients were shown to be capable of killing autologous leukemia cells and were on average at higher frequencies than those in healthy individuals, with no difference in respective responses to influenza reference control [20]. BCR-ABL-specific CD8 and CD4 T cells are enriched in bone marrow relative to peripheral blood in CML patients. They are also inversely correlated with MRD status, and absent at relapse [80]. Additional studies demonstrate that in CML patients T cells specific for common BCR-ABL neoantigens, such as the HLA-A3-presented E255K mutation, are dependent on the patient's tumor harboring the mutation. Furthermore, they are seen at high frequency post-HCT [79]. Peptide vaccination has been shown to reliably induce BCR-ABL T cell responses in CML patients but without clear association with clinical response [81–83]. When administered to post-HCT CML patients, donor T cells sensitized in vitro to BCR-ABL appear to be safe, persist up to at least 6 months post-infusion, and are potentially capable of eliminating residual blasts [42, 55]. Ultimately, BCR-ABL peptides are ideal neoantigen targets for adoptive cell therapy of leukemia due to their exclusive expression by leukemic cells.

4.3.3.2 ETV6-RUNX1

Similar to BCR-ABL translocations, ETV6-RUNX1 translocations are recurring leukemia mutations. ETV6-RUNX1 mutations frequently occurring in pediatric B cell ALL were recently shown to give rise to neoantigens presented by multiple HLA molecules including HLA-A2, -B35, and -B15 [84]. Such HLA class I-presented neoantigens were comprehensively shown to be recognized by bone marrow CD8 T cells directly ex vivo in 7 of 9 B-ALL patients expressing the mutated gene [84]. The aforementioned study did not find such responses detectable directly ex vivo in healthy donors, but as has been shown with other neoantigens, they may be induced via in vitro sensitization [85–88]. Such an approach would prove useful for generating donor T cells to treat ETV6-RUNX1+tumors expressing a range of common HLA molecules.

4.3.3.3 Microsatellite-Unstable Tumor-Derived Frameshift Mutations

Defective DNA mismatch repair mechanisms in tumors give rise to microsatellite instability (MSI), in turn producing frameshift mutations which can be presented as neoantigens by both liquid and solid tumors. Tumors displaying high MSI (MSI-H) express frameshift mutations that are often shared between patients due to the limited number of MSI loci, which bottlenecks and genetically focuses the process of accumulating somatic mutations during carcinogenesis [89]. This focused accumulation of mutations and resultant neoantigens that may be expressed may in part explain the comparatively higher clinical response rate to checkpoint blockade therapy in MSI-H tumors [90, 91]. Indeed, several groups have reported immunogenic T cell epitopes derived from recurring MSI-H frameshift mutations, including TGFbRII, MSH3, and RNF43 [92–95]. Recently, Roudko et al. performed a systematic analysis of MSI-H mutations and found that compared to missense mutations, frameshift mutations more frequently produce neoantigens that are less similar to both the human proteome and common virus antigens [95]. They further demonstrated that their allelic frequency is on average 30–40% in MSI-H endometrial carcinomas, indicating such frameshift mutations represent a substantial clonal fraction of the tumor. Lastly, they verified by mass spectrometry that frameshift neoantigens are presented on the surface of cancer cells and that they are able to induce effector T cell responses in healthy donors and endometrial cancer patients after in vitro sensitization. Immunotherapy approaches targeting recurrent frameshift mutation neoantigens derived from AIM2, TAF1B, and HT001 have thus far been limited to peptide vaccines which were shown to produce T cell responses in MSI-H colorectal cancer patients, but tumor regression was not observed [96]. In this study, neoantigen-specific CD4 T cell responses were preferentially produced in patients compared to neoantigen-specific CD8 T cell responses. An adoptive cell therapy approach would enable ex vivo selection of neoantigen-specific T cell subsets with vetted antitumor activity prior to infusion to potentially enhance therapeutic benefit. Given their tumor-exclusive expression, high allelic frequency, consistent immunogenicity, and recurring nature, frameshift neoantigens could be a useful class of targets for donor-derived adoptive T cell therapy.

4.3.4 Use of Multi Tumor-Antigen Specific T Cells

The well-documented, functional T cell responses to the differentially expressed antigens WT1 and PRAME in normal donors has led to several ongoing efforts to develop allogeneic T cells recognizing a mix of WT1, PRAME, and other tumor antigens to treat or prevent disease relapse [55, 97, 98]. Defining features of these multi-tumor antigen specific T cells (TAA-T) are their safety due to lack of alloreactivity, multi-tumor antigen specificity, which renders them capable of addressing a wide range of tumors, and their capacity to be generated from healthy donors enabling production of an off-the-shelf T cell product. As new tumor antigens are discovered and validated as targets, the production of such T cells can be easily modified to include peptides from other antigens during the sensitization cycles.

Clinical trials conducted by the group at Children's National Hospital (Washington, DC, USA) have demonstrated safe infusion of autologous or allogeneic TAA-T in patients presenting with pediatric solid tumors [99, 100]. A phase I study by Hont et al. described use of autologous TAA-T targeting WT1, PRAME, and Survivin to treat 15 patients with high-risk solid tumors, including Ewing sarcoma, neuroblastoma, and Wilms' tumor [100]. 14 of 15 treated patients were pediatric or young adults. Notably, no dose-limiting toxicities were observed. Of 12 with detectable disease, 10 had stable disease with progression-free survivals of >4.1 to >19.9 months. Response correlated with epitope spread, and clinical responses were observed despite the absence of lymphodepleting chemotherapy prior to T cell infusion. This group also reported use of allogeneic TAA-T recognizing WT1, PRAME, and Survivin to treat two Hodgkin's lymphoma patients without prior lymphodepletion while still achieving some degree of clonal expansion and one of two patients experiencing continued complete remission [101]. In an ongoing trial, the group at Baylor College (Houston, TX, USA) treated 25 AML/MDS patients with HCT donor-derived TAA-T cells reactive to PRAME, WT1, Survivin, and NYESO1 [45]. 17 patients were treated in CR, of whom 6 relapsed. 17 patients were treated in CR, of whom 11 remained leukemia-free at median follow-up of 1.9 years. 8 patients were treated with active disease, 2 of whom experienced objective responses (1 CR and 1 PR). These studies demonstrate the ability of TAA-T to induce clinical responses and, when generated from allogeneic sources, can be a safe off-the-shelf cell therapy. Multiple clinical trials of TAA-T are ongoing (Table 4.1).

4.3.5 Tumor Antigens Recognized by Unconventional T Cells

T cells other than those acting through $\alpha\beta$ T cell receptors and are not MHC-restricted may serve as potent effectors for donor-derived cancer immunotherapy via recognition of nonpeptidic ligands on tumor cells. This category includes $\gamma\delta$ T cells and CD1-restricted T cells. Compared to conventional MHC-restricted $\alpha\beta$ T cells, however, these unconventional T cells still have an unclear role in host immunity.

γδ T cells are a class of T cells expressing a γδ, rather than αβ, TCR. They were initially found to control mycobacterium tuberculosis and CMV infection in an MHC-independent manner [102–105]. Vγ9Vδ2 T cells represent a subset of γδ T cells that respond to mycobacteria that is minor at birth and expands to account for 70–90% of peripheral γδ T cells in adults [106]. Vγ9Vδ2 recognize and display cytolytic activity toward phosphorylated nonpeptidic antigens derived from multiple parasitic species including plasmodium and mycobacterium [107]. Butyrophilins BTN2A1 and BTN3A1 are the antigen presenting molecules essential to phosphoantigen presentation exclusively to Vγ9Vδ2 T cells [108, 109]. Allogeneic Vδ1+T cells are currently being investigated in clinical trials for MRD+AML on the basis of their expansion post-HCT in leukemia patients [110], as well as their proliferative and cytotoxic response to leukemic blasts in the absence of activity toward allogeneic healthy cells in vitro [111]. That they display antitumor activity in an MHC-unrestricted manner without alloreactivity suggests Vδ1+T cells may be a broadly reactive T cell therapy not constrained by HLA-restriction; however, their cognate ligand and efficacy when adoptively transferred are yet to be determined. While evidence indicates that γδ T cells are also capable of killing many epithelial tumors expressing MICA/MICB in vitro via engagement of the MICA/MICB receptor, NKG2D [112]. Furthermore, γδ T cells may also promote tumor growth via an IL-17-mediated mechanism [113] or via suppression of αβ T cells [114]. Nevertheless, Park et al. recently found a favorable prognostic significance to the presence of tumor-infiltrating γδ T cells in brain tumors further and demonstrated that the antitumor activity of such γδ T cells could be enhanced by alleviating tumor hypoxia-induced T cell dysfunction [115].

An evolutionarily conserved subset of immune effectors of emerging significance are T cells with αβ TCRs that are specific for lipids presented by CD1 [116]. Similar to MHC molecules, CD1 molecules present antigens between a pair of alpha helices, albeit from different sources. CD1a, CD1b, CD1c, and CD1d present lipids which are much smaller molecules than MHC peptide ligands. CD1 molecules are expressed on myeloid cells depending on their differentiation stage and display limited polymorphism depending on the isoform [117, 118]. CD1b and CD1c do not appear to display amino acid substitutions between individuals [118], rendering them potentials targets for broadly applicable immunotherapy. Further details of the antigen presentation pathways for CD1 molecules are reviewed elsewhere [119, 120].

CD1-restricted T cells have been reported to comprise up to approximately 10% of the CD4+or CD4-CD8- double negative T cell repertoire in peripheral blood and umbilical cord blood [121]. Most of the reactivity of these CD1-restricted T cells in adult peripheral blood is restricted by CD1a [121]. Lepore et al. described a double negative donor-derived T cell clone, DN4.99, bearing a TCR specific for methyl-lysophosphatidic acids presented by CD1c. This TCR specifically recognized acute leukemias in vitro [122]. Further, it was recently shown that the DN4.99 TCR when transduced into donor T cells could prolong survival in AML xenografted mice [123]. The high frequency of CD1-restricted T cells in peripheral blood, the differential expression of CD1 on myeloid cells, limited CD1 poly-morphism, and in vivo efficacy against AML xenografts warrant further study of donor-derived CD1-restricted T cells for allogeneic cell therapy of myeloid leukemias.

4.4 Technological Advances Facilitating Tumor Antigen Discovery and Evolution of More Effective and Persistent Tumor-Selective T Cells

Advances in technologies used to screen antigenic peptides and identify T cells recognizing them can now be utilized to streamline identification, generation, and expansion of tumor antigen-specific donor T cells.

Reverse immunology approaches, which aim to predict immunogenic peptides for a given HLA allele from protein sequence data, now benefit from multiple independently-developed user-friendly prediction algorithms for HLA class I and II such as the NetMHC suite, MHCflurry, MARIA, MixMHCpred, and HLAthena. The prediction power of these tools has been significantly enhanced in recent years due to the emergence of sophisticated machine learning algorithms and their being trained on mass spectrometry-identified peptides eluted from peptide-HLA complexes on tumor cell lines. The quality of training data for an algorithm is critical to its performance. Early models were trained on HLA binding affinity of synthetic peptides; however, these data do not account for determinants of peptide presentation beyond competition binding, such as proteasomal cleavage, protein abundance, or physicochemical properties. To this end, mass spectrometry-identified peptides eluted from peptide-HLA complexes on cells provide a picture of the immunopeptidome with unprecedented throughput and verisimilitude. Prediction tools have thus benefitted from being trained on monoallelic cell lines stably expressing a given HLA allele to enable resolution of the immunopeptidome for a particular HLA allele. Mass spectrometry-based HLA peptide identification can be further improved via interfacing with other systems biology approaches. For example, whole exome sequencing (WES) of tumor DNA has been used by multiple groups to construct a database of tumor-specific nonsynonymous muta-tions, against which mass spectra of tumor-eluted HLA peptides can be matched and identified [124–126]. Beyond nonsynonymous mutations, Chong et al. com-bined WES, RNAseq, and ribosome profiling to construct a database to enable high accuracy mass spectrometric identification of tumor-specific HLA peptides derived from noncanonical proteins, such as those produced by translation of alternative open reading frames, long non-coding RNAs, or transposable elements [127].

Tumor antigen-specific T cell induction in normal donors typically requires repeated in vitro sensitization of T cells with peptide-pulsed autologous antigen presenting cells over several weeks to promote outgrowth of low frequency pre-cursors, usually originating from the naïve T cell pool, and deplete alloreactive T cells [14, 85, 128, 129]. Optimizations in culture and assay conditions have recently reduced this process to a single sensitization cycle lasting only 10–14 days

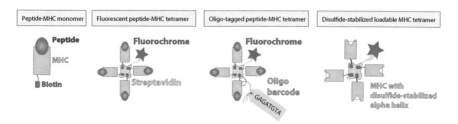

Fig. 4.1 Illustration of peptide-MHC tetramer reagents. Conventional fluorescent tetramers comprise biotinylated peptide-MHC monomers tetramerized via a fluorescent streptavidin backbone for analysis by flow cytometry. Oligo-tagged peptide-MHC tetramers enable detection of antigenic specificity by PCR in addition to flow cytometry, further enhancing multiplexed analysis. Disulfide-stabilized loadable MHC tetramers are a shelf-stable tetramer reagent that can be readily loaded with an antigenic peptide of interest

[129], after which tumor antigen-specific T cells with effector function can be detected in antigen recall assays which can be performed without compromising T cell viability via staining for antigen-dependent upregulation of activation markers (CD137/4-1BB, CD69, CD154/CD40L, CD134/OX40), surface-captured or –trapped cytokines (IFNg, IL-2, TNFa), or degranulation markers (CD107a/b). All of the aforementioned markers can be combined with magnetic enrichment or flow cytometric sorting to permit direct capture for downstream culture or transcriptome applications. Furthermore, multiplexed flow cytometric assessment of T cell reactivities in parallel on the basis of these markers is feasible via the use of cell membrane dye barcoding in which combinatorial titrations of fluorescent succinimidyl ester-based dyes are used to barcode different T cell lines or clones prior to coculture with target antigen in a single culture tube [85]. Using this technique, up to 1024 different cultures can theoretically be barcoded and analyzed at once with available dyes and a capable flow cytometer; in practice, 96-wells have been feasibly analyzed simultaneously [130, 131].

Fluorescent peptide-HLA multimers (Fig. 4.1) remain the gold standard for direct ex vivo identification of antigen-specific T cells by flow cytometry. The use of multimers, typically tetramers, to detect low frequency tumor antigen-specific T cells has been enabled by recent methodological and reagent improvements. A methodological improvement of broad applicability to multimer staining protocols is the use of protein kinase inhibitors to prevent multimer internalization and subsequent signal quenching, coreceptor-specific antibodies to augment multimer staining, and multimer fluorochrome-specific antibodies to reduce multimer off-rate [132, 133]. A key improvement to the production of multimer reagents is the use of conditional ligand-based peptide-HLA reagents, which comprise HLA molecules refolded around a placeholder peptide which can later be replaced. The ability to assemble multimers at will with any HLA-binding peptide of interest further facilitates the throughput with which different antigenic specificities can be analyzed [134, 135]. Rather than refolding soluble HLA monomers in the presence of the peptide antigen of interest, an allele-specific UV-sensitive conditional ligand is used to initially refold the monomer but is eventually exchanged for a peptide of interest via brief exposure to long wavelength UV irradiation. This technology allows multiple peptide-HLA complexes for a given allele to be generated in parallel based on a single batch of conditional ligand-bound HLA protein. An alternative approach to tetramer generation improves the shelf-life and facilitates peptide loading by use of disulfide-stabilized HLA refolded by a short dipeptide so that it is functionally "empty", which is then tetramerized and thereafter can be readily loaded with peptides of interest [136, 137] (Fig. 4.1). Impressively, disulfide-stabilized HLA-A2 molecules have been stored as empty-loadable tetramers for use with a peptide of interest on-demand and have boasted a higher staining index than their wildtype counterparts [137]. Furthermore, peptide-HLA complexes can be barcoded with combinations of different fluorescent conjugates or oligonucleotides, enabling multiple antigenic specificities to be analyzed in a single tube by flow cytometry or PCR, respectively [138, 139].

4.5   Towards a Broadly Applicable Cell Therapy for Cancer

While technological advances have radically improved our capacity to identify and characterize tumor-unique or differentially expressed tumor-associated antigens, and to expand TAA-specific T cells for adoptive therapy, the logistics of doing so from a tumor bearing host or even a healthy HCT donor are daunting and costly enough to prevent its application. Some tumor-bearing patients may generate TAA-specific T cells with the capacity to recognize and then lyse their own tumors or to expand and persist sufficiently in vivo to eradicate their malignancy. For patients receiving allogeneic, HLA-matched HCT for a hematologic malignancy, TAA-specific T cells may have significant advantage in that the TAA-specific T cells are derived from a healthy donor and can expand and persist post-transplant in the engrafted host. They can also be readily cloned and selected for their quality, effector functions, and potential for persistence. However, the logistics and time required to create HCT donor-derived TAA-specific T cells remain an issue. For patients not receiving an allogeneic HCT, and particularly those being treated for treatment-resistant solid tumors, allogeneic TAA-specific T cells have not been a practicable option.

Recently, this situation has begun to change as a result of clinical trials demonstrating the potential of banked 3rd party donor-derived virus-specific T cells, selected on the basis of partial HLA compatibility and restriction by an HLA allele shared by the patient's disease, to induce durable remission of EBV+ lymphomas as well as CMV and adenoviral infections.

In a multicenter trial reported by Leen et al. [140], 3rd party T cells specific for EBV, CMV pp65, and adenovirus 5 hexon induced a CR or PR in 6/9 (67%) HCT patients with EBV-PTLD, 17/19 (73.9%) with CMV and 14/17 (77.8%) with adenoviral infections. In two subsequent trials employing 3rd party T cells sensitized in vitro with antigen presenting cells loaded with immunogenic peptides from the viruses, the largest cohorts of HCT patients were treated for recurrent or persistent CMV infection or viremia that failed to respond to antiviral drugs. In these two trials, 28/30 (93%) and 16/17 (94%) of patients treated achieved a CR or PR, respectively [141, 142].

In our own trial of EBV-specific 3rd party T cells for the treatment of rituximab-refractory EBV+lymphomas in 46 HCT and 13 organ allograft recipients, CR or durable PR was achieved in 68% and 54% respectively. As with the allogeneic TAA-specific T cell experience, these "off-the-shelf" T cells were associated with minimal or no toxicity or GvHD. Thus, these cryopreserved, immediately accessible allogeneic 3rd party T cells are safe and can induce durable responses, even in patients with advanced, otherwise refractory viral infections and EBV+lymphomas.

The use of allogeneic cells for off-the-shelf cancer immunotherapy hinges on generating a bank of T cell lines with defined HLA-restriction and antigen-specificity. Such a bank has been generated by us and others by repeated in vitro sensitization of HCT donor or 3rd party donor T cells with autologous stimulator cells. Autologous EBV-BLCL are used to generate EBV-specific T cells. Autologous WT1-peptide pulsed dendritic cells or EBV-BLCL can be used to generate WT1-specific T cells without or with co-specificity for EBV, respectively. The resulting cell lines can be characterized as to their specificity and HLA restriction, cryopreserved, and administered to a tumor-bearing patient sharing the HLA alleles by which the T cells are restricted. We and others have repeatedly shown that these cells elicit in vivo antitumor activity and do not cause GvHD. As more tumor antigens are recognized, autologous EBV-BLCL pulsed with peptides derived from these antigens can be used to stimulate donor T cells. Other types of stimulator cells include autologous dendritic cells or artificial antigen presenting cells pulsed with peptides of interest.

An alternative approach to the generation of potent effectors with defined tumor antigen specificity is the use of EBV- or other viral-specific T cells transduced with a TCR recognizing an antigen of interest. As human TCRs recognizing different tumor antigens are discovered, they can be transduced into preexisting EBV-specific T cell lines and expanded to treatment numbers for adoptive therapy. The benefits of this approach are (1) EBV-specific T cells and their autologous EBV-BLCL can be feasibly expanded from as little as 10 mL of blood from an EBV seropositive individual; (2) EBV-specific T cells exhibit favorable in vivo persistence by virtue of their derivation from the central memory pool as well as their continuous stimulation by endogenous EBV-transformed B cells in seropositive patients; and (3) banked EBV-specific T cell lines of defined antigenic specificity can be transduced with a TCR or multiple TCRs of interest to readily endow them with the capacity to recognize new tumor antigens. The study by Chapuis et al. serves as a heuristic for this approach; a vetted WT1-specific TCR was transduced into HCT donor EBV-specific T cells, administered without prior lymphodepletion, and persisted in 4/12 patients for >1 year at frequencies of 11–51% of total peripheral CD8 T cells [41]. The use of 3rd party cells raises the issue of T cell persistence in vivo. In immunosuppressed transplant recipients, 3rd party cells can last for 2–6 months and in some instances up to 2 years post-infusion. However, in the absence of lymphodepletion, one would expect these cells to be rejected relatively rapidly. However, several studies have demonstrated that infusions of these allogeneic T cells also induce endogenous T cells to react against other tumor antigens, also known as "epitope spreading". These cells are thought to result from cross presentation of tumor antigens released from tumor cells lysed by the infused T cells via endogenous dendritic cells or other antigen-presenting cells that stimulate endogenous T cells, as well as from the activation of endogenous T cells induced by cytokines released by patient T cells responding to the allogeneic stimulus of the 3rd party T cells infused. As a result, these endogenous T cells, may contribute to the maintenance of responses to the patient's tumor.

References

1. Fefer A, Buckner CD, Thomas ED, Cheever MA, Clift RA, Glucksberg H et al (1977) Cure of hematologic neoplasia with transplantation of marrow from identical twins. N Engl J Med 297(3):146–148
2. Weiden PL, Flournoy N, Thomas ED, Prentice R, Fefer A, Buckner CD et al (2010) Antileukemic effect of graft-versus-host disease in human recipients of allogeneic-marrow grafts. 300(19):1068–73. https://doi.org/10.1056/NEJM197905103001902
3. Apperley JF, Jones L, Hale G, Waldmann H, Hows J, Rombos Y et al (1986) Bone marrow transplantation for patients with chronic myeloid leukaemia: T-cell depletion with Campath-1 reduces the incidence of graft-versus-host disease but may increase the risk of leukaemic relapse. Bone Marrow Transplant 1(1):53–66
4. Kolb HJ, Mittermuller J, Clemm C, Holler E, Ledderose G, Brehm G et al (1990) Donor leukocyte transfusions for treatment of recurrent chronic myelogenous leukemia in marrow transplant patients. Blood 76(12):2462–2465
5. Mackinnon S, Papadopoulos E, Carabasi M, Reich L, Collins N, Boulad F et al (1995) Adoptive immunotherapy evaluating escalating doses of donor leukocytes for relapse of chronic myeloid leukemia after bone marrow transplantation: separation of graft-versus-leukemia responses from graft-versus-host disease. Blood 86(4):1261–1268
6. Collins RH, Shpilberg O, Drobyski WR, Porter DL, Giralt S, Champlin S et al (1997) Donor leukocyte infusions in 140 patients with relapsed malignancy after allogeneic bone marrow transplantation. J Clin Oncol 15(2):433–44
7. Smit WM, Rijnbeek M, Van Bergen CAM, Fibbe WE, Willemze R, Frederik Falkenburg JH (1998) T cells recognizing leukemic CD34+ progenitor cells mediate the antileukemic effect of donor lymphocyte infusions for relapsed chronic myeloid leukemia after allogeneic stem cell transplantation. Proc Natl Acad Sci USA 95(17):10152–10157
8. Marijt WAE, Heemskerk MHM, Kloosterboer FM, Goulmy E, Kester MGD, Van der Hoorn MAWG et al (2003) Hematopoiesis-restricted minor histocompatibility antigens HA-1- or HA-2-specific T cells can induce complete remissions of relapsed leukemia. Proc Natl Acad Sci USA 100(5):2742–2747
9. de Rijke B, van Horssen-Zoetbrood A, Beekman JM, Otterud B, Maas F, Woestenenk R et al (2005) A frameshift polymorphism in P2X5 elicits an allogeneic cytotoxic T lymphocyte response associated with remission of chronic myeloid leukemia. J Clin Invest 115(12):3506–3516
10. Griffioen M, Kessler JH, Borghi M, van Soest RA, van der Minne CE, Nouta J et al (2006) Detection and functional analysis of CD8+ T cells specific for PRAME: a target for T-cell therapy. Clin Cancer Res 12(10):3130–3136
11. Molldrem JJ, Lee PP, Wang C, Felio K, Kantarjian HM, Champlin RE et al (2000) Evidence that specific T lymphocytes may participate in the elimination of chronic myelogenous leukemia. Nat Med 6(9):1018–23

12. Molldrem J, Dermime S, Parker K, Jiang Y, Mavroudis D, Hensel N et al (1996) Targeted T-cell therapy for human leukemia: cytotoxic T lymphocytes specific for a peptide derived from proteinase 3 preferentially lyse human myeloid leukemia cells. Blood 88(7):2450–2457

13. Molldrem JJ, Lee PP, Wang C, Champlin RE, Davis MM (1999) A PR1-human leukocyte antigen-A2 tetramer can be used to isolate low-frequency cytotoxic T lymphocytes from healthy donors that selectively lyse chronic myelogenous leukemia 1. Cancer Res 59:2675–2681

14. Doubrovina ES, Doubrovin MM, Lee S, Shieh JH, Heller G, Pamer E et al (2004) In vitro stimulation with WT1 peptide-loaded epstein-barr virus-positive B cells elicits high frequencies of WT1 peptide-specific T cells with in vitro and in vivo tumoricidal activity. Clin Cancer Res 10(21):7207–7219

15. Doubrovina E, Carpenter T, Pankov D, Selvakumar A, Hasan A, O'Reilly RJ (2012) Mapping of novel peptides of WT-1 and presenting HLA alleles that induce epitope-specific HLA-restricted T cells with cytotoxic activity against WT-1+ leukemias. Blood 120(8):1633–1646

16. Ohminami H, Yasukawa M, Fujita S (2000) HLA class I-restricted lysis of leukemia cells by a CD8+ cytotoxic T-lymphocyte clone specific for WT1 peptide. Blood 95(1):286–293

17. Tyler EM, Jungbluth AA, O'Reilly RJ, Koehne G (2013) WT1-specific T-cell responses in high-risk multiple myeloma patients undergoing allogeneic T cell–depleted hematopoietic stem cell transplantation and donor lymphocyte infusions. Blood 121(2):308–317

18. Ishikawa T, Fujii N, Imada M, Aoe M, Meguri Y, Inomata T et al (2017) Graft-versus-leukemia effect with a WT1-specific T-cell response induced by azacitidine and donor lymphocyte infusions after allogeneic hematopoietic stem cell transplantation. Cytotherapy 19(4):514–520

19. Tyler EM, Jungbluth AA, Gnjatic S, O'Reilly RJ, Koehne G (2014) Cancer-testis antigen 7 expression and immune responses following allogeneic stem cell transplantation for multiple myeloma. Cancer Immunol Res 2(6):547–558

20. Yotnda P, Firat H, Garcia-Pons F, Garcia Z, Gourru G, Vernant JP et al (1998) Cytotoxic T cell response against the chimeric p210 BCR-ABL protein in patients with chronic myelogenous leukemia. J Clin Invest 101(10):2290

21. Bocchia M, Korontsvit T, Xu Q, Mackinnon S, Yang SY, Sette A et al (1996) Specific human cellular immunity to bcr-abl oncogene-derived peptides. Blood 87(9):3587–92

22. Clark RE, Dodi IA, Hill SC, Lill JR, Aubert G, Macintyre AR et al (2001) Direct evidence that leukemic cells present HLA-associated immunogenic peptides derived from the BCR-ABL b3a2 fusion protein. Blood 98(10):2887–2893

23. Butt NM, Rojas JM, Wang L, Christmas SE, Abu-Eisha HM, Clark RE (2005) Circulating bcr-abl-specific CD8+ T cells in chronic myeloid leukemia patients and healthy subjects. Haematologica 90(10 SE-Comparative Studies):1315–23

24. Wu CJ, Biernacki M, Kutok JL, Rogers S, Chen L, Yang X-F et al (2005) Graft-versus-leukemia target antigens in chronic myelogenous leukemia are expressed on myeloid progenitor cells. Clin Cancer Res 11(12):4504–4511

25. Wang Q, Li M, Wang Y, Zhang Y, Jin S, Xie G et al (2008) RNA interference targeting CML66, a novel tumor antigen, inhibits proliferation, invasion and metastasis of HeLa cells. Cancer Lett 269(1):127–138

26. Zhang W, Choi J, Zeng W, Rogers SA, Alyea EP, Rheinwald JG et al (2010) Graft-versus-leukemia antigen CML66 elicits coordinated B-cell and T-cell immunity after donor lympho-cyte infusion. Clin Cancer Res 16(10):2729–2739

27. Suemori K, Fujiwara H, Ochi T, Azuma T, Yamanouchi J, Narumi H et al (2008) Identification of an epitope derived from CML66, a novel tumor-associated antigen expressed broadly in human leukemia, recognized by human leukocyte antigen-A*2402-restricted cytotoxic T lymphocytes. Cancer Sci 99(7):1414–1419

28. van Baren N, Chambost H, Ferrant A, Michaux L, Ikeda H, Millard I et al (1998) PRAME, a gene encoding an antigen recognized on a human melanoma by cytolytic T cells, is expressed in acute leukaemia cells. Br J Haematol 102(5):1376–9

29. Oehler VG, Guthrie KA, Cummings CL, Sabo K, Wood BL, Gooley T et al (2009) The preferentially expressed antigen in melanoma (PRAME) inhibits myeloid differentiation in normal hematopoietic and leukemic progenitor cells. Blood 114(15):3299

30. De Carvalho DD, Binato R, Pereira WO, Leroy JMG, Colassanti MD, Proto-Siqueira R et al (2011) BCR–ABL-mediated upregulation of PRAME is responsible for knocking down TRAIL in CML patients. Oncogene 30(2):223–33

31. Kessler JH, Beekman NJ, Bres-Vloemans SA, Verdijk P, Van Veelen PA, Kloosterman-Joosten AM et al (2001) Efficient identification of novel HLA-A * 0201-presented cytotoxic T lymphocyte epitopes in the widely expressed tumor antigen PRAME by proteasome-mediated digestion analysis. J Exp Med 193(1):73–88

32. Sommermeyer D, Conrad H, Krönig H, Gelfort H, Bernhard H, Uckert W (2013) NY-ESO-1 antigen-reactive T cell receptors exhibit diverse therapeutic capability. Int J Cancer 132(6):1360–1367

33. Zarour HM, Storkus WJ, Brusic V, Williams E, Kirkwood JM (2000) NY-ESO-1 Encodes DRB1*0401-restricted epitopes recognized by melanoma-reactive CD4+ T cells. Cancer Res 60(17):4946–4952

34. Zarour HM, Maillere B, Brusic V, Coval K, Williams E, Pouvelle-Moratille S et al (2002) NY-ESO-1 119–143 is a promiscuous major histocompatibility complex class II T-helper epitope recognized by Th1- and Th2-type tumor-reactive CD4+ T cells. Cancer Res 62(1):213–218

35. Kayser S, Boβ C, Feucht J, Witte K-E, Scheu A, Bülow H-J et al (2015) Rapid generation of NY-ESO-1-specific CD4+ THELPER1 cells for adoptive T-cell therapy. Oncoimmunology 4(5):e1002723

36. Valmori D, Souleimanian NE, Hesdorffer CS, Old LJ, Ayyoub M (2005) Quantitative and qualitative assessment of circulating NY-ESO-1 specific CD4+ T cells in cancer-free individuals. Clin Immunol 117(2):161–167

37. Hofmann S, Schmitt M, Götz M, Döhner H, Wiesneth M, Bunjes D et al (2019) Donor lymphocyte infusion leads to diversity of specific T cell responses and reduces regulatory T cell frequency in clinical responders. Int J Cancer 144(5):1135–1146

38. van Balen P, Jedema I, van Loenen MM, de Boer R, van Egmond HM, Hagedoorn RS et al (2020) HA-1H T-cell receptor gene transfer to redirect virus-specific T cells for treatment of hematologic malignancies after allogeneic stem cell transplantation: a phase 1 clinical study. Front Immunol 20:1804

39. Warren EH, Fujii N, Akatsuka Y, Chaney CN, Mito JK, Loeb KR et al (2010) Therapy of relapsed leukemia after allogeneic hematopoietic cell transplantation with T cells specific for minor histocompatibility antigens. Blood 115(19):3869–3878

40. Meij P, Jedema I, van der Hoorn MAWG, Bongaerts R, Cox L, Wafelman AR et al (2012) Generation and administration of HA-1-specific T-cell lines for the treatment of patients with relapsed leukemia after allogeneic stem cell transplantation: a pilot study. Haematologica 97(8):1205

41. Chapuis AG, Egan DN, Bar M, Schmitt TM, Mcafee MS, Paulson KG et al (2019) T cell receptor gene therapy targeting WT1 prevents acute myeloid leukemia relapse post-transplant. Nat Med 25

42. Comoli P, Basso S, Riva G, Barozzi P, Guido I, Gurrado A et al (2017) BCR-ABL–specific T-cell therapy in Ph+ ALL patients on tyrosine-kinase inhibitors. Blood 129(5):582–586

43. Chapuis AG, Ragnarsson GB, Nguyen HN, Chaney CN, Pufnock JS, Schmitt TM et al (2013) Transferred WT1-reactive CD8+ T cells can mediate antileukemic activity and persist in post-transplant patients. Sci Transl Med 5(174):174ra27–174ra27

44. Prockop SE, Doubrovina E, Adams R, Boulad F, Kernan NA, O'Reilly RJ (2013) Adoptive transfer of WT-1 specific HLA class 2 restricted donor-derived T-cells induces sustained remission of AML relapse post transplant presenting as leukemia cutis. Blood 122(21):2085–2085

45. Lulla PD, Naik S, Vasileiou S, Tzannou I, Watanabe A, Kuvalekar M et al (2021) Clinical effects of administering leukemia-specific donor T cells to patients with AML/MDS after allogeneic transplant. Blood 137(19):2585–97

46. van Bergen CAM, van Luxemburg-Heijs SAP, de Wreede LC, Eefting M, von dem Borne PA, van Balen P et al (2017) Selective graft-versus-leukemia depends on magnitude and diversity of the alloreactive T cell response. J Clin Invest 127(2):517–529
47. Bleakley M, Riddell SR (2011) Exploiting T cells specific for human minor histocompatibility antigens for therapy of leukemia. Immunol Cell Biol 89(3):396
48. Bykova NA, Malko DB, Efimov GA (2018) In silico analysis of the minor histocompatibility antigen landscape based on the 1000 genomes project. Front Immunol 1819
49. Kloosterboer FM, van Luxemburg-Heijs SA, van Soest RA, Barbui AM, van Egmond HM, Strijbosch MP et al (2004) Direct cloning of leukemia-reactive T cells from patients treated with donor lymphocyte infusion shows a relative dominance of hematopoiesis-restricted minor histocompatibility antigen HA-1 and HA-2 specific T cells. Leukemia 18(4):798–808
50. Molldrem JJ, Clave E, Jiang YZ, Mavroudis D, Raptis A, Hensel N et al (1997) Cytotoxic T lymphocytes specific for a nonpolymorphic proteinase 3 peptide preferentially inhibit chronic myeloid leukemia colony-forming units. Blood 90(7):2529–2534
51. Rao NV, Wehner NG, Marshall BC, Gray WR, Gray BH, Hoidal JR (1991) Characterization of proteinase-3 (PR-3), a neutrophil serine proteinase. Structural and functional properties. J Biol Chem 266(15):9540–8
52. Yang T-H, John LS St, Garber HR, Kerros C, Ruisaard KE, Clise-Dwyer K et al (2018) Membrane-associated proteinase 3 on granulocytes and acute myeloid leukemia inhibits T cell proliferation. J Immunol Author Choice 201(5):1389
53. Qazilbash MH, Wieder E, Thall PF, Wang X, Rios R, Lu S et al (2017) PR1 peptide vaccine induces specific immunity with clinical responses in myeloid malignancies. Leukemia 31(3):697–704
54. Alatrash G, Molldrem JJ, Qazilbash MH (2017) Targeting PR1 in myeloid leukemia. Oncotarget 9(4):4280–4281
55. Bornhäuser M, Thiede C, Platzbecker U, Kiani A, Oelschlaegel U, Babatz J et al (2011) Prophylactic transfer of BCR-ABL–, PR1-, and WT1-reactive donor T cells after T cell–depleted allogeneic hematopoietic cell transplantation in patients with chronic myeloid leukemia. Blood 117(26):7174–7184
56. Buckler AJ, Pelletier J, Haber DA, Glaser T, Housman DE (1991) Isolation, characterization, and expression of the murine Wilms' tumor gene (WT1) during kidney development. Mol Cell Biol 11(3):1707–1712
57. Mundlos S, Pelletier J, Darveau A, Bachmann M, Winterpacht A, Zabel B (1993) Nuclear localization of the protein encoded by the Wilms' tumor gene WT1 in embryonic and adult tissues. Development 119(4):1329–1341
58. Keilholz U, Menssen HD, Gaiger A, Menke A, Oji Y, Oka Y et al (2005) Wilms' tumour gene 1 (WT1) in human neoplasia. Leuk 19(8):1318–23
59. Tatsumi N, Oji Y, Tsuji N, Tsuda A, Higashio M, Aoyagi S et al (2008) Wilms' tumor gene WT1-shRNA as a potent apoptosis-inducing agent for solid tumors. Int J Oncol 32(3):701–711
60. Glienke W, Maute L, Koehl U, Esser R, Milz E, Bergmann L (2007) Effective treatment of leukemic cell lines with wt1 siRNA. Leuk 21(10):2164–70
61. Doubrovina ES, Doubrovin MM, Lee S, Shieh J-H, Heller G, Pamer E et al (2004) In vitro stimulation with WT1 peptide-loaded epstein-barr virus-positive B cells elicits high frequencies of WT1 peptide-specific T cells with in vitro and in vivo tumoricidal activity. Clin Cancer Res 10(21):7207–7219
62. Gao L, Xue SA, Hasserjian R, Cotter F, Kaeda J, Goldman JM et al (2003) Human cytotoxic T lymphocytes specific for Wilms' tumor antigen-1 inhibit engraftment of leukemia-initiating stem cells in non-obese diabetic-severe combined immunodeficient recipients. Transplantation 75(9):1429–1436
63. Xue S-A, Gao L, Hart D, Gillmore R, Qasim W, Thrasher A et al (2005) Elimination of human leukemia cells in NOD/SCID mice by WT1-TCR gene–transduced human T cells. Blood 106(9):3062–3067

64. Maslak PG, Dao T, Bernal Y, Chanel SM, Zhang R, Frattini M et al (2018) Phase 2 trial of a multivalent WT1 peptide vaccine (galinpepimut-S) in acute myeloid leukemia. Blood Adv 2(3):224–234

65. Koehne G, Devlin S, Chung DJ, Landau HJ, Korde N, Mailankody S et al (2017) WT1 heteroclitic epitope immunization following autologous stem cell transplantation in patients with high-risk multiple myeloma (MM). 35(15_suppl):8016–8016. https://doi.org/10.1200/JCO20173515_suppl8016

66. O'Cearbhaill RE, Gnjatic S, Aghajanian C, Iasonos A, Konner JA, Losada N et al (2018) A phase I study of concomitant galinpepimut-s (GPS) in combination with nivolumab (nivo) in patients (pts) with WT1+ ovarian cancer (OC) in second or third remission. 36(15_suppl):5553–5553. https://doi.org/10.1200/JCO20183615_suppl5553

67. Ikeda H, Lethé B, Lehmann F, Van BN, Baurain J-F, De SC et al (1997) Characterization of an antigen that is recognized on a melanoma showing partial HLA loss by CTL expressing an NK inhibitory receptor. Immunity 6(2):199–208

68. Chang AY, Dao T, Gejman RS, Jarvis CA, Scott A, Dubrovsky L et al (2017) A therapeutic T cell receptor mimic antibody targets tumor-associated PRAME peptide/HLA-I antigens. J Clin Invest 127(7):1–14

69. Chang AY, Dao T, Gejman RS, Jarvis CA, Scott A, Dubrovsky L et al (2017) A therapeutic T cell receptor mimic antibody targets tumor-associated PRAME peptide/HLA-I antigens. J Clin Invest 127(7):2705

70. Kalaora S, Sang Lee J, Barnea E, Levy R, Greenberg P, Alon M et al Immunoproteasome expression is associated with better prognosis and response to checkpoint therapies in melanoma

71. Keller M, Ebstein F, Bürger E, Textoris-Taube K, Gorny X, Urban S et al (2015) The proteasome immunosubunits, PA28 and ER-aminopeptidase 1 protect melanoma cells from efficient MART-126-35-specific T-cell recognition. Eur J Immunol 45(12):3257–3268

72. Stanojevic M, Hont AB, Geiger A, O'Brien S, Ulrey R, Grant M et al (2021) Identification of novel HLA-restricted preferentially expressed antigen in melanoma peptides to facilitate off-the-shelf tumor-associated antigen-specific T-cell therapies. Cytotherapy 23(8):694–703

73. Mahoney KE, Shabanowitz J, Hunt DF (2021) MHC phosphopeptides: promising targets for immunotherapy of cancer and other chronic diseases. Mol Cell Proteomics 20:100112

74. Cobbold M, De La Peña H, Norris A, Polefrone JM, Qian J, English AM et al (2013) MHC class I-associated phosphopeptides are the targets of memory-like immunity in leukemia. Sci Transl Med 5(203):1–11

75. Zarling AL, Polefrone JM, Evans AM, Mikesh LM, Shabanowitz J, Lewis ST et al (2006) Identification of class I MHC-associated phosphopeptides as targets for cancer immunotherapy. Proc Natl Acad Sci 103(40):14889–14894

76. Engelhard VH, Obeng RC, Cummings KL, Petroni GR, Ambakhutwala AL, Chianese-Bullock KA et al (2020) MHC-restricted phosphopeptide antigens: preclinical validation and first-in-humans clinical trial in participants with high-risk melanoma. J Immunother Cancer 8(1):e000262

77. Mohammed F, Cobbold M, Zarling AL, Salim M, Barrett-Wilt GA, Shabanowitz J et al (2008) Phosphorylation-dependent interaction between antigenic peptides and MHC class I: a molecular basis for the presentation of transformed self. Nat Immunol 9(11):1236–1243

78. Kessler JH, Bres-Vloemans SA, van Veelen PA, de Ru A, Huijbers IJG, Camps M et al (2006) BCR-ABL fusion regions as a source of multiple leukemia-specific CD8+ T-cell epitopes. Leuk 20(10):1738–50.

79. Cai A, Keskin DB, DeLuca DS, Alonso A, Zhang W, Zhang GL et al (2012) Mutated BCR-ABL generates immunogenic T-cell epitopes in CML patients. Clin Cancer Res 18(20):5761–5772

80. Riva G, Luppi M, Barozzi P, Quadrelli C, Basso S, Vallerini D et al (2010) Emergence of BCR-ABL–specific cytotoxic T cells in the bone marrow of patients with Ph+ acute lymphoblastic leukemia during long-term imatinib mesylate treatment. Blood 115(8):1512–1518

81. Rojas JM, Knight K, Wang L, Clark RE (2007) Clinical evaluation of BCR-ABL peptide immunisation in chronic myeloid leukaemia: results of the EPIC study. Leuk 21(11):2287–95
82. Cathcart K, Pinilla-Ibarz J, Korontsvit T, Schwartz J, Zakhaleva V, Papadopoulos EB et al (2004) A multivalent bcr-abl fusion peptide vaccination trial in patients with chronic myeloid leukemia. Blood 103(3):1037–1042
83. Pinilla-Ibarz J, Cathcart K, Korontsvit T, Soignet S, Bocchia M, Caggiano J et al (2000) Vaccination of patients with chronic myelogenous leukemia with bcr-abl oncogene breakpoint fusion peptides generates specific immune responses. Blood 95(5):1781–1787
84. Zamora AE, Crawford JC, Allen EK, Guo XJ, Bakke J, Carter RA et al (2019) Pediatric patients with acute lymphoblastic leukemia generate abundant and functional neoantigen-specific CD8+ T cell responses. Sci Transl Med 11(498):eaat8549
85. Ali M, Foldvari Z, Giannakopoulou E, Böschen ML, Strønen E, Yang W et al (2019) Induction of neoantigen-reactive T cells from healthy donors. Nat Protoc 14(6):1926–1943
86. Kato T, Matsuda T, Ikeda Y, Park J-H, Leisegang M, Yoshimura S et al (2018) Effective screening of T cells recognizing neoantigens and construction of T-cell receptor-engineered T cells. Oncotarget 9(13):11009
87. Matsuda T, Leisegang M, Park J-H, Ren L, Kato T, Ikeda Y et al (2018) Induction of neoantigen-specific cytotoxic T cells and construction of T-cell receptor-engineered T cells for ovarian cancer. Clin Cancer Res 24(21):5357–67
88. Strønen E, Toebes M, Kelderman S, Van Buuren MM, Yang W, Van Rooij N et al (2016) Targeting of cancer neoantigens with donor-derived T cell receptor repertoires. Science (80-) 352(6291):1337–41
89. Duval A, Reperant M, Hamelin R (2002) Comparative analysis of mutation frequency of coding and non coding short mononucleotide repeats in mismatch repair deficient colorectal cancers. Oncogene 21(52):8062–6
90. Mardis ER (2019) Neoantigens and genome instability: impact on immunogenomic phenotypes and immunotherapy response. Genome Med 11(1):1–12
91. Sahin IH, Akce M, Alese O, Shaib W, Lesinski GB, El-Rayes B et al (2019) Immune checkpoint inhibitors for the treatment of MSI-H/MMR-D colorectal cancer and a perspective on resistance mechanisms. Br J Cancer 121(10):809–18
92. Schwitalle Y, Linnebacher M, Ripberger E, Gebert J, Doeberitz MVK (2004) Immunogenic peptides generated by frameshift mutations in DNA mismatch repair-deficient cancer cells. 4:1–10
93. Linnebacher M, Gebert J, Rudy W, Woerner S, Yuan YP, Bork P et al (2001) Frameshift peptide-derived T-cell epitopes: a source of novel tumor-specific antigens. Int J Cancer 93(1):6–11
94. Garbe Y, Maletzki C, Linnebacher M (2011) An MSI tumor specific frameshift mutation in a coding microsatellite of MSH3 encodes for HLA-A0201-restricted CD8+ cytotoxic T cell epitopes. PLoS One 6(11)
95. Roudko V, Bozkus CC, Orfanelli T, McClain CB, Carr C, O'Donnell T et al (2020) Shared immunogenic poly-epitope frameshift mutations in microsatellite unstable tumors. Cell 183(6):1634-1649.e17
96. Kloor M, Reuschenbach M, Pauligk C, Karbach J, Rafiyan M-R, Al-Batran S-E et al (2020) A frameshift peptide neoantigen-based vaccine for mismatch repair-deficient cancers: a phase I/IIa clinical trial. Clin Cancer Res 26(17):4503–4510
97. Mohamed YS, Bashawri LA, Vatte C, Abu-Rish EY, Cyrus C, Khalaf WS et al (2016) The in vitro generation of multi-tumor antigen-specific cytotoxic T cell clones: candidates for leukemia adoptive immunotherapy following allogeneic stem cell transplantation. Mol Immunol 77:79–88
98. Gerdemann U, Katari U, Christin AS, Cruz CR, Tripic T, Rousseau A et al (2011) Cytotoxic T lymphocytes simultaneously targeting multiple tumor-associated antigens to treat EBV negative lymphoma. Mol Ther 19(12):2258–68

99. Vasileiou S, Lulla PD, Tzannou I, Watanabe A, Kuvalekar M, Callejas WL et al (2021) T-cell therapy for lymphoma using nonengineered multiantigen-targeted T cells is safe and produces durable clinical effects. J Clin Oncol 39(13):1415–1425

100. Hont AB, Cruz CR, Ulrey R, O'Brien B, Stanojevic M, Datar A et al (2019) Immunotherapy of relapsed and refractory solid tumors with ex vivo expanded multi-tumor associated antigen specific cytotoxic T lymphocytes: a phase I study. J Clin Oncol 37(26):2349

101. Dave H, Terpilowski M, Mai M, Toner K, Grant M, Stanojevic M et al (2021) Tumor associated antigen specific T cells with nivolumab are safe and persist in vivo in rel/ref Hodgkin Lymphoma. Blood Adv

102. Em J, Sh K, Rh S, Dm P (1989) Activation of gamma delta T cells in the primary immune response to mycobacterium tuberculosis. Science 244(4905):713–716

103. Ravens S, Schultze-Florey C, Raha S, Sandrock I, Drenker M, Oberdörfer L et al (2017) Human γδ T cells are quickly reconstituted after stem-cell transplantation and show adaptive clonal expansion in response to viral infection. Nat Immunol 18(4):393–401

104. Knight A, Madrigal AJ, Grace S, Sivakumaran J, Kottaridis P, Mackinnon S et al (2010) The role of Vδ2-negative γδ T cells during cytomegalovirus reactivation in recipients of allogeneic stem cell transplantation. Blood 116(12):2164–2172

105. Scheper W, van Dorp S, Kersting S, Pietersma F, Lindemans C, Hol S et al (2013) γδT cells elicited by CMV reactivation after allo-SCT cross-recognize CMV and leukemia. Leuk 27(6):1328–38

106. Vila LM, Haftel HM, Park HS, Lin MS, Romzek NC, Hanash SM et al (1995) Expansion of mycobacterium-reactive gamma delta T cells by a subset of memory helper T cells. Infect Immun 63(4):1211

107. Behr C, Poupot R, Peyrat MA, Poquet Y, Constant P, Dubois P et al (1996) Plasmodium falciparum stimuli for human gammadelta T cells are related to phosphorylated antigens of mycobacteria. Infect Immun 64(8):2892–6

108. Vavassori S, Kumar A, Wan GS, Ramanjaneyulu GS, Cavallari M, El Daker S et al (2013) Butyrophilin 3A1 binds phosphorylated antigens and stimulates human γδ T cells. Nat Immunol 14(9):908–16

109. Rigau M, Ostrouska S, Fulford TS, Johnson DN, Woods K, Ruan Z et al (2020) Butyrophilin 2A1 is essential for phosphoantigen reactivity by γδ T cells. Science (80-) 367(6478)

110. Godder KT, Henslee-Downey PJ, Mehta J, Park BS, Chiang K-Y, Abhyankar S et al (2007) Long term disease-free survival in acute leukemia patients recovering with increased γδ T cells after partially mismatched related donor bone marrow transplantation. Bone Marrow Transplant 39(12):751–7

111. Lamb L, Musk P, Ye Z, van Rhee F, Geier S, Tong J-J et al (2001) Human γδ+ T lymphocytes have in vitro graft vs leukemia activity in the absence of an allogeneic response. Bone Marrow Transplant 27(6):601–6

112. Bauer S, Groh V, Wu J, Steinle A, Phillips JH, Lanier LL et al (1999) Activation of NK cells and T cells by NKG2D, a receptor for stress-inducible MICA. Science (80-) 285(5428):727–9

113. Jin C, Lagoudas GK, Zhao C, Bullman S, Bhutkar A, Hu B et al (2019) Commensal microbiota promote lung cancer development via γδ T cells. Cell 176(5):998-1013.e16

114. Payne KK, Mine JA, Biswas S, Chaurio RA, Perales-Puchalt A, Anadon CM, et al (2020) BTN3A1 governs antitumor responses by coordinating αβ and γδ T cells. Science (80-) 369(6506):942–9

115. Park JH, Kim H-J, Kim CW, Kim HC, Jung Y, Lee H-S et al (2021) Tumor hypoxia represses γδ T cell-mediated antitumor immunity against brain tumors. Nat Immunol 22(3):336–46

116. Treiner E, Lantz O (2006) CD1d- and MR1-restricted invariant T cells: of mice and men. Curr Opin Immunol 18(5):519–526

117. Brigl M, Brenner MB (2004) CD1: antigen presentation and T cell function. 22:817–90. https://doi.org/10.1146/annurev.immunol22012703104608

118. Han M, Hannick LI, DiBrino M, Robinson MA (1999) Polymorphism of human CD1 genes. Tissue Antigens 54(2):122–127

119. Barral DC, Brenner MB (2007) CD1 antigen presentation: how it works. Nat Rev Immunol 7(12):929–41
120. Mori L, Lepore M, De Libero G (2016) The Immunology of CD1- and MR1-restricted T cells. Annu Rev Immunol 34(1):479–510
121. de Lalla C, Lepore M, Piccolo FM, Rinaldi A, Scelfo A, Garavaglia C et al (2011) High-frequency and adaptive-like dynamics of human CD1 self-reactive T cells. Eur J Immunol 41(3):602–10
122. Lepore M, de Lalla C, Gundimeda SR, Gsellinger H, Consonni M, Garavaglia C et al (2014) A novel self-lipid antigen targets human T cells against CD1c+ leukemias. J Exp Med 211(7):1363–1377
123. Consonni M, Garavaglia C, Grilli A, de Lalla C, Mancino A, Mori L et al (2021) Human T cells engineered with a leukemia lipid-specific TCR enables donor-unrestricted recognition of CD1c-expressing leukemia. Nat Commun 12(1):1–14
124. Bassani-Sternberg M, Bräunlein E, Klar R, Engleitner T, Sinitcyn P, Audehm S et al (2016) Direct identification of clinically relevant neoepitopes presented on native human melanoma tissue by mass spectrometry. Nat Commun 7(1):13404
125. Yadav M, Jhunjhunwala S, Phung QT, Lupardus P, Tanguay J, Bumbaca S et al (2014) Predicting immunogenic tumour mutations by combining mass spectrometry and exome sequencing. Nature 515(7528):572–576
126. Kalaora S, Barnea E, Merhavi-Shoham E, Qutob N, Teer JK, Shimony N et al (2016) Use of HLA peptidomics and whole exome sequencing to identify human immunogenic neo-antigens. Oncotarget 7(5):5110–5117
127. Chong C, Müller M, Pak H, Harnett D, Huber F, Grun D et al (2020) Integrated proteoge-nomic deep sequencing and analytics accurately identify non-canonical peptides in tumor immunopeptidomes. Nat Commun 11(1):1–21
128. Ho WY, Nguyen HN, Wölfl M, Kuball J, Greenberg PD (2006) In vitro methods for gener-ating CD8+ T-cell clones for immunotherapy from the naïve repertoire. J Immunol Methods 310(1–2):40–52
129. Wölfl M, Greenberg PD (2014) Antigen-specific activation and cytokine-facilitated expansion of naive, human CD8+ T cells. Nat Protoc 9(4):950–966
130. Krutzik PO, Nolan GP (2006) Fluorescent cell barcoding in flow cytometry allows high-throughput drug screening and signaling profiling. Nat Methods 3(5):361–8
131. Krutzik PO, Clutter MR, Trejo A, Nolan GP (2011) Fluorescent cell barcoding for multiplex flow cytometry. Curr Protoc Cytom 55(1):6.31.1–6.31.15
132. Dolton G, Tungatt K, Lloyd A, Bianchi V, Theaker SM, Trimby A et al (2015) More tricks with tetramers: a practical guide to staining T cells with peptide-MHC multimers. Immunol-ogy 146(1):11–22
133. Wooldridge L, Lissina A, Cole DK, Van Den Berg HA, Price DA, Sewell AK (2009) Tricks with tetramers: how to get the most from multimeric peptide-MHC. Immunology 126(2):147–164
134. Toebes M, Coccoris M, Bins A, Rodenko B, Gomez R, Nieuwkoop NJ et al (2006) Design and use of conditional MHC class I ligands. Nat Med 12(2):246–51
135. Bakker AH, Hoppes R, Linnemann C, Toebes M, Rodenko B, Berkers CR et al (2008) Con-ditional MHC class I ligands and peptide exchange technology for the human MHC gene products HLA-A1, -A3, -A11, and -B7. Proc Natl Acad Sci 105(10):3825–3830
136. Anjanappa R, Garcia-Alai M, Kopicki J-D, Lockhauserbäumer J, Aboelmagd M, Hinrichs J et al (2020) Structures of peptide-free and partially loaded MHC class I molecules reveal mechanisms of peptide selection. Nat Commun 11(1):1–11
137. Saini SK, Tamhane T, Anjanappa R, Saikia A, Ramskov S, Donia M et al (2019) Empty peptide-receptive MHC class I molecules for efficient detection of antigen-specific T cells. Sci Immunol 4(37)
138. Hadrup SR, Bakker AH, Shu CJ, Andersen RS, van Veluw J, Hombrink P et al (2009) Par-allel detection of antigen-specific T-cell responses by multidimensional encoding of MHC multimers. Nat Methods 6(7):520–6

139. Bentzen AK, Marquard AM, Lyngaa R, Saini SK, Ramskov S, Donia M et al (2016) Large-scale detection of antigen-specific T cells using peptide-MHC-I multimers labeled with DNA barcodes. Nat Biotechnol 34(10):1037–45
140. Leen AM, Bollard CM, Mendizabal AM, Shpall EJ, Szabolcs P, Antin JH et al (2013) Multicenter study of banked third-party virus-specific T cells to treat severe viral infections after hematopoietic stem cell transplantation. Blood 121(26):5113–5123
141. Tzannou I, Papadopoulou A, Naik S, Leung K, Martinez CA, Ramos CA et al (2017) Off-the-shelf virus-specific T cells to treat BK virus, human herpesvirus 6, cytomegalovirus, epstein-barr virus, and adenovirus infections after allogeneic hematopoietic stem-cell transplantation. J Clin Oncol 35(31):3547
142. Withers B, Blyth E, Clancy LE, Yong A, Fraser C, Burgess J et al (2017) Long-term control of recurrent or refractory viral infections after allogeneic HSCT with third-party virus-specific T cells. Blood Adv 1(24):2193–2205

Chimeric Antigen Receptor (CAR) T Cell Therapy for Glioblastoma

5

Amitesh Verma and Sarwish Rafiq

5.1 CAR T Cells: A Novel Cell Therapy for Cancer

5.1.1 Structure and Manufacturing

Chimeric antigen receptors (CARs) are synthetic molecules that can redirect T cells to target cancer cells. T cells that are genetically engineered to express CARs have demonstrated high therapeutic responses in the treatment of B cell malignancies [1–8].

The CAR is composed of four primary components that redirect T cell function specificity toward tumor-associated antigens (TAA). These are an extracellular targeting moiety, a hinge domain, a transmembrane domain, and an intracellular signaling domain (Fig. 5.1). Traditionally, the extracellular targeting moiety consists of the variable heavy (V_H) and variable light (V_L) chains of monoclonal antibodies connected via a linker to form a single-chain variable fragment (scFv) [9, 10]. The targeting domain of a CAR characteristically operates in a human leukocyte antigen (HLA)-independent manner, directly recognizing tumor cells without the necessity of antigen presentation. The hinge or spacer component links the extracellular ectodomain to the intracellular signaling domain and provides flexibility against steric hindrance [11]. Additionally, the CAR hinge provides

A. Verma
Winship Summer Scholars Research Program, Emory University, Atlanta, GA, USA

A. Verma · S. Rafiq (✉)
Department of Hematology and Medical Oncology, Emory University School of Medicine, Atlanta, GA, USA
e-mail: Sarwish.Rafiq@emory.edu

S. Rafiq
Winship Cancer Institute, Emory University, Atlanta, GA, USA

© The Author(s), under exclusive license to Springer Nature Switzerland AG 2022   161
P. Hays (ed.), *Cancer Immunotherapies*, Cancer Treatment and Research 183,
https://doi.org/10.1007/978-3-030-96376-7_5

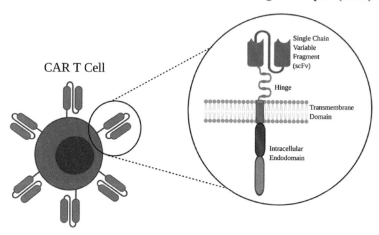

Fig. 5.1 **Chimeric antigen receptor (CAR) construction.** Chimeric antigen receptors consist of four principal components, including a single-chain variable fragment (scFv), a CAR hinge, a transmembrane domain, and an intracellular ectodomain. An extracellular targeting moiety, usually in the form of an scFv, allows a CAR construct to operate HLA-independently and pinpoint tumor-associated antigens. A hinge component, securing the extracellular ectodomain to the intracellular signaling domain, provides ample flexibility against steric effects. Consisting of a type I protein, a transmembrane domain provides anchorage for the extracellular receptor to the T cell counterparts. An intracellular endodomain consists of a primary activation component, allowing CAR T cell persistence and function. Figure created with BioRender.com

necessary length for the extracellular portion of the receptor, allowing CAR interactions with the targeted tumor antigen. The transmembrane domain of the CAR consists of a type I protein and provides anchorage for the extracellular CAR domain to intracellular signaling domains. The intracellular signaling domains consist of a principal activation component, commonly CD3ζ, and at least one costimulatory domain to maintain T cell persistence and optimal function. Most commonly, these costimulatory domains are derived from either CD28 or 4-1BB. These intracellular domains allow for CAR T cell activation, proliferation, and the cytokine release and enable anti-tumor function [9, 10, 12].

Currently, most CAR T cells are made as an autologous product and are manufactured specifically for the patient. Briefly, the patient undergoes leukapheresis, and T cells are isolated, activated, and genetically engineered to express CAR by means of lenti- or retroviral transduction. The engineered CAR T cells are then expanded and reinfused into the patient as a living drug after conditioning chemotherapy [13].

5.1.2   Clinical CAR T Cell Success in Hematologic Malignancies

Numerous clinical trials have highlighted the efficacy of CAR T cell use in hematologic malignancies. For example, axicabtagene ciloleucel, a CAR T cell therapy for refractory or relapsed large B cell lymphoma yielded an 82% objective response rate in a phase I trial [5]. Another phase II clinical trial demonstrated high rates of durable response rates utilizing tisagenlecleucel, a CD19-directed CAR T cell therapy for relapsed diffuse large B cell lymphoma [6]. Additional seminal trials have been conducted in acute lymphoblastic leukemia, diffuse large B cell lymphoma, mantle cell lymphoma, and multiple myeloma [1–4, 6–8] and have led to the Food and Drug Administration's approval of six CAR T cell products. Four of these products are targeted to the B cell marker CD19 and two products are directed toward B cell maturation antigen (BCMA). Such advances have naturally fostered the investigation of CAR T cell use in solid tumors, such as "difficult-to-treat" malignancies like glioblastoma.

5.2   Glioblastoma: Current Barriers and the Promise for CAR T Cells

Glioblastoma (GBM) is an aggressive central nervous system malignancy that develops from astrocytic glial cells. It is the most common primary malignant brain tumor in adults, constituting 48.3% of all central nervous system malignancies [14]. To date, treating GBM has been a rigorous and challenging process. The current standard-of-care for GBM includes maximum safe surgical resection followed by a combination of radiation and temozolomide (TMZ), a deoxyribonucleic acid methylator that triggers tumor cell death [15]. However, various issues persist with the current standard-of-care, including TMZ-resistant gliomas that produce O^6-methylguanine DNA methyltransferase (MGMT) and limit the efficacy of chemotherapy [16, 17]. As a result, the current prognosis for patients with GBM remains poor, with a median survival rate of roughly 15 months and a five-year relative survival rate of 6.8% [14, 18]. Therefore, new treatment modalities are needed for this malignancy. Given the recent successes of CAR T cells in hematologic malignancies, an expanding number of researchers have focused on translating CAR T cell therapy to solid tumors [19]. The next sections will highlight the primary challenges in treating GBM and how CAR T cells can overcome these issues.

5.2.1   The Blood-Brain-Barrier

The blood-brain-barrier (BBB) is a selective semipermeable interface that manages the influx and efflux of substances to the central nervous system (CNS) and protects CNS parenchyma from toxins, pathogens, and disease. Composed of various cell types such as endothelial cells, pericytes, and astrocytic glial cells, the

BBB tightly regulates the homeostasis of the CNS parenchyma (Fig. 5.2). However, the BBB also greatly diminishes the trafficking of drugs to the GBM tumor microenvironment [20], blocking 100% of all large neurotherapeutics and 98% of all small neurotherapeutics [21, 22]. Small-molecule drug treatments must therefore meet two major criteria in order to surpass the BBB in significant amounts: a molecular mass under a 500 Dalton threshold and a high lipid solubility [21]. Several attempts to produce BBB penetrable drugs have been made for GBM, the most successful being TMZ [23]. However, because only limited levels of TMZ penetrate into the brain parenchyma (roughly 20% of plasma levels), TMZ must be given in high doses to be effective and may result in several off-tumor toxicities [24]. Despite the disruption of the integrity of the BBB through the course of GBM progression, a significant tumor burden with an intact BBB still remains, limiting drug delivery to these regions [25]. As a result, CNS drug delivery may be suboptimal, thus reducing efficacy of chemotherapy regimens in GBM [26].

Immune cell therapies may be able to overcome this physical barrier. The discovery of lymphatic vessels in the meningeal membrane of the brain challenged the notion of the CNS as an immune-privileged site [27]. Naturally occurring autoimmune disorders, such as multiple sclerosis, provide supplementary evidence of

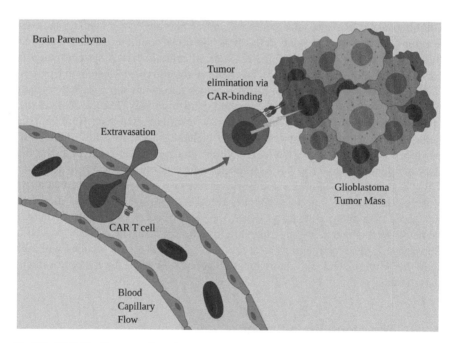

Fig. 5.2 CAR T cell penetration of the blood-brain-barrier. Recent trials involving peripherally administered CAR T cells and GBM have elucidated the ability of CAR T cells to surpass the restrictive blood-brain-barrier into the brain parenchyma. Cellular extravasation through epithelial cells allows T lymphocytes to successfully reach antigen sites for tumor elimination. Figure created with BioRender.com

immune system trafficking to parts of the brain [28, 29]. Furthermore, clinical trials have demonstrated the presence of various mechanisms that enable T cells to traffic across the BBB and into the brain parenchyma to elicit an immune response [30, 31].

Neurotoxicities associated with CAR T cell treatment may also provide insight into the trafficking of CAR T cells across the BBB. Immune effector cell-associated neurotoxicity syndrome (ICANS) presents as delirium, encephalopathy, seizures, and other neurological symptoms in patients receiving CD19-directed CAR T cell treatment and may result in fatal cerebral edema [32]. Rates of grade 3 ICANS can range from 12 to 30% for patients with lymphoma and 13–42% for patients with leukemia [1, 2, 5, 6, 33, 34]. These toxicities have been associated with high serum levels of IL-1, IL-6, and IL-15, indicating the passive diffusion of cytokines in the brain induced by CAR T cell treatment [35, 36]. Clinically, high-grade ICANS is managed with the anti-IL-6 Receptor antibody tocilizumab and steroids. Additionally, the detection of CAR T cells in the cerebrospinal fluid in patients with ICANS further indicates that T cell trafficking across the BBB into the CNS parenchyma may play a role in the development of ICANS [37, 38]. Finally, neurotoxicity could also be caused by on-target killing of CD19-expressing brain mural cells which provides further evidence for CAR T cell infiltration into the brain [39]. Overall, the pathophysiology of CAR T cell-associated neurotoxicities provides further evidence of the ability of CAR T cells to both traffic across the BBB and influence processes associated with the CNS. These characteristics suggest that CAR T cells may be able to cross the BBB and mediate anti-tumor responses to GBM (Fig. 5.2).

5.2.2 Low Tumor Mutational Burden and Antigen-Processing Machinery Defects

Inherent characteristics of GBM influence how these tumors interact with the immune system and have limited the therapies that can be utilized for this disease. Tumor mutational burden (TMB) is a measure of mutations expressed by tumor cells and correlates with expression of neoantigens that the endogenous immune system can target. Melanoma, which can have high TMBs and high levels of neoantigens, has benefited from therapies that involve endogenous T cell responses [40]. However, GBM tumors generally have a lower average TMB than other tumor types, which limits the endogenous immune response [41]. Additionally, GBM tumors often have defects in antigen-processing and presentation machinery which further hinders recognition of the cells by the immune system [42, 43]. Decreased expression of HLA due to genotypic mutations leads to decreased activation of cytotoxic T cells [44] and overall evasion of adaptive immune system responses [45]. As a result, glioblastoma cells can remain undetected by the immune system, increasing the risk of "stealth invasion" and infiltration across the brain parenchyma [46].

CAR T cells can overcome the issues of low TMB and decreased antigen presentation. CAR T cells are generally engineered to target a tumor-associated surface antigen, recognizing target antigen independent of HLA presentation [47, 48]. While a low TMB may limit the abundance of available and targetable TAA, numerous mutated or GBM-associated antigens have been identified for CAR T cell targeting. The most studied of these include epidermal growth factor receptor variant III (EGFRvIII), IL-13 receptor subunit alpha 2 (IL-13Rα2), and human epidermal growth factor receptor 2 (HER2) [49] and will be discussed in depth elsewhere in this review.

5.2.3 Immunosuppressive Tumor Microenvironment

The GBM tumor microenvironment is characterized as an immunosuppressive one that enhances chemoresistance and tumor evasion (Fig. 5.3). Tumor-associated endothelial cells are instrumental in sustaining tumor tissue and appear morphologically different from normal brain parenchyma [50]. Additional cytogenetic

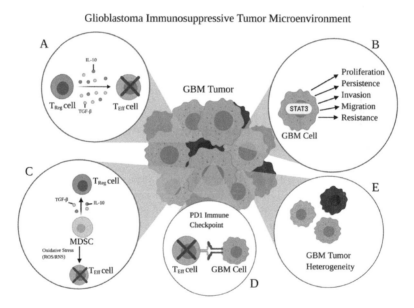

Fig. 5.3 **Features of the glioblastoma microenvironment**. Several features of the GBM microenvironment make implementing CAR T cells challenging. **a** Regulatory T cell (Treg) secretion of cytokines (Il-10, TGF-β) inhibits effector T cell activity. **b** Expression of STAT3 in GBM cells promotes tumor progression and invasion. **c** Myeloid-derived suppressor cells (MDSC) secretion of cytokines to promote Treg activity, ligand binding of PD1 checkpoint inhibitor in effector T cells, and induction of oxidative stress. **d** Ligand binding of PD1/PDL1 induce effector T cell inhibition. **e** GBM morphological variation and heterogeneity make it difficult to target a single antigen using a CAR construct. Figure created with BioRender.com

abnormalities, including abnormal centromeres, have enhanced GBM-associated endothelial cells' chemoresistance, allowing them to migrate faster and produce higher levels of growth factors [50, 51]. Moreover, overexpression of drug efflux transporter pumps including P-glycoprotein, and the upregulation of DNA repair mechanisms in GBM cells have limited the effects of TMZ treatment [52]. Several extracellular matrix-related factors including the CCN1 protein and the promotion of pro-inflammatory cytokines have further facilitated glioma cell migration within the tumor microenvironment [53, 54]. Non-cellular elements like a reduction in oxygen concentration and pH, as hypothesized by the Warburg effect, increase glioma cell migration, invasion, and chemoresistance [55, 56]. Tumor microenvironment-induced hypoxia also upregulates chemokine ligand 28 (CCL28), assisting in sustained angiogenesis [57]. Lastly, glioma stem cells within the microenvironment upregulate anti-apoptotic pathways and contribute to persistent tumorigenesis [58, 59]. These challenges associated with the complex GBM milieu can hinder the effectiveness of immunotherapy for glioblastoma.

Several genomic mutations within malignant glioma cells hinder the also persistence of T lymphocytes in the GBM microenvironment [60]. Signal transducer and activator of transcription 3 (STAT3) is phosphorylated in GBM cells [61]. This can lead to the expression of factors such as increased transcription of IL-10, TGF-β, and FOX3P and promotion of dendritic cell maturation that suppress T cell proliferation and persistence and contribute to GBM tumor progression [62–64]. However, recent studies have demonstrated that STAT3 inhibition can greatly enhance the activity of immune cells, including T cells in the immunosuppressive tumor microenvironment [65].

Additionally, the presence of suppressive cells in the tumor microenvironment adversely counter an immune response by CAR T cells. Regulatory T cells (Tregs) secrete immunosuppressive cytokines, including transforming growth factor-beta (TGF-β), that limit the efficacy of CAR T cells [66–68]. Additionally, myeloid-derived suppressor cells (MDSC), a heterogeneous group of myeloid progenitor and precursor cells, negatively regulate immune responses and are a major contributor to T cell inhibition and gliomagenesis in the MDSC populous tumor microenvironment [69]. MDSC secrete reactive oxygen species (ROS) and nitrogen species (RNS) and induces oxidative stress in T cells [70].

While inherent characteristics of glioblastoma hinder the effects of traditional therapies, CAR T may be able to overcome these hurdles. Therefore, preclinical and clinical research has studied the potential of this therapy in GBM.

5.3   CAR T Cell Therapy for Glioblastoma

Many preclinical studies have tested the efficacy of CAR T cells against GBM. Below we will highlight the antigens that have been most widely studied as targets and have published clinical trial results (Table 5.1).

Table 5.1 Published data of CAR T cells in GBM

Institution/NCT#	Phase	Target	Study treatment/features	Results
Perelman School of Medicine at the University of Pennsylvania, Philadelphia, PA NCT02209376	I	EGFRvIII	Patients with EGFRvIII + GBM were treated with EGFRvIII-directed CAR T cells	CAR T cell persistence and trafficking were observed with presence of T cells in tumor environment after peripheral infusion. A median overall survival of ~8 months was reported. One patient remains alive >18 months after peripheral CAR T cell infusion [71]
National Cancer Institute, Bethesda, MD NCT01454596	I/II	EGFRvIII	Patients with EGFRvIII + GBM receive intravenous lymphodepleting therapy (cyclophosphamide and fludarabine) prior to administration of anti-EGFRvIII CAR gene-transduced peripheral blood mononuclear cells, plus intravenous aldesleukin	Median progression-free survival was 1.3 months. 2 patients experienced severe hypoxia, including a single treatment-related mortality. Median overall survival was 6.9 months with two patients surviving over a year and another patient surviving >59 months. No objective responses were reported [72]
Perelman School of Medicine at the University of Pennsylvania, Philadelphia, PA	–	EGFRvIII	A 59-year -old woman received a single peripheral infusion of 9.2×10^7 autologous EGFRvIII-targeted CAR T cells	Survival 36 months after disease recurrence was observed. Histopathological analysis of tissue obtained during a second stage surgical resection showed decreased expression of EGFRvIII on tumor tissue. CAR T cells persisted in peripheral circulation at 29 months follow-up [73]

(continued)

Table 5.1 (continued)

Institution/NCT#	Phase	Target	Study treatment/features	Results
Baylor College of Medicine, Houston, TX NCT01109095	I	HER2	Patients treated intravenously with cytomegalovirus (CMV)-HER2-specific CARs	HER2-specific CARs were successfully manufactured and infusions of $1 \times 10^6/m^2$ to $1 \times 10^8/m^2$ were well tolerated without severe adverse events. Of the 17 subjects, 8 clinically benefited from CAR T cell therapy as shown by a partial response ($n = 1$) and a stable disease ($n = 7$). Median survival for adult patients was 30 months from diagnosis [74]
City of Hope National Medical Center, Duarte, CA NCT00730613	I	IL13Rα2	Patients with recurrent GBM were locally infused with IL13Rα2-directed CD8+CAR T cells, receiving up to 12 infusions with a maximum dosage of 10^8 engineered CAR T lymphocytes	IL13Rα2-directed CAR T cells were manufactured and intracranial administration was tolerated with no severe brain inflammation. 1 of 3 patients showed reduced overall IL13Rα2 expression of GBM cells. Increase in GBM necrotic volume was seen in intracranial sites in 1 patient [75]
Xuanwu Hospital, Capital Medical University, Beijing, China NCT03423992	I	EphA2	Three patients with EphA2-positive recurrent GBM received 1×10^6 EphA2-directed CAR T cells/kilogram, intravenously after lymphodepletion	Expansion of CAR T cells was observed in both the peripheral blood and cerebral-spinal-fluid, persisting for more than four weeks. One patient had transit diminishment of the tumor with an overall survival ranging from 86 to 181 days for the three patients [76]

(continued)

Table 5.1 (continued)

Institution/NCT#	Phase	Target	Study treatment/features	Results
West China Hospital, West China Medical School, Sichuan University, Chengdu, Sichuan Province, China	–	B7-H3	A 56-year-old woman with recurrent GBM received weekly intracavitary infusions of B7-H3-targeted CAR T cells	After the first infusion, dramatic reduction of recurrent tumor tissue was observed. Clinical responses were sustained for 50 days after CAR T cell infusion with the patient opting out of the study after seven cycles of infusion [77]

5.3.1 Epidermal Growth Factor Receptor Variant III

EGFRvIII is a mutant protein that is the result of the deletion of exons 2–7 of EGFR. This deletion creates a new glycine residue at the junction of exons 1 and 8 and results in an abbreviated extracellular domain of the protein. It is the most common mutant of EGFR observed in GBM cells and is expressed in 24–67% of cases. Expression of EGFRvIII in GBM has been associated with a poor prognosis in patients surviving greater than a year [78].

Targeting EGFRvIII on GBM cells has demonstrated promise in preclinical studies. Similar to studies in hematologic malignancies, lymphodepletive therapy prior to CAR T cell administration enhances efficacy of treatment [79]. This effect is improved when CAR T cells were studied in combination with TMZ [80]. Dose-intensified TMZ pre-treatment with EGFRvII-directed CARs induced complete regression of 21-day established brain tumors in murine models, signaling enhanced CAR proliferation and persistence in microenvironment [80]. In addition, the binding of the CAR to EGFRvIII could be fine-tuned by varying the length of the spacer domain [81] or increasing the affinity of the scFv [82]. Finally, preclinical studies have demonstrated that intravenous infusion of CAR T cells can traffic through the BBB. EGFRvIII-specific CAR T cells were intravenously administered to orthotopic and subcutaneous xenograft models of human EGFRvIII+ glioblastoma [83]. The results of this study demonstrated efficacy of EGFRvIII-directed CAR T cells in eliciting an anti-tumor response and suggested that peripheral administration of CAR T cells results in penetration through the BBB and trafficking to the GBM tumor microenvironment [83].

Johnson et al. developed and tested anti-EGFRvIII-CAR T cells in in vivo xenograft models of glioblastoma. EGFRvIII-directed CAR T cells were capable of targeting EGFRvIII-expressing cells and controlled tumor progression in subcutaneous and orthotopic models of disease [83]. Based on these results, a phase I clinical trial was opened to test the efficacy of anti-EGFRvIII CAR T cells in patients with recurrent GBM. Although heterogeneric EGFRvIII expression on glioblastoma remained a hurdle, decreased levels of EGFRvIII expression were observed in tumor tissues after intravenous infusion of CAR T cells. This indicates antigen-directed cytolysis and CAR T cell trafficking [71]. A second case report reported prolonged survival of a patient with recurrent glioblastoma following EGFRvIII CAR T cell treatment [73].

5.3.2 Human Epidermal Growth Factor Receptor 2

HER2 is a receptor tyrosine-protein that is expressed by up to 80% of GBM cases [84]. Preclinical studies demonstrated that HER2-directed CAR T cells can target HER2-positive GBM cells in orthotopic xenograft models of GBMs. Additionally, HER2-directed CAR T cells recognized CD133-positive glioma stem cells, denoting the possibility of targeting glioma-initiating cells and mitigating GBM recurrence. Overall, these results suggested that HER2-directed CAR T cells could be an effective treatment for glioblastoma [85]. This data was used to initiate a phase I clinical trial utilizing HER2-directed CAR T cells manufactured from virus-specific T cells (NCT01109095) [86]. Infusion of HER2-directed CAR T cells was well tolerated by 17 patients with progressive recurrent GBM, with no significant off-tumor toxicities. Eight patients clinically benefited from CAR T cell therapy, with one patient achieving a partial response and seven remaining with stable disease. The poor clinical responses were seen in most patients on this trial highlight the need to improve anti-HER2-CAR T cells to enhance in vivo persistence and expansion [86].

5.3.3 IL-13 Receptor Subunit Alpha 2

IL-13Rα2 is overexpressed in 44% of GBM tumor samples [87] and not significantly expressed on normal brain tissue. Recent research has demonstrated that IL-13Rα2 is a plausible antigen target for CAR T cell therapy in GBM. Pituch et al. demonstrated that intratumoral injection of IL-13Rα2-targeted CARs promoted a pro-inflammatory microenvironment, with increased pro-inflammatory cytokine production, T cell proliferation, and a decrease in MDSC [88]. These changes resulted in sustained long-term survival in murine models of malignant glioma [88]. Similarly, intracranial administration of IL-13Rα2-directed CD8+cytotoxic T cells targeting glioma stem-like cells resulted in robust anti-tumor activity and elimination of GBM-initiating activity in an orthotopic mouse tumor model [89].

Based on preclinical results, researchers conducted a phase I study utilizing IL-13Rα2-directed CD8+cytotoxic T cells (NCT00730613) [75]. Although the time to manufacture CAR T cells was lengthy (3–4 months), the trial demonstrated the feasibility and efficacy for IL-13Rα2-directed CD8+cytotoxic T lymphocytes against recurrent GBM. Three patients were treated and had a median survival of 11 months after relapse. There was significant reduction in the expression of

IL-13Rα2 + cells in GBM samples and increased tumor necrotic volume following infusion, suggesting that IL-13Rα2-directed CAR T cells may be a promising therapy for this disease [75].

5.3.4 Alternative GBM-Associated Targets for CAR T Cell Therapy

Preclinical studies have studied other glioblastoma-associated antigens as targets for CAR T cells. For example, chlorotoxin (CLTX) is found in the venom of scorpions and binds with high specificity to GBM tumor cells [90]. CARs constructed with CLTX as the targeting domain were tested in orthotopic xenograft models of human GBM. GBM tumor regression was reported in vivo after CLTX-CAR T cell infusion and no off-tumor immune response was observed [90]. Another group identified P32 to be highly expressed on malignant gliomas and tumor endothelial cells [91]. P32-specific CAR T cells were found to be effective in orthotopic syngeneic and xenograft mouse models of GBM [91].

Other CAR T cell target antigens have been validated preclinically to led to clinical testing in patients with glioblastoma. Ephrin type-A receptor 2 (EphA2) is selectively overexpressed on GBM cells and CAR T cells targeted to EphA2 improved survival of mice inoculated with glioblastoma [92]. A first-human-trial of EphA2-targeted CAR T cells demonstrated in vivo expansion of CAR T cells with one out of three patients treated achieving stable disease [76].

B7-H3 is a type I transmembrane protein that is overexpressed on many cancer types, including glioblastoma [93]. B7-H3-targeted CAR T cells demonstrated cytotoxic activity against primary glioblastoma cells and GBM cell lines [93]. A case study has been published on a single patient with recurrent GBM that was treated weekly with B7-H3-specific CAR T cells infusions [77]. Dramatic reduction of tumor was observed after the first infusion and the patient responded for 50 days.

Building upon preclinical data and the success in B cell malignancies, CAR T cells have been tested in patients with GBM (Table 5.1). These studies have established the safety of the CAR T cell approach in glioblastoma and have targeted a number of GBM-associated antigens. However, given the poor clinical responses seen in trials thus far, more research is needed to improve the efficacy of CAR T cells in this malignancy.

5.4 Optimizing CAR T Cell Therapy for Glioblastoma

As evident in the clinical responses discussed above, several barriers limit the efficacy of CAR T cells in malignant gliomas. For example, the harsh, immunosuppressive GBM microenvironment can hinder CAR T cell persistence. Additionally, because GBM exhibits a broad spectrum of tumor heterogeneity, targeting a single neoantigen may lead to immune escape. Furthermore, although T cells may penetrate the restrictive blood-brain-barrier, challenges associated with off-tumor toxicities and T cell persistence have been encountered. Recent advances in the field have elucidated combinations and engineering methods that can be used to surmount such barriers to optimize and strengthen the clinical application of CAR T cells for GBM.

5.4.1 Overcoming the Immunosuppressive Glioblastoma Microenvironment

The immunosuppressive glioblastoma tumor microenvironment has limited the clinical success of GBM therapies. Combination therapy with immunosuppression inhibitors has been studied to overcome this. Celecoxib is a non-steroidal anti-inflammatory drug that inhibits the build-up of MDSCs in the GBM tumor microenvironment via the inhibition of prostaglandin E2 production [94]. However, administration of celecoxib in combination with CAR T cell limited CAR T cell function [95]. More promising results were demonstrated with an inhibitor of BRD4, an epigenetic modifier of immunosuppressive genes [96]. Combination therapy of CAR T cells and a BRD4 inhibitor suppressed metastasis of glioblastoma in mouse models and improved survival [96].

Improving CAR T cell function through combination therapy with cytokines has been widely reported [97–99]. IL-12 is a pro-inflammatory cytokine that improves T cell function. Combination therapy of intratumoral delivery of IL-12 and infusion of EGFRvIII-directed CAR T cells achieves durable anti-tumor responses through reshaping the GBM microenvironment by increasing CD4 T cells, decreasing Tregs, and activating the myeloid compartment [100]. IL-7 is an important cytokine needed for T cell development and function. Delivery of IL-7 to GBM cells by oncolytic adenoviruses improved the proliferation, persistence, and anti-tumor efficacy of B7-H3 CAR T cells [101]. Finally, TGFβ is found in the GBM microenvironment and decreases immune response. EGFRvIII-specific CAR T cells engineered to express only the TGFRII ectodomain worked as a "TGFβ cytokine trap" that helped polarize GBM-infiltrated microglia toward a pro-inflammatory phenotype and improved CAR T cell function [102].

Combination therapy with immune checkpoint inhibitors has also been found to aid with CAR T persistence in glioblastoma. GBM cells express a multitude of suppressive immune checkpoint molecules that limit T cell function, including programmed cell death ligand 1 (PD-L1) [71]. Recent advances have demonstrated that checkpoint blockades of the PD-1/PD-L1 axis can be enhancing T cell function at the tumor site and are efficacious when applied to GBM [103, 104]. CRISPR-Cas9 knockout of PD-1 in CAR T cells resulted in cells that were resistant to PD-L1 inhibition. These CAR T cells demonstrated enhanced anti-tumor activity in preclinical glioma models with prolonged survival in mice through intracerebral administration [105, 106]. Given the success in combination checkpoint blockade and CAR T cell therapy in other diseases [107, 108], this approach is currently being studied currently in clinic for glioblastoma (Table 5.2).

Table 5.2 Ongoing trials of CAR T cells in GBM (results not obtained)

Institution/NCT#	Clinical trial and NCT#	Phase	Target	Study treatment/features
Beijing Sanbo Brain Hospital, Beijing, China NCT02844062	A safety and efficacy study of autologous chimeric antigen receptor engineered T cells redirected to EGFRvIII in patients with recurrent glioblastoma	I	EGFRvIII	Patients with recurrent GBM receive intravenous administration of lentiviral-transfected EGFRvIII-targeted CAR T cells (doses given in a split-day regimen in the range of 5×10^4/kg to 1×10^7/kg) Lymphodepleting therapy prior to T cell infusion includes 250 mg/m^2 of cyclophosphamide and 25 mg/m^2 of fludarabine on days 1–3
Perelman School of Medicine at the University of Pennsylvania, Philadelphia, PA NCT03726515	Phase 1 study of EGFRvIII-directed CAR T Cells combined with PD-1 inhibition in patients with newly diagnosed, MGMT-unmethylated glioblastoma	I	EGFRvIII	Patients with newly diagnosed, EGFRvIII +, MGMT-unmethylated GBM infused with EGFRvIII-targeted CAR T cells consisting of a tandem signaling domain of the 4-1BB and TCRζ signaling modules Administered in combination with pembrolizumab
Shenzhen Geno-Immune Medical Institute, Shenzhen, China NCT03170141	Immunogene-modified antigen-specific T (IgT) cells for the treatment of glioblastoma	I	EGFRvIII	GBM patients receive non-myeloablative chemotherapy (25 mg/m^2 of fludarabine and 250 mg/m^2 of cyclophosphamide on days 1–3) prior to infusion of EGFRvIII-IgT cells Dosage of IgT cells will be given on a 3 + 3 escalation approach, ranging from 1×10^5/kg to 1×10^7/kg
City of Hope National Medical Center, Duarte, CA NCT04003649	A phase 1 study to evaluate IL13Rα2-targeted chimeric antigen receptor (CAR) T cells combined with checkpoint inhibition for patients with resectable recurrent glioblastoma	I	IL13Rα2	Arm I patients with recurrent or refractory GBM receive nivolumab or ipilimumab intravenously prior to intracerebroventricular infusion of IL13Rα2-directed CAR T cells co-stimulated with 4-1BB. Arm II patients receive nivolumab in combination with CAR T cells

(continued)

Table 5.2 (continued)

Institution/NCT#	Clinical trial and NCT#	Phase	Target	Study treatment/features
City of Hope National Medical Center, Duarte, CA NCT02208362	Phase I study of cellular immunotherapy using memory-enriched T cells lentivirally transduced to express an IL13Rα2-specific, hinge-optimized, 41BB-costimulatory chimeric receptor and a truncated CD19 for recurrent/refractory malignant glioma patients	I	IL13Rα2	Patients randomized into 5 strata receive: 1. Intatumoral administration of IL13Rα2-directed CAR T cells co-stimulated with 4-1BB 2. Intracavitary administration of IL13Rα2-directed CAR T cells co-stimulated with 4-1BB 3. Intraventricular administration of IL13Rα2-directed CAR T cells co-stimulated with 4-1BB 4. Intratumoral/intraventricular administration of IL13Rα2-directed CAR T cells co-stimulated with 4-1BB 5. Intratumoral/intraventricular administration of IL13Rα2-directed BBzeta/truncated CD19[t] + Tn/mem
City of Hope National Medical Center, Duarte, CA, NCT04661384	Brain tumor-specific immune cells for the treatment of leptomeningeal glioblastoma, ependymoma, or medulloblastoma	I	IL13Rα2	Patients with IL13Rα2-positive leptomeningeal disease from glioblastoma, ependymoma, and medulloblastoma receive IL13Rα2-specific hinge-optimized 41BB costimulatory CAR truncated CD19-expressing autologous T lymphocytes
Xijing Hospital, Xi'an, China NCT04045847	A clinical study to investigate the safety, tolerance and efficacy evaluation of single-center, open-label of local treatment of CD147-CART in recurrent glioblastoma	I	CD147	Patients with recurrent malignant glioma receive intracavitary administration of CD147-directed CAR T cells in a 3 + 3 dose escalation design
Second Affiliated Hospital, School of Medicine, Zhejiang University, Hangzhou, China NCT04077866	B7-H3-targeted Chimeric antigen receptor (CAR) T cells in treating patients with recurrent or refractory glioblastoma	I/II	B7-H3	Patients with recurrent/refractory GBM are assigned to receive oral temozolomide (150 mg/m^2 followed by 200 ^{mg}/m^2 after the first day) or doses of temozolomide given in combination with intratumoral/intracerebroventricular administration of B7-H3-directed CAR T cells. Treatments will be given every 5 days with an interval of 23 days

(continued)

Table 5.2 (continued)

Institution/NCT#	Clinical trial and NCT#	Phase	Target	Study treatment/features
Second Affiliated Hospital, School of Medicine, Zhejiang University, Hangzhou, China NCT04385173	A pilot study of chimeric antigen receptor (CAR) T cells targeting B7-H3 antigen in treating patients with recurrent and refractory glioblastoma	I	B7-H3	Patients with recurrent/refractory GBM receive regular cycles of oral temozolomide (150 mg/m^2 followed by 200 ^{mg}/m^2 after the first day) in combination with intratumoral/intracerebroventricular administration of B7-H3-directed CAR T cells between cycles
City of Hope National Medical Center, Duarte, CA NCT03389230	Phase I study of cellular immunotherapy using memory-enriched T cells lentivirally transduced to express a HER2-specific, hinge-optimized, 41BB-costimulatory chimeric receptor and a truncated CD19 for patients with recurrent/refractory malignant glioma	I	HER2	Patients with recurrent/refractory GBM are assigned: Intratumoral/intracavitary infusion of HER2 BBzeta/CD19t+ Tcm cells Intratumoral/intracavitary and intraventricular catheter infusion of HER2 BBzeta/CD19t+ Tcm cells Intratumoral/intracavitary and intraventricular catheter infusion of HER2 BBzeta/CD19t+ Tn/mem cells
Baylor College of Medicine, Houston, TX NCT02442297	Phase I study of intracranial injection of T cells expressing HER2-specific chimeric antigen receptors (CAR) in subjects with HER2-positive tumors of the central nervous system (iCAR)	I	HER2	Patients with recurrent/refractory brain tumors labeled as high-risk (HER2 staining of grade 3 and intensity scores of 3+) or standard risk (not meeting high-risk description) receive intracavitary infusion of HER2-directed CAR T cells on a 3-cell dosing schedule consisting of three cell doses
Johann Wolfgang Goethe University Hospital, Frankfurt, Germany, NCT03383978	Intracranial injection of NK-92/5.28.z cells in patients with recurrent HER2-positive glioblastoma (CAR2BRAIN)	I	HER2	Patients with recurrent/refractory HER2-positive glioblastoma intracranially receive NK-92 and/or CAR 5.28 cells to determine maximum tolerated dose or maximum feasible dose
City of Hope National Medical Center, Duarte, CA NCT04214392	A phase 1 study to evaluate chimeric antigen receptor (CAR) T cells with a chlorotoxin tumor-targeting domain for patients with MMP2+ recurrent or progressive glioblastoma	I	Chlorotoxin	Patients with recurrent or progressive MMP2+ GBM receive chlorotoxin-directed CD28-CD3 zeta-CD19t-expressing CAR T cells via dual delivery on a 28-day cycle

(continued)

Table 5.2 (continued)

Institution/NCT#	Clinical trial and NCT#	Phase	Target	Study treatment/features
Beijing Sanbo Brain Hospital, Beijing, China NCT02937844	A safety and efficacy study of autologous chimeric switch receptor engineered T Cells redirected to PD-L1 in patients with recurrent glioblastoma	I	PD-L1	Patients with recurrent GBM receive lymphodepletion therapy consisting of 250 mg/m^2 of cyclophosphamide and 25 ^{mg}/m^2 of fludarabine on days 1–3 prior to intravenous infusion of PD-L1 CAR T cells. Infusion will occur in a standard 3 + 3 escalation approach in dosage ranges of 5 × 104/kg to 1 × 107/kg
UWELL Biopharma, NCT04717999	Pilot study of NKG2D CAR T in treating patients with recurrent glioblastoma	I	NKG2D	Patients with refractory/relapsed glioblastoma receive NKG2D CAR T cells via intracerebroventricular injection

5.4.2 Battling Tumor Heterogeneity

Tumor heterogeneity presents a formidable barrier in the application of CAR T cells for GBM. Challenges posed by tumor heterogeneity are not unique to GBM; however, numerous trials have observed antigen escape after CAR T cell use in GBM [109]. Epidermal growth factor receptor variant III (EGFRvIII), was found to be expansively rearranged and amplified in various parts of the tumor, including EGFRvIII-positive and negative GBM masses. EGFRvIII-positive cells were able to give rise to EGFRvIII-negative glioma cells as well [110]. In a genomic profile of glioblastoma, widespread tumor heterogeneity was observed, including the loss/partial loss of chromosome 10, polysomy of chromosome 7, focal deletion of the CDKN2A/B, and focal high-level amplification of EGFR [111]. Because CAR T cells are traditionally directed to one TAA, antigen heterogeneity may lead to antigen escape and GBM recurrence.

Investigators have developed numerous approaches to surmount GBM heterogeneity using combinatorial techniques that target multiple TAA using a single CAR construct. Bispecific CAR T cells engineered to co-express HER2 and IL-13Rα2 moieties successfully minimized antigen escape and enhanced T cell effector function for GBM—0.18% of input tumor cells survived the bispecific CAR T cells in vivo and improved tumor control with 80% of mice surviving greater than 160 days [112]. Tandem CAR T cells that link a HER2-binding scFv and an IL-13Rα2-binding IL-13 mutein more effectively mitigated antigen escape and improved in vivo survival in mouse models [113]. Trivalent CAR T cells constructed to target HER2, IL-13Rα2, and EphA2, were able to overcome interpatient GBM heterogeneity, eliminating 100% of tumor cells in patient-derived xenograft (PDX) models [114]. Finally, T cells can be engineered with biological circuits that enable CAR activation through recognition of multiple antigens [115]. These synNotch-CAR T cells are primed to express a CAR against EGFRvII only after recognizing the CNS-specific antigen myelin oligodendrocyte glycoprotein (MOS) and demonstrated precise and potent control of PDX models of glioblastoma [115].

The use of bispecific T cell engagers (BiTEs) has also shown clinical efficacy in targeting TAA in various solid tumors, including EGFR. BiTEs form a link between tumor cells and T cells to enhance cytotoxicity of tumors [116]. Wing and colleagues demonstrated efficacy of a combinatory approach of folate receptor alpha-specific CAR T cells with oncolytic adenoviruses encoding EGFR-targeting BiTEs [117]. This approach resulted in increased cytokine production, improved anti-tumor response, and overall survival in heterogeneous solid tumor murine models [117]. Furthermore, Choi et al. demonstrated that EGFRvIII-targeted CAR T cells modified to secrete EGFR-specific BiTEs were able to successfully redirect CAR T cells and activate bystander effector T cells in the GBM tumor microenvironment. Although EGFRvIII-targeted CAR T cells were unable to thoroughly treat glioblastoma tumors heterogenous for EGFRvIII expression, CAR T cells secreting BiTE molecules were able to effectively eliminate heterogeneous glioblastoma tumor cells in murine models after intracranial administration [118].

5.4.3 Enhancing CAR T Cell Trafficking

Another obstacle to effective therapy for solid tumors is effective CAR T cell trafficking to the site of tumor. Hence, numerous studies have investigated approaches to improve the migration of CAR T cells to the tumor. Brown and colleagues overcame this hurdle by injecting CAR T cells at the tumor site. Intracranial administration of IL-13Rα2-targeted CAR T cells to the resected tumor cavity and the cerebral ventricular system demonstrated no severe safety issues or > Grade 3 toxicities. Furthermore, local delivery of CAR T cells resulted in regression of all spinal and intracranial GBM tumors [119].

Additionally, the prospective promise of utilizing exosomes, nanovesicles excreted by cells into the extracellular space expressing uniform receptors, in concert with CAR T cells has been highlighted in recent literature. Because of their nanoscale size and ability to surpass the BBB, CAR exosomes may facilitate in targeting difficult-to-traffic tumor sites of glioblastomas [120, 121].

5.5 Concluding Remarks and the Path Forward

In conclusion, though implementing CAR T cell therapy for solid tumors has seen slowed development, recent clinical trials have highlighted the prospective promise for this therapy against GBM. Nevertheless, the path ahead poses various challenges for optimal CAR T cell use in GBM, including the immunosuppressive GBM microenvironment, rampant tumor heterogeneity, and inability to traffic T cells to antigen sites. Recent development of innovative solutions, including multi-specific or armored CAR T cells, have proven to be more effective in treating GBM. Though the Food and Drug Administration has not hitherto approved a

CAR T cell therapy for solid tumors, an expanding number of clinical trials (Table 5.2) continue to highlight the promising prospects of the therapy's use in GBM.

References

1. Park JH et al (2018) Long-term follow-up of CD19 CAR therapy in acute lymphoblastic leukemia. N Engl J Med 378:449–459
2. Maude SL et al (2018) Tisagenlecleucel in children and young adults with B-cell lymphoblastic leukemia. N Engl J Med 378:439–448
3. Brentjens RJ et al (2013) CD19-targeted T cells rapidly induce molecular remissions in adults with chemotherapy-refractory acute lymphoblastic leukemia. Sci Transl Med 5:177ra38–177ra38
4. Schuster SJ et al (2017) Chimeric antigen receptor T cells in refractory B-cell lymphomas. N Engl J Med 377:2545–2554
5. Neelapu SS et al (2017) Axicabtagene ciloleucel CAR T-cell therapy in refractory large B-cell lymphoma. N Engl J Med 377:2531–2544
6. Schuster SJ et al (2019) Tisagenlecleucel in adult relapsed or refractory diffuse large B-cell lymphoma. N Engl J Med 380:45–56
7. Wang M et al (2020) KTE-X19 CAR T-cell therapy in relapsed or refractory mantle-cell lymphoma. N Engl J Med 382:1331–1342
8. Raje N et al (2019) Anti-BCMA CAR T-cell therapy bb2121 in relapsed or refractory multiple myeloma. N Engl J Med 380:1726–1737
9. Sadelain M, Brentjens R, Rivière I (2013) The basic principles of chimeric antigen receptor design. Cancer Discov 3:388–398
10. Srivastava S, Riddell SR (2015) Engineering CAR-T cells: design concepts. Trends Immunol 36:494–502
11. Stoiber S et al (2019) Limitations in the design of chimeric antigen receptors for cancer therapy. Cells 8:472
12. Whilding LM, Maher J (2015) CAR T-cell immunotherapy: the path from the by-road to the freeway? Mol Oncol 9:1994–2018
13. Zhang C, Liu J, Zhong JF, Zhang X (2017) Engineering CAR-T cells. Biomark Res 5
14. Ostrom QT et al (2019) CBTRUS Statistical report: primary brain and other central nervous system tumors diagnosed in the United States in 2012–2016. Neuro Oncol 21:V1–V100
15. Silantyev AS et al (2019) Current and future trends on diagnosis and prognosis of glioblastoma: from molecular biology to proteomics. Cells 8
16. Hegi ME et al (2005) MGMT Gene silencing and benefit from temozolomide in glioblastoma. N Engl J Med 352:997–1003
17. Oldrini B et al (2020) MGMT genomic rearrangements contribute to chemotherapy resistance in gliomas. Nat Commun 11:1–10
18. Tamimi AF, Juweid M (2017) Epidemiology and outcome of glioblastoma. In: Glioblastoma, Codon Publications, pp 143–153. https://doi.org/10.15586/codon.glioblastoma.2017.ch8
19. Rafiq S, Hackett CS, Brentjens RJ (2020) Engineering strategies to overcome the current roadblocks in CAR T cell therapy. Nat Rev Clin Oncol 17:147–167
20. Theodorakis PE, Müller EA, Craster RV, Matar OK (2016) Physical insights into the blood-brain barrier translocation mechanisms
21. Pardridge WM (2005) The blood-brain barrier: bottleneck in brain drug development. NeuroRx 2:3–14
22. Dufresne RL (2002) Brain drug targeting: the future of brain drug development. Ann Pharmacother 36:733–734
23. Schreck KC, Grossman SA (2018) Role of temozolomide in the treatment of cancers involving the central nervous system. Oncol (Williston Park, N.Y.) 32
24. Bae SH et al (2014) Toxicity profile of temozolomide in the treatment of 300 malignant glioma patients in Korea. J Korean Med Sci 29:980–984

25. Sarkaria JN et al (2018) Is the blood-brain barrier really disrupted in all glioblastomas? a critical assessment of existing clinical data. Neuro Oncol 20:184–191
26. Agarwal S, Manchanda P, Vogelbaum MA, Ohlfest JR, Elmquist WF (2013) Function of the blood-brain barrier and restriction of drug delivery to invasive glioma cells: findings in an orthotopic rat xenograft model of glioma. Drug Metab Dispos 41:33–39
27. Mäkinen T (2019) Lymphatic vessels at the base of the mouse brain provide direct drainage to the periphery. Nat 572:34–35
28. Babbe H et al (2000) Clonal expansions of CD8+ T cells dominate the T cell infiltrate in active multiple sclerosis lesions as shown by micromanipulation and single cell polymerase chain reaction. J Exp Med 192:393–404
29. Jacobsen M et al (2002) Oligoclonal expansion of memory CD8+ T cells in cerebrospinal fluid from multiple sclerosis patients. Brain 125:538–550
30. Galea I et al (2007) An antigen-specific pathway for CD8 T cells across the blood-brain barrier. J Exp Med 204:2023–2030
31. Wilson EH, Weninger W, Hunter CA (2010) Trafficking of immune cells in the central nervous system. J Clin Invest 120:1368–1379
32. Santomasso BD et al (2018) Clinical and biological correlates of neurotoxicity associated with CAR T-cell therapy in patients with B-cell acute lymphoblastic leukemia. Cancer Discov 8:958–971
33. Abramson JS et al (2018) Updated safety and long term clinical outcomes in TRANSCEND NHL 001, pivotal trial of lisocabtagene maraleucel (JCAR017) in R/R aggressive NHL. J Clin Oncol 36:7505–7505
34. Gardner RA et al (2017) Intent-to-treat leukemia remission by CD19 CAR T cells of defined formulation and dose in children and young adults. Blood 129:3322–3331
35. Lee DW et al (2015) T cells expressing CD19 chimeric antigen receptors for acute lymphoblastic leukaemia in children and young adults: a phase 1 dose-escalation trial. Lancet 385:517–528
36. Norelli M et al (2018) Monocyte-derived IL-1 and IL-6 are differentially required for cytokine-release syndrome and neurotoxicity due to CAR T cells. Nat Med 24:739–748
37. Grupp SA et al (2013) Chimeric antigen receptor-modified T cells for acute lymphoid leukemia. N Engl J Med 368:1509–1518
38. Maude SL et al (2014) Chimeric antigen receptor T cells for sustained remissions in leukemia. N Engl J Med 371:1507–1517
39. Parker KR et al (2020) Single-cell analyses identify brain mural cells expressing CD19 as potential off-tumor targets for CAR-T immunotherapies. Cell 183:126-142.e17
40. Van Rooij N et al (2013) Tumor exome analysis reveals neoantigen-specific T-cell reactivity in an ipilimumab-responsive melanoma. J Clin Oncol 31
41. McGranahan T, Therkelsen KE, Ahmad S, Nagpal S (2019) Current state of immunotherapy for treatment of glioblastoma. Curr Treat Options Oncol 20
42. Schartner JM et al (2005) Impaired capacity for upregulation of MHC class II in tumor-associated microglia. Glia 51:279–285
43. Badie B, Bartley B, Schartner J (2002) Differential expression of MHC class II and B7 costimulatory molecules by microglia in rodent gliomas. J Neuroimmunol 133:39–45
44. Leone P et al (2013) MHC class I antigen processing and presenting machinery: organization, function, and defects in tumor cells. J Nat Cancer Inst 105:1172–1187
45. Zagzag D et al (2005) Downregulation of major histocompatibility complex antigens in invading glioma cells: stealth invasion of the brain. Lab Invest 85:328–341
46. Gomez GG, Kruse CA (2006) Mechanisms of malignant glioma immune resistance and sources of immunosuppression. Gene Ther Mol Biol 10:133–146
47. Chmielewski M, Hombach AA, Abken H (2013) Antigen-specific T-cell activation independently of the MHC: chimeric antigen receptor-redirected T cells. Front Immunol 4
48. Hombach A et al (2001) CD4+ T cells engrafted with a recombinant immunoreceptor efficiently lyse target cells in a MHC antigen- and fas-independent fashion. J Immunol 167:1090–1096

49. Yu S et al (2017) Chimeric antigen receptor T cells: a novel therapy for solid tumors. J Hematol Oncol 10
50. Charalambous C, Hofman FM, Chen TC (2005) Functional and phenotypic differences between glioblastoma multiforme-derived and normal human brain endothelial cells. J Neurosurg 102:699–705
51. Hida K et al (2004) Tumor-associated endothelial cells with cytogenetic abnormalities. Cancer Res 64:8249–8255
52. Munoz JL, Walker ND, Scotto KW, Rameshwar P (2015) Temozolomide competes for P-glycoprotein and contributes to chemoresistance in glioblastoma cells. Cancer Lett 367:69–75
53. Haseley A et al (2012) Extracellular matrix protein CCN1 limits oncolytic efficacy in glioma. Cancer Res 72:1353–1362
54. Yeh W-L, Lu D-Y, Liou H-C, Fu W-M (2012) A forward loop between glioma and microglia: glioma-derived extracellular matrix-activated microglia secrete IL-18 to enhance the migration of glioma cells. J Cell Physiol 227:558–568
55. Adams DJ, Morgan LR (2011) Tumor physiology and charge dynamics of anticancer drugs: implications for camptothecin-based drug development. Curr Med Chem 18:1367–1372
56. Joseph JV et al (2015) Hypoxia enhances migration and invasion in glioblastoma by promoting a mesenchymal shift mediated by the HIF1α-ZEB1 axis. Cancer Lett 359:107–116
57. Facciabene A et al (2011) Tumour hypoxia promotes tolerance and angiogenesis via CCL28 and T reg cells. Nature 475:226–230
58. Hjelmeland AB et al (2011) Acidic stress promotes a glioma stem cell phenotype. Cell Death Differ 18:829–840
59. Eyler CE, Rich JN (2008) Survival of the fittest: cancer stem cells in therapeutic resistance and angiogenesis. J Clin Oncol 26:2839–2845
60. Liu A, Hou C, Chen H, Zong X, Zong P (2016) Genetics and epigenetics of glioblastoma: applications and overall incidence of IDH1 mutation. Front Oncol 6:1
61. de la Iglesia N et al (2008) Identification of a PTEN-regulated STAT3 brain tumor suppressor pathway. Genes Dev 22:449–462
62. Piperi C, Papavassiliou KA, Papavassiliou AG (2019) Pivotal role of STAT3 in shaping glioblastoma immune microenvironment. Cells 8:1398
63. Ferguson SD, Srinivasan VM, Heimberger AB (2015) The role of STAT3 in tumor-mediated immune suppression. J Neuro-Oncol 123(3):385–394
64. Krawczyk CM et al (2010) Toll-like receptor–induced changes in glycolytic metabolism regulate dendritic cell activation. Blood 115:4742
65. Kortylewski M et al (2005) Inhibiting Stat3 signaling in the hematopoietic system elicits multicomponent antitumor immunity. Nat Med 11:1314–1321
66. Wing K, Sakaguchi S (2010) Regulatory T cells exert checks and balances on self tolerance and autoimmunity. Nat Immunol 11:7–13
67. Humphries W, Wei J, Sampson JH, Heimberger AB (2010) The role of tregs in glioma-mediated immunosuppression: potential target for intervention. Neurosurg Clin N Am 21:125–137
68. Zhang E, Gu J, Xu H (2018) Prospects for chimeric antigen receptor-modified T cell therapy for solid tumors. Mol Cancer 17
69. Mi Y et al (2020) The emerging role of myeloid-derived suppressor cells in the glioma immune suppressive microenvironment. Front Immunol 11:737
70. Liu Y, Wei J, Guo G, Zhou J (2015) Norepinephrine-induced myeloid-derived suppressor cells block T-cell responses via generation of reactive oxygen species. Immunopharmacol Immunotoxicol 37:359–365
71. O'Rourke DM et al (2017) A single dose of peripherally infused EGFRvIII-directed CAR T cells mediates antigen loss and induces adaptive resistance in patients with recurrent glioblastoma. Sci Transl Med 9
72. Goff SL et al (2019) Pilot trial of adoptive transfer of chimeric antigen receptor-Transduced t cells targeting EGFRVIII in patients with glioblastoma. J Immunother 42:126–135

73. Durgin JS et al (2021) Case report: prolonged survival following EGFRvIII CAR T cell treatment for recurrent glioblastoma. Front Oncol 11
74. Ahmed N et al (2015) Autologous HER2 CMV bispecific CAR T cells are safe and demonstrate clinical benefit for glioblastoma in a Phase I trial. J Immunother Cancer 3:4–8
75. Brown CE et al (2015) Bioactivity and safety of IL13Rα2-redirected chimeric antigen receptor CD8+ T cells in patients with recurrent glioblastoma. Clin Cancer Res 21:4062–4072
76. Lin Q et al (2021) First-in-human trial of EphA2-redirected CAR T-cells in patients With recurrent glioblastoma: a preliminary report of three cases at the starting dose. Front Oncol 11
77. Tang X et al (2021) Administration of B7-H3 targeted chimeric antigen receptor-T cells induce regression of glioblastoma. Signal Transduct Target Ther 6
78. Heimberger AB et al (2005) Prognostic effect of epidermal growth factor receptor and EGFRvIII in glioblastoma multiforme patients. Clin Cancer Res 11:1462–1466
79. Sampson JH et al (2014) EGFRvIII mCAR-modified T-cell therapy cures mice with established intracerebral glioma and generates host immunity against tumor-antigen loss. Clin Cancer Res 20:972–984
80. Suryadevara CM et al (2018) Temozolomide lymphodepletion enhances CAR abundance and correlates with antitumor efficacy against established glioblastoma. 7:1434464. https://doi.org/10.1080/2162402X.2018
81. Ravanpay AC et al (2019) EGFR806-CAR T cells selectively target a tumor-restricted EGFR epitope in glioblastoma. Oncotarget 10:7080–7095
82. Abbott RC et al (2021) Novel high-affinity EGFRvIII-specific chimeric antigen receptor T cells effectively eliminate human glioblastoma. Clin Transl Immunol 10
83. Johnson LA et al (2015) Rational development and characterization of humanized anti-EGFR variant III chimeric antigen receptor T cells for glioblastoma. Sci Transl Med 7:275ra22
84. Jian GZ et al (2007) Antigenic profiling of glioma cells to generate allogeneic vaccines or dendritic cell-based therapeutics. Clin Cancer Res 13:566–575
85. Ahmed N et al (2010) HER2-specific T cells target primary glioblastoma stem cells and induce regression of autologous experimental tumors. Clin Cancer Res 16:474–485
86. Ahmed N et al (2017) HER2-specific chimeric antigen receptor–modified virus-specific T cells for progressive glioblastoma: a phase 1 dose-escalation trial. JAMA Oncol 3:1094–1101
87. Jarboe JS, Johnson KR, Choi Y, Lonser RR, Park JK (2007) Expression of IL-13 receptor α2 in glioblastoma multiforme: implications for targeted therapies. Cancer Res 67:7983–7986
88. Pituch KC et al (2018) Adoptive transfer of IL13Rα2-specific chimeric antigen receptor T cells creates a pro-inflammatory environment in glioblastoma. Mol Ther. https://doi.org/10.1016/j.ymthe.2018.02.001
89. Brown CE et al (2012) Stem-like tumor-initiating cells isolated from IL13Rα2 expressing gliomas are targeted and killed by IL13-zetakine-redirected T cells. Clin Cancer Res 18:2199–2209
90. Wang D et al (2020) Chlorotoxin-directed CAR T cells for specific and effective targeting of glioblastoma. Sci Transl Med 12
91. Rousso-Noori L et al (2021) P32-specific CAR T cells with dual antitumor and antiangiogenic therapeutic potential in gliomas. Nat Commun 12
92. An Z et al (2021) Antitumor activity of the third generation EphA2 CAR-T cells against glioblastoma is associated with interferon gamma induced PD-L1. Oncoimmunology 10
93. Tang X et al (2019) B7-H3 as a novel CAR-T therapeutic target for glioblastoma. Mol Ther Oncolytics 14:279–287
94. Fujita M et al (2011) COX-2 blockade suppresses gliomagenesis by inhibiting myeloid-derived suppressor cells. Cancer Res 71:2664–2674
95. Yang M et al (2021) Dual effects of cyclooxygenase inhibitors in combination with CD19.CAR-T cell immunotherapy. Front Immunol 0:1895
96. Xia L et al (2021) BRD4 inhibition boosts the therapeutic effects of epidermal growth factor receptor-targeted chimeric antigen receptor T cells in glioblastoma. Mol Ther. https://doi.org/10.1016/J.YMTHE.2021.05.019

97. Uricoli B et al (2021) Engineered cytokines for cancer and autoimmune disease immunotherapy. Adv Healthc Mater 10
98. Jin J, Cheng J, Huang M, Luo H, Zhou J (2020) Fueling chimeric antigen receptor T cells with cytokines. Am J Cancer Res 10:4038
99. Evans AN, Lin HK, Hossian AKMN, Rafiq S (2021) Using Adoptive cellular therapy for localized protein secretion. Cancer J 27:159–167
100. Agliardi G et al (2021) Intratumoral IL-12 delivery empowers CAR-T cell immunotherapy in a pre-clinical model of glioblastoma. Nat Commun 12
101. Huang J et al (2021) IL-7-loaded oncolytic adenovirus improves CAR-T cell therapy for glioblastoma. Cancer Immunol Immunother 70:2453–2465
102. Li Y et al (2020) Arming anti-EGFRvIII CAR-T with TGFβ trap improves antitumor efficacy in glioma mouse models. Front Oncol 10
103. Peng W et al (2012) PD-1 blockade enhances T-cell migration to tumors by elevating IFN-γ inducible chemokines. Cancer Res 72:5209–5218
104. Bouffet E et al (2016) Immune checkpoint inhibition for hypermutant glioblastoma multiforme resulting from germline biallelic mismatch repair deficiency 34:2206–2211. https://doi.org/10.1200/JCO.2016.66.6552
105. Choi BD et al (2019) CRISPR-Cas9 disruption of PD-1 enhances activity of universal EGFRvIII CAR T cells in a preclinical model of human glioblastoma. J Immunother Cancer 7(1):1–8
106. Zhu H, You Y, Shen Z, Shi L (2020) EGFRvIII-CAR-T cells with PD-1 knockout have improved anti-glioma activity. Pathol Oncol Res 26:2135–2141
107. Adusumilli PS et al (2021) A phase I trial of regional mesothelin-targeted CAR T-cell therapy in patients with malignant pleural disease, in combination with the anti-PD-1 agent pembrolizumab. Cancer Discov Candisc 0407.2021. https://doi.org/10.1158/2159-8290.CD-21-0407
108. Chong EA et al (2021) Pembrolizumab for B-cell lymphomas relapsing after or refractory to CD19-directed CAR T-cell therapy. Blood. https://doi.org/10.1182/BLOOD.2021012634
109. Akhavan D et al (2019) CAR T cells for brain tumors: lessons learned and road ahead. Immunol Rev 290:60–84
110. Del Vecchio CA et al (2013) EGFRvIII gene rearrangement is an early event in glioblastoma tumorigenesis and expression defines a hierarchy modulated by epigenetic mechanisms. Oncogene 32:2670–2681
111. Aubry M et al (2015) From the core to beyond the margin: a genomic picture of glioblastoma intratumor heterogeneity. Oncotarget 6:12094–12109
112. Hegde M et al (2013) Combinational targeting offsets antigen escape and enhances effector functions of adoptively transferred T cells in glioblastoma. Mol Ther 21:2087–2101
113. Hegde M et al (2016) Tandem CAR T cells targeting HER2 and IL13Rα2 mitigate tumor antigen escape. J Clin Invest 126:3036–3052
114. Bielamowicz K et al (2018) Trivalent CAR T cells overcome interpatient antigenic variability in glioblastoma. Neuro Oncol 20:506–518
115. Choe JH et al (2021) SynNotch-CAR T cells overcome challenges of specificity, heterogeneity, and persistence in treating glioblastoma. Sci Transl Med 13
116. Huehls AM, Coupet TA, Sentman CL (2015) Bispecific T-cell engagers for cancer immunotherapy. Immunol Cell Biol 93:290–296
117. Wing A et al (2018) Improving CART-cell therapy of solid tumors with oncolytic virus–driven production of a bispecific T-cell engager. Cancer Immunol Res 6:605–616
118. Choi BD et al (2019) CAR-T cells secreting BiTEs circumvent antigen escape without detectable toxicity. Nat Biotechnol 37:1049–1058
119. Brown CE et al (2016) Regression of glioblastoma after chimeric antigen receptor T-cell therapy. N Engl J Med 375:2561–2569
120. Fu W et al (2019) CAR exosomes derived from effector CAR-T cells have potent antitumour effects and low toxicity. Nat Commun 10:1–12

121. Alvarez-Erviti L et al (2011) Delivery of siRNA to the mouse brain by systemic injection of targeted exosomes. Nat Biotechnol 29:341–345

Lag3: From Bench to Bedside

6

Francesca Aroldi, Reem Saleh, Insiya Jafferji, Carmelia Barreto, Chantal Saberian, and Mark R. Middleton

6.1 Introduction

The hypothesis of immune surveillance and the confirmation that immune system plays a role in controlling cancer growth has completely revolutionized the treatment of cancer and spawned extensive research into novel strategies to control cancer progression. The approval of the monoclonal antibody (mAb) targeting the inhibitory immune checkpoint cytotoxic T-lymphocyte-associated protein 4 (CTLA-4), Ipilimumab, for the treatment of melanoma, represented a great success of immunotherapy in the treatment of cancer. Shortly, this was followed by the agents targeting another inhibitory immune checkpoint, programmed cell

F. Aroldi (✉) · M. R. Middleton
Department of Oncology, The University of Oxford, OX 37LE Oxford, England
e-mail: dssafrancesca.aroldi@gmail.com

R. Saleh
Peter MacCallum Cancer Centre, Tumor Suppression and Cancer Sex Disparity Laboratory, Melbourne, VIC 3000, Australia

Department of Oncology, The University of Melbourne, The Sir Peter MacCallum Cancer Centre, Melbourne, VIC 3000, Australia

I. Jafferji
Department of Immunology, The University of Texas MD Anderson Cancer Centre, Houston, TX 77030, USA

C. Barreto
Investigational Cancer Therapeutics (A Phase I Program), The University of Texas MD Anderson Cancer Centre, Houston, TX 77030, USA

C. Saberian
Melanoma Medical Oncology, The University of Texas MD Anderson Cancer Centre, Houston, TX 77030, USA

death protein 1 (PD-1), and the approval of anti-PD-1 inhibitors, nivolumab, and pembrolizumab. Even though the response rate is high in tumor types with high infiltration of immune cells, it is often suboptimal in eradicating cancer cells resulting in primary and secondary resistance [1, 2]. Many other immunomodulatory pathways have been investigated in clinical trials. LAG-3 has recently emerged as a promising target as it is expressed on both CD4$^+$ and CD8$^+$ T cells and is linked to the suppressive function of regulatory T cells (Tregs). LAG-3 acts as a co-receptor on activated cytotoxic CD8$^+$ T cells, B cells, natural killer (NK) cells, Tr1 (CD4$^+$FOXP3$^+$ and IL-10-secreting type 1 regulatory T cells), dendritic cells, and activated or exhausted effector CD4$^+$ T cells. It has a high binding affinity to the major histocompatibility complex (MHC) Class II and leads to T cell inactivation or exhaustion [3–5] (Table 6.1, Fig. 6.1a). LAG-3 can also bind to a newly identified ligand, fibrinogen-like protein 1 (FGL-1) secreted by tumor cells, for example in hepatic carcinoma, and promote tumor progression and T cell inactivation by unknown mechanisms [6] (Fig. 6.1b). Based on that, the inhibition of LAG-3 is relevant and could have promising clinical benefits in treating several solid tumor types, particularly those that are dense with T cell infiltration.

6.2   Preclinical Evidence Supporting the Benefit of LAG-3 Inhibition

There is mounting evidence suggesting a promising therapeutic potential for LAG-3 inhibition in a wide range of tumor types ranging from melanoma, head, and neck, breast, colon, lung, pancreatic, ovarian, and renal tumors [7]. Increased expression levels of inhibitory immune checkpoints (ICs), such as LAG-3, CTLA-4, and PD-1, have been detected on dysfunctional T effector cells (cytotoxic CD8$^+$ T cells and CD4$^+$ T helper cells) and activated Tregs present in the tumor microenvironment [8, 9]. Upregulated levels of LAG-3 on Tregs can indirectly inhibit the activation of T effector (Teff) cells by negatively influencing the function of antigen-presenting cells (APCs), which is essential for Teff activation [10, 11]. LAG-3 on Tregs induces inhibitory signals upon its interaction with its ligands, leading to the inhibition of APC function, Teff activation, and proliferation within tumor sites [10, 11]. The overexpression of LAG-3 on Tregs can also positively regulate Treg stability/differentiation and enhance their suppressive activity [10]. Grosso et al. demonstrated that the combinatorial therapy incorporating anti-LAG-3 mAb and vaccination with tumor-associated antigen increased the number of activated CD8$^+$ T cells in the tumor and disrupted the tumor parenchyma in the tumor-tolerance model of prostate cancer [12]. A study of mouse head and neck cancer models showed that the administration of anti-LAG-3 mAb halts tumor growth and is associated with enhanced anti-tumor CD8$^+$ immune responses and reduced numbers of tumor-infiltrating Tregs [13]. Similarly, the efficacy of anti-LAG-3 mAb in inducing anti-tumor activity associated with reduced tumorigenesis was observed in a mouse model of fibrosarcoma [14].

Table 6.1 Most known immune checkpoints

Receptor	LAG-3	TIM-3	TIGIT	PD-1	CTLA-4
Synonyms	CD223	HAVCR2	WUCAM	PDCD1 and CD279	CD152
Gene location	Chromosome 12p13.32	Chromosome 5q33.3	Chromosome 3q13.31	Chromosome 2q37.3	Chromosome 2q33
Number of amino acids	498 amino acids	301 amino acids	244 amino acids	288 amino acids	223 amino acids
Signaling motif	KIEELE motif in the cytoplasmic tail	Tyrosine residues in the cytoplasmic tail	ITT-ITIM	ITSM	Cytoplasmic tail
Ligands	MHC II class, LSECtin, Galectin-3	Galectin-9, Ceacam-1, HMGB-1 and phosphatidylserine	PVR/CD155 and CD112	PD-L1 and PD-L2	CD80 and CD86
Cells expressing receptor	Activated CD4+ T cells, Tregs, Tr1 cells, activated CD8+ T cells, NK cells, DCs, B cells and exhausted effector T cell	Activated T cells, TH17 cells, Tregs, DCs, NK cells, monocytes, and exhausted effector T cell	Activated T cells, memory T cells, TFH cells (Follicular Helper T Cells), TRegs, Tr1 cells, NK cells, NKT cells and exhausted effector T cells	Activated T cells, Tregs, NK cells, macrophages, and exhausted effector T cells	Activated T cells, Tregs, exhausted effector T cells
Cells expressing ligand	APCs	APCs, tumor cells	APCs, fibroblasts, endothelial cells, and tumor cells	APCs, hematopoietic and non-hemato-poietic cells, and tumor cells	APCs

Adapted from Rotte et al. [54]

6.3   Preclinical Evidence Supporting the Benefit of Dual Inhibition of LAG-3 and Other Immune Checkpoints

Despite the therapeutic efficacy of the single blockade of LAG-3 in cancer models (as described above), there is a high potential for limited response rates and resistance development upon the administration of anti-LAG-3 mAb alone. It has been reported that LAG-3 acts in a synergistic manner with PD-1 to suppress anti-tumor immunity and autoimmunity [15]. The co-expression of PD-1 and LAG-3 on tumor-infiltrating CD8+ and CD4+ T cells has been shown in multiple cancer

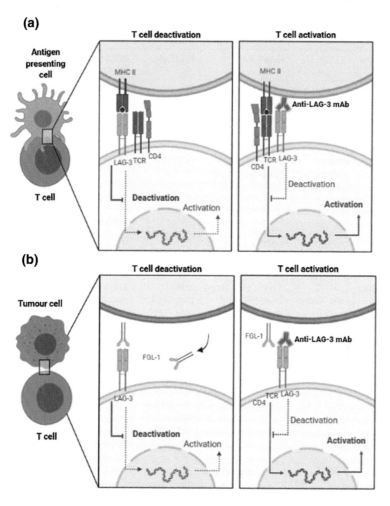

Fig. 6.1 Inhibition of T cell activation via LAG-3 signaling. Interaction between LAG-3 (on T cells) and the major histocompatibility complex class II-MHCII (on antigen-presenting cells) **a** or with its ligand secreted by tumor cells, fibrinogen-like protein 1 (FGL-1) **b** abrogates T cell activation and inhibits gene transcription associated with anti-tumor immunity

types, suggesting a potential cooperative role between the two immune checkpoint pathways in the induction of T cell dysfunction or exhaustion [16]. For example, Matsuzaki et al. reported that approximately 80% of LAG-3$^+$ and 50% of LAG-3$^-$ tumor-infiltrating CD8$^+$ T cells expressed PD-1 in patients with ovarian cancer [17]. The co-expression of LAG-3 and PD-1 on CD4$^+$ and CD8$^+$ T cells has also been demonstrated in tumors from mice with ovarian cancer and chronic lymphocytic leukemia [18]. These, and related observations, provide the basis for the investigation of the safety and therapeutic efficacy of dual inhibition of LAG-3 and PD-1 or PD-L1 in several cancer models [17, 19].

Matsuzaki et al. demonstrated *ex vivo* that the co-inhibition of LAG-3 and PD-1 enhanced the proliferation of tumor-infiltrating CD8$^+$ T cells and cytokine release following their stimulation with the tumor-associated antigen, NY-ESO-1 [17]. The genetic deletion of LAG-3 and PD-1 or their co-inhibition in mouse ovarian cancer model and chronic lymphocytic leukemia showed a synergistic effect causing a marked delay in tumor growth and enhancing anti-tumor T cell immune responses [18]. Although LAG-3 blockade is expected to activate tumor-specific CD4$^+$ and/or CD8$^+$ T cells in such studies, it may also increase the number of Tregs since LAG-3 blockade may have a significant impact on Treg numbers within the tumor microenvironment [20]. This is evident in findings by Goding et al. who demonstrated that the dual blockade of PD-L1 and LAG-3, via mAbs, in the B16 melanoma mouse model is sufficient to restore anti-tumor immunity [20].

More recently, Kraman et al. showed that the dual inhibition of LAG-3 and PD-1, via mAbs or a bispecific antibody, in mouse colon carcinoma models was more effective in the activation of anti-tumor immunity and reducing tumor progression than the single inhibition of either molecule [21]. Huang et al. showed that anti-CTLA-4 or anti-PD-1 mAb in a mouse ovarian cancer model resulted in the up-regulation of LAG-3 on CD8$^+$ T cells, suggesting a potential resistance mechanism [22]. In the same study, authors showed that the co-blockade of CTLA-4 and LAG-3 or PD-1 and LAG-3 had better therapeutic efficacy, associated with increased numbers of CD8$^+$ T cell and/or reduced numbers of Tregs in the tumor, compared to targeting CTLA-4 or PD-1 alone [22].

Other studies using mouse models of triple-negative breast cancer (TNBC), B16-F10 melanoma, and MC38 colon adenocarcinoma and fibrosarcoma (Sa1N) tumors demonstrated superior anti-tumor effects upon the co-administration of anti-LAG-3/anti-PD-1 mAbs, compared to the single administration of either treatment [19, 23]. Woo et al. showed that B16-F10, MC38 and Sa1N tumor-bearing mice lacking both LAG-3 and PD-1 had markedly reduced tumor weights and increased survival, compared to those lacking only one immune checkpoint [19]. However, it should be noted that minimal immune-adverse related events were observed in PD-1 and LAG-3 single knockout mice, while a lethal phenotype was observed in dual knockout mice associated with reduced self-tolerance and increased autoimmune immune cell infiltrates in multiple organs [19]. Collectively, these findings implicate the strong synergy between LAG-3 and PD-1 inhibitory pathways in self-tolerance and tumor reactivity and provide a solid basis for the combined use of these inhibitors in clinical studies, suggesting potential risks of adverse immune-related events upon the dual inhibition of LAG-3 and PD-1 in cancer patients.

6.4   Agents Targeting LAG-3 Used in Preclinical Models

Different drugs against the LAG-3 have been developed due to the rationale of targeting both LAG-3 and PD-1 and the strong preclinical data supporting the efficacy of dual blockade of the LAG-3 and PD-1 axes (Table 6.2).

Monoclonal antibodies (mAbs) against LAG-3 in preclinical studies have shown promising anti-tumor effects. For example, anti-LAG-3 mAb P13BO2-3 (IgG1) and LBL-007 (scFv-IgG4 fusion from human phage library) have both been shown to elicit reduced tumor growth in preclinical studies [24, 25]. In MC38 colon mouse cancer model, treatment with either anti-PD-1 mAb or LBL-007 resulted in a significant delay in tumor growth compared with control IgG treatment, and their combination was even more effective [26]. Additionally, REGN3767, a fully human IgG4 Ab targeting LAG-3, showed favorable pharmacokinetics in non-human primates, and when it was used in combination with anti-PD-1 mAb in mice with syngeneic colorectal carcinomas, REGN3767 enhanced the anti-tumor efficacy of anti-PD-1 mAb [27].

Interestingly, some anti-LAG-3 drugs were generated to be bispecific and target both LAG-3 and PD-1/PD-L1 axes, such as PRS-332 (LAG-3 × PD-1 bispecific fusion protein) and FS118 (LAG-3 × PD-1 tetravalent bispecific IgG1 antibody) [21]. The therapeutic efficacy of PRS-332 in preclinical settings has not yet been reported. However, FS118 has demonstrated simultaneous binding to LAG-3 and PD-L1 with a high affinity and better efficacy than the combined blockade of LAG-3 and PD-L1 using neutralizing mAbs and was associated with enhanced T cell activation [21]. Together, these findings supported the development of clinical trials to target LAG-3 alone or in combination with PD-1/PD-L1 in various cancer patients.

Alternatively, the use of a dimeric recombinant form of LAG-3, such as IMP321, was proven to induce a sustained immune response via the stimulation

Table 6.2 Summary of preclinical models targeting LAG-3 alone or in combination with other immune checkpoints

Preclinical summary of LAG-3 inhibitors					
Name	Company	Format	Tumor type	Combined with	References
PRS-332	Pieris Pharmaceuticals	LAG-3 × PD-1 bispecific protein	Solid		[28]
PI3B02-3	Agenus	IgG1	Solid	PD-1	[58]
LBL-007	Nanjing Leads Biolabs	scFv-IgG4 fusion	Solid	PD-1	[26]
IMP321	Immutep	Soluble LAG-3	Solid	PD-1	[30]
REGN3767	Regeneron	LAG-3 × PD-1 bispecific protein	Solid		[27]
FS118		LAG-3 × PD-L1 bispecific protein	Solid		[21]

of APCs, including dendritic cells [28, 29]. Additionally, it is currently being used in combination with other therapeutic approaches in clinical studies in patients with metastatic breast cancer, advanced pancreatic adenocarcinoma, and metastatic melanoma [26, 30–32].

6.5 Clinical Evidence Demonstrating the Efficacy of LAG-3 Targeting in Cancer

Several preclinical studies confirmed the immune-suppressive activity of LAG-3 and its involvement in modulating cancer immune response. Based on these results, antibodies targeting LAG-3 such as Relatlimab (BMS-986016), LAG525, INCAGN2385-101, REGN3767, MK-4280 Sym022, BI754111, TSR-033, and IMP321 are in clinical development in monotherapy or in combination and have been tested in different tumor types (Table 6.3). Most of these compounds are still in early stages of clinical development and strong data on anti-tumor activity are available only for Relatlimab in melanoma.

Recently, an interim analysis of RELATIVITY-047 trial, a phase III study that investigates the combination of nivolumab and relatlimab versus nivolumab in treatment of naive advanced melanoma, confirmed the superiority of the experimental arm over the standard treatment. Almost 6 months advantage in median progression-free survival (PFS) was reported with the addition of anti-LAG-3 therapy to nivolumab (10.1 months versus 4.6 months, respectively), HR 0.75 [95% CI, 0.6–0.9]; $P = 0.0055$). Interestingly, high expression of LAG-3 was correlated with a positive trend in PFS [33].

Although the PFS could be considered a surrogate outcome for overall survival (OS) in metastatic melanoma, combination with Ipilimumab-nivolumab reported increasing of OS and durable response in the same setting. Therefore, until data from RELATIVITY-047 trial are mature, the doublet Ipilimumab and nivolumb remains the recommended therapy, especially in patients with aggressive disease, brain, bone, and liver involvement [33–35]. However, with a toxicity rate of 18.9% the combination relatlimab and nivolumab is better tolerated than Ipilimumab plus Nivolumab.

Since a strong synergy between anti-LAG-3 therapy and other checkpoint inhibitors was observed in preclinical models, Novartis investigated the combination of LAG525, an anti-LAG3, and Spartalizumab, an anti-PD-1 in a phase I/II study. This study showed a durable response in 9.9% of patients with advanced solid tumors, including triple-negative breast cancer and mesothelioma (NCT02460224).

Conversely, IMP321, a soluble recombinant fusion protein with the Fc region of IgG and the extracellular region of LAG-3, showed an acceptable safety profile but limited efficacy (ORR was 48.3% with IMP321 versus 38.4% in the placebo group), despite the proven biological activity [32, 36, 37].

Following the enthusiasm for bispecific antibodies, some companies have recently produced compounds that simultaneously target LAG-3 and PD-1 like

Table 6.3 LAG-3 inhibitors current under investigation

Name	Therapy	Tumor type	Phase	Estimated number of patients	Trial number
Sym022	Monotherpy	Solid tumor or lymphoma	I	15	NCT03489369
Sym022	+ Anti-PD-1 or anti-PD-1 + anti-TIM-3	Solid tumor	I	200	NCT04641871
LAG525	+ anti-PD-1	TNBC	I	220	NCT03742349
REGN3767	± anti-PD-1	Solid tumor	I	669	NCT03005782
IMP321	+ gemcitabine	Advanced Pancreas	I	18	NCT00732082
IMP321	± Nivo	HNSCC	II	60	NCT04080804
IMP321	+ Pembrolizumab	NSCLC, HNSCC	II	183	NCT03625323
IMP321	+ weekly paclitaxel	HR$^+$ breast	I	24	NCT04252768
Relatlimab	Nivo ± relatlimab	Prior CT-RT Gastric, esophageal cancer	I	32	NCT03044613
Relatlimab	+ Nivo and IDO inhibitor versus Nivo and Ipi	Solid tumor	I/II	230	NCT03459222
Relatlimab	± Nivo	Solid tumor	I	45	NCT02966548
Relatlimab	Nivo versus Nivo + ipi versus Nivo + Relatlimab versus Nivo + Daratumumab	Solid tumor	I/II	584	NCT02488759
Relatlimab	Nivo + Relatlimab	HCC	I/II	20	NCT04658147
Relatlimab	Nivo + Relatlimab	HCC, liver cancer	II	250	NCT04567615
INCAGN02385	+ anti-PD-1 and anti-TIM-3	Melanoma	I/II	52	NCT04370704
TSR033	± anti-PD-1	Solid tumor	I	111	NCT03250832

TNBC: triple-negative breast cancer, Ipi: Ipilimumab, Nivo: Nivolumab, NSCLC: non-small cell lung cancer, HNSCC: head and neck squamous cell carcinoma, HCC: hepatocarcinoma

FS118, MGD013, IBI323, or CTLA-4 like XmAb22841 (Table 6.4). These drugs aimed to prevent the interaction between MHCII and LAG-3 but most of these trials are still recruiting, and the result is awaited.

Table 6.4 Bispecific antibodies targeting LAG-3 under investigation

Name trial number	Sponsor	Agent	Therapy	Tumor type	Study phase	Estimated n of pts
FS118 NCT03440437	F-star Delta Limited	PD-1-LAG-3	Monotherapy	Solid tumor HNSCC	I/II	80
IBI323 NCT04916119	Innovent Biologics	PD-1-LAG-3	Monotherapy	Solid tumor	I	332
RO7247669 NCT04140500	Hoffman-La Roche	PD-1-LAG-3	Monotherapy	Solid tumor, Melanoma NSCLC, Esophageal	I	320
MDG013 NCT03219268	MacroGenics	PD-1-LAG3	Monotherapy	Solid tumor	I	353
MDG013 NCT04082364	MacroGenics	PD-1-LAG-3	+ chemotherapy	Her2⁺ gastric, gastroesophageal	II/III	860
MDG013 NCT04178460	Zai Lab	PD-1-LAG3	+ niraparib	Gastric, TNBC, biliary, endometrial	I	164
XmAb22841 NCT03849469	Xencor	CTLA-4-LAG-3	+ Pembrolizumab	Solid tumor	I	242

TNBC: triple-negative breast cancer

However, FS118 reported encouraging clinical activity in heavily treated patients with advanced solid tumors. The phase I trial showed disease stabilization > 6 months in patients with secondary resistance. Retrospective analysis of LAG-3 and PD-L1 expression in the tumor was assessed as a potential biomarker of response, but still unpublished [38].

6.6   LAG-3 as a Prognostic Biomarker in Solid Tumors

A meta-analysis of fifteen studies showed that higher LAG-3 expression is associated with better OS in many solid tumors, especially in the early stages of disease [39]. However, this result is controversial because LAG- 3 is implicated in immune suppression and many studies have identified it as a negative prognostic factor and novel potential target for immunotherapy [13, 40–43].

A prospective analysis of 430 samples from patients with advanced solid tumors investigated LAG-3 expression and confirmed that LAG-3 is well represented across tumor types (33.3% of the LAG-3$^+$ were pancreatic cancer, 24.7% gastric, 23.6% colorectal, 12.5% melanoma, 9.5% genitourinary, 6.3% biliary tract, and 5.4% sarcoma) [42].

In immunotherapy naive patients, Edwards et al. identified a phenotype characterized by increased number of CD69$^+$CD103$^+$ CD8$^+$ T cells that correlated to better melanoma-specific survival. The authors suggested that this phenotype expresses high PD-1 and LAG-3, expands early during anti-PD-1 treatment, and may drive the response to anti-PD1 and anti-LAG-3 therapies [44, 45].

Based on the data from micro-RNA database: gene expression analyses, clinical-pathological features, and clinical outcomes, LAG-3 seems to have potential negative prognostic value in triple-negative breast cancer as LAG-3 expression was higher in grade III than grade I and II tumors [40].

In colorectal cancer, whether the LAG-3 expression is predominant in tumor or stroma plays an orchestral importance [41]. In a cohort of stage I-III colorectal adenocarcinoma, high expression of LAG-3 in tumor compartment was associated with poor prognosis, whereas high expression of LAG-3, TIM-3, and PD-1 in the stroma compartment was associated with improved survival [46].

In muscle-invasive bladder cancer (MIBC) the abundance of stromal LAG-3 cells resulted in immunoevasive path with dysfunctional CD8$^+$ T cells [47]. Recently, Kates et al. observed that increased Tregs and CD8$^+$ T cells expressing CTLA-4, TIM-3, and LAG-3 correlated to resistance to anti-PD-1 therapy [48]. This suggests that the blockade of these checkpoints should be explored for the treatment of urothelial carcinoma [48].

In a cohort of 38 patients with advanced platinum-resistant non-small cell lung cancer (NSCLC) LAG-3 appeared to be an independent positive prognostic factor for stage I-IIIB [49]. In this study, the co-expression of LAG-3 and PD-L1 was associated with increased OS and PFS after anti-PD-1 therapy [49]. In another study, LAG-3 was a predictive biomarker of response to PD-1inhibition [50].

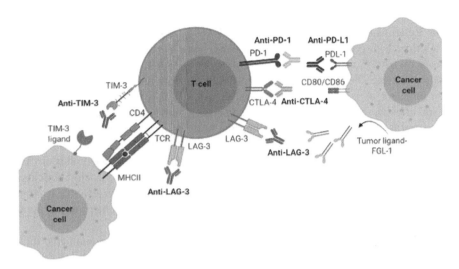

Fig. 6.2 Combined blockade of multiple immune checkpoints could maximize T cell activation and boost anti-tumor immunity. The application of monoclonal antibodies (mAbs) targeting multiple inhibitory immune checkpoint pathways, CTLA-4, PD-1/PD-L1 axis, TIM-3, and LAG-3 could alleviate tumor resistance to therapy against one target and enhance anti-tumor immune responses

In clear cell renal carcinoma (ccRCC), LAG-3 hypomethylation correlated to phenotype expressing CD3, CD8, CD45, and gamma-interferon which seems to be predictive of better outcome and response to immunotherapy [51].

Some data supported LAG-3 as negative prognostic factor and potential target for the treatment of bone and soft tissue sarcomas (STS). Indeed, in comparison with healthy volunteers, patients with STS reported higher CD4$^+$ and CD8$^+$ in peripheral blood, LAG-3 was independently associated with high grade and advanced pathological stages, and the expression of LAG-3 and PD-1 in TILs of STS was associated with poor prognosis [27, 52].

LAG-3 may play a role also in pediatric solid tumors, where the complex MHC-II/LAG-3 showed to regulate the immune response and might be investigated as target for future novel immunotherapies [53]. Additionally, targeting multiple immune checkpoints, such as CTLA-4, PD-1/PD-L1, TIM-3, and LAG-3 could be beneficial in maximizing the clinical outcomes of cancer patients, if the combined treatment was proven to be well-tolerated and safe (Fig. 6.2).

6.7   Conclusions and Future Directions

LAG-3 is a transmembrane protein involved in the inhibitory signaling of T cells but also fosters the activity of Tregs and it is expressed on NK cells, T cells, B cells, and TILs. LAG-3 expression could reflect the infiltration of T cells in tumors, similarly to PD-1 and CTLA-4. LAG-3 inhibition has reported promising results in preclinical studies, which can be translated into clinical development. Most of the trials are still in early phase, but encouraging results have been reported for Relatlimab in a phase III trial. Anti-LAG-3 therapy could have a synergistic effect with other immunotherapy agents, and clinical trials should explore the level of LAG-3 and PD-1 expression together with CTLA-4, TIM-3, as well as other immune checkpoints. They also should examine the potential for LAG-3 as an indicator of poor disease prognosis across various solid tumor types. Therefore, effective combination treatment could be tumor type-specific and dependent on several important factors, including the host immune response, molecular and cellular composition of the tumor microenvironment, resistance and immune suppression mechanisms, and the expression levels of immune checkpoints. In the future, this information could be obtained by next-generation sequencing, immunohistochemistry, and immune profiling and would provide a rational for appropriate immune checkpoint inhibitors [2, 54–56]. Ultimately, this may help in improving response rates to cancer immunotherapy.

Apart from this, safety profiles of combination therapies are another critical clinical aspect, which should be considered. The combination of TIM-3 and LAG-3 blockade along with anti-PD-1/PD-L1 therapy could be safer and well-tolerated, compared to the combination of anti-CTLA-4 and anti-PD-1/PD-L1 therapies which has been associated with immune-related adverse events [57]. In fact, in Relativity -047 trial, the toxicity rate of the combination anti-LAG3 and anti-PD1 was sufficiently low to be considered as potential backbone for triplet therapy.

References

1. Restifo NP, Smyth MJ, Snyder A (2016) Acquired resistance to immunotherapy and future challenges. Nat Rev Cancer 16(2):121–126
2. Saleh R, Elkord E (2020) Acquired resistance to cancer immunotherapy: role of tumor-mediated immunosuppression. Semin Cancer Biol 65:13–27
3. Lythgoe MP, Liu DSK, Annels NE, Krell J, Frampton AE (2021) Gene of the month: lymphocyte-activation gene 3 (LAG-3). J Clin Pathol 74(9):543–547
4. He Y, Rivard CJ, Rozeboom L, Yu H, Ellison K, Kowalewski A et al (2016) Lymphocyte-activation gene-3, an important immune checkpoint in cancer. Cancer Sci 107(9):1193–1197
5. Anderson AC, Joller N, Kuchroo VK (2016) Lag-3, Tim-3, and TIGIT: co-inhibitory receptors with specialized functions in immune regulation. Immunity 44(5):989–1004
6. Wang J, Sanmamed MF, Datar I, Su TT, Ji L, Sun J et al (2019) Fibrinogen-like protein 1 is a major immune inhibitory ligand of LAG-3. Cell 176(1–2):334–47.e12
7. Long L, Zhang X, Chen F, Pan Q, Phiphatwatchara P, Zeng Y et al (2018) The promising immune checkpoint LAG-3: from tumor microenvironment to cancer immunotherapy. Genes Cancer 9(5–6):176–189
8. Saleh R, Elkord E (2019) Treg-mediated acquired resistance to immune checkpoint inhibitors. Cancer Lett 457:168–179
9. Le Mercier I, Chen W, Lines JL, Day M, Li J, Sergent P et al (2014) VISTA regulates the development of protective antitumor immunity. Cancer Res 74(7):1933–1944
10. Camisaschi C, Casati C, Rini F, Perego M, De Filippo A, Triebel F et al (2010) LAG-3 expression defines a subset of CD4(+)CD25(high)Foxp3(+) regulatory T cells that are expanded at tumor sites. J Immunol 184(11):6545–6551
11. Workman CJ, Vignali DA (2003) The CD4-related molecule, LAG-3 (CD223), regulates the expansion of activated T cells. Eur J Immunol 33(4):970–979

12. Grosso JF, Kelleher CC, Harris TJ, Maris CH, Hipkiss EL, De Marzo A et al (2007) LAG-3 regulates CD8+ T cell accumulation and effector function in murine self- and tumor-tolerance systems. J Clin Invest 117(11):3383–3392
13. Deng WW, Mao L, Yu GT, Bu LL, Ma SR, Liu B et al. (2016) LAG-3 confers poor prognosis and its blockade reshapes antitumor response in head and neck squamous cell carcinoma. Oncoimmunology 5(11):e1239005
14. Que Y, Fang Z, Guan Y, Xiao W, Xu B, Zhao J et al (2019) LAG-3 expression on tumor-infiltrating T cells in soft tissue sarcoma correlates with poor survival. Cancer Biol Med 16(2):331–340
15. Maruhashi T, Sugiura D, Okazaki IM, Okazaki T (2020) LAG-3: from molecular functions to clinical applications. J Immunother Cancer 8(2)
16. Grosso JF, Goldberg MV, Getnet D, Bruno TC, Yen HR, Pyle KJ et al (2009) Functionally distinct LAG-3 and PD-1 subsets on activated and chronically stimulated CD8 T cells. J Immunol 182(11):6659–6669
17. Matsuzaki J, Gnjatic S, Mhawech-Fauceglia P, Beck A, Miller A, Tsuji T et al (2010) Tumor-infiltrating NY-ESO-1-specific CD8+ T cells are negatively regulated by LAG-3 and PD-1 in human ovarian cancer. Proc Natl Acad Sci U S A 107(17):7875–7880
18. Wierz M, Pierson S, Guyonnet L, Viry E, Lequeux A, Oudin A et al (2018) Dual PD1/LAG3 immune checkpoint blockade limits tumor development in a murine model of chronic lymphocytic leukemia. Blood 131(14):1617–1621
19. Woo SR, Turnis ME, Goldberg MV, Bankoti J, Selby M, Nirschl CJ et al (2012) Immune inhibitory molecules LAG-3 and PD-1 synergistically regulate T-cell function to promote tumoral immune escape. Cancer Res 72(4):917–927
20. Goding SR, Wilson KA, Xie Y, Harris KM, Baxi A, Akpinarli A et al (2013) Restoring immune function of tumor-specific CD4+ T cells during recurrence of melanoma. J Immunol 190(9):4899–4909
21. Kraman M, Faroudi M, Allen NL, Kmiecik K, Gliddon D, Seal C et al (2020) FS118, a bispecific antibody targeting LAG-3 and PD-L1, enhances T-Cell activation resulting in potent antitumor activity. Clin Cancer Res 26(13):3333–3344
22. Huang RY, Francois A, McGray AR, Miliotto A, Odunsi K (2017) Compensatory upregulation of PD-1, LAG-3, and CTLA-4 limits the efficacy of single-agent checkpoint blockade in metastatic ovarian cancer. Oncoimmunology 6(1):e1249561
23. Du H, Yi Z, Wang L, Li Z, Niu B, Ren G (2020) The co-expression characteristics of LAG3 and PD-1 on the T cells of patients with breast cancer reveal a new therapeutic strategy. Int Immunopharmacol 78:106113
24. Everett KL, Kraman M, Wollerton FPG, Zimarino C, Kmiecik K, Gaspar M et al (2019) Generation of Fcabs targeting human and murine LAG-3 as building blocks for novel bispecific antibody therapeutics. Methods 154:60–69
25. Lecocq Q, Keyaerts M, Devoogdt N, Breckpot K (2020) The next-generation immune checkpoint LAG-3 and its therapeutic potential in oncology: third time's a charm. Int J Mol Sci 22(1)
26. Yu X, Huang X, Chen X, Liu J, Wu C, Pu Q et al (2019) Characterization of a novel anti-human lymphocyte activation gene 3 (LAG-3) antibody for cancer immunotherapy. MAbs 11(6):1139–1148
27. Burova E, Hermann A, Dai J, Ullman E, Halasz G, Potocky T et al (2019) Preclinical development of the anti-LAG-3 antibody REGN3767: characterization and activity in combination with the anti-PD-1 antibody cemiplimab in human PD-1xLAG-3-knockin mice. Mol Cancer Ther 18(11):2051–2062
28. Avice MN, Sarfati M, Triebel F, Delespesse G, Demeure CE (1999) Lymphocyte activation gene-3, a MHC class II ligand expressed on activated T cells, stimulates TNF-alpha and IL-12 production by monocytes and dendritic cells. J Immunol 162(5):2748–2753
29. Buisson S, Triebel F (2003) MHC class II engagement by its ligand LAG-3 (CD223) leads to a distinct pattern of chemokine and chemokine receptor expression by human dendritic cells. Vaccine 21(9–10):862–868

30. Dirix L, Triebel F (2019) AIPAC: a phase IIb study of eftilagimod alpha (IMP321 or LAG-3Ig) added to weekly paclitaxel in patients with metastatic breast cancer. Future Oncol 15(17):1963–1973
31. Brignone C, Gutierrez M, Mefti F, Brain E, Jarcau R, Cvitkovic F et al (2010) First-line chemoimmunotherapy in metastatic breast carcinoma: combination of paclitaxel and IMP321 (LAG-3Ig) enhances immune responses and antitumor activity. J Transl Med 8:71
32. Legat A, Maby-El Hajjami H, Baumgaertner P, Cagnon L, Abed Maillard S, Geldhof C et al (2016) Vaccination with LAG-3Ig (IMP321) and peptides induces specific CD4 and CD8 T-cell responses in metastatic melanoma patients-report of a phase I/IIa clinical trial. Clin Cancer Res 22(6):1330–1340
33. Lipson EJ (ed) Relatlimab (RELA) plus nivolumab (NIVO) versus NIVO in first-line advanced melanoma: primary phase III results from RELATIVITY-047 (CA224–047). JCO
34. Flaherty KT, Hennig M, Lee SJ, Ascierto PA, Dummer R, Eggermont AM et al (2014) Surrogate endpoints for overall survival in metastatic melanoma: a meta-analysis of randomised controlled trials. Lancet Oncol 15(3):297–304
35. Larkin J, Chiarion-Sileni V, Gonzalez R, Grob JJ, Rutkowski P, Lao CD et al (2019) Five-year survival with combined nivolumab and ipilimumab in advanced melanoma. N Engl J Med 381(16):1535–1546
36. Romano E, Michielin O, Voelter V, Laurent J, Bichat H, Stravodimou A et al (2014) MART-1 peptide vaccination plus IMP321 (LAG-3Ig fusion protein) in patients receiving autologous PBMCs after lymphodepletion: results of a Phase I trial. J Transl Med 12:97
37. Wildiers (2021) Abstract PD14–08: primary efficacy results from AIPAC: a double-blinded, placebo controlled, randomized multinational phase IIb trial comparing weekly paclitaxel plus eftilagimod alpha (soluble LAG-3 protein) versus weekly paclitaxel plus placebo in HR-positive metastatic breast cancer patients
38. Yap T (ed) (2020) A first-in-human study of FS118, a tetravalent bispecific antibody targeting LAG-3 and PD-L1, in patients with advanced cancer and resistance to PD-(L)1 therapy. BMJ
39. Saleh RR, Peinado P, Fuentes-Antras J, Perez-Segura P, Pandiella A, Amir E et al (2019) Prognostic value of lymphocyte-activation gene 3 (LAG3) in cancer: a meta-analysis. Front Oncol 9:1040
40. Fang J, Chen F, Liu D, Gu F, Chen Z, Wang Y (2020) Prognostic value of immune checkpoint molecules in breast cancer. Biosci Rep 40(7)
41. Al-Badran SS, Grant L, Campo MV, Inthagard J, Pennel K, Quinn J et al (2021) Relationship between immune checkpoint proteins, tumour microenvironment characteristics, and prognosis in primary operable colorectal cancer. J Pathol Clin Res. 7(2):121–134
42. Lee SJ, Byeon SJ, Lee J, Park SH, Park JO, Park YS et al (2019) LAG3 in solid tumors as a potential novel immunotherapy target. J Immunother 42(8):279–283
43. Wu S, Shi X, Wang J, Wang X, Liu Y, Luo Y et al. (2021) Triple-negative breast cancer: intact mismatch repair and partial co-expression of PD-L1 and LAG-3. Front Immunol 12:561793
44. Edwards J, Wilmott JS, Madore J, Gide TN, Quek C, Tasker A et al (2018) CD103(+) Tumor-resident CD8(+) T cells are associated with improved survival in immunotherapy-naive melanoma patients and expand significantly during anti-PD-1 treatment. Clin Cancer Res 24(13):3036–3045
45. Wei SC, Levine JH, Cogdill AP, Zhao Y, Anang NAS, Andrews MC et al (2017) Distinct cellular mechanisms underlie anti-CTLA-4 and anti-PD-1 checkpoint blockade. Cell 170(6):1120–33.e17
46. Richards CH, Roxburgh CS, Powell AG, Foulis AK, Horgan PG, McMillan DC (2014) The clinical utility of the local inflammatory response in colorectal cancer. Eur J Cancer 50(2):309–319
47. Zeng H, Zhou Q, Wang Z, Zhang H, Liu Z, Huang Q et al. (2020) Stromal LAG-3(+) cells infiltration defines poor prognosis subtype muscle-invasive bladder cancer with immunoevasive contexture. J Immunother Cancer 8(1)
48. Kates M, Nirschl TR, Baras AS, Sopko NA, Hahn NM, Su X et al (2021) Combined next-generation sequencing and flow cytometry analysis for an anti-PD-L1 partial responder over

time: an exploration of mechanisms of PD-L1 activity and resistance in bladder cancer. Eur Urol Oncol 4(1):117–120

49. Hald SM, Rakaee M, Martinez I, Richardsen E, Al-Saad S, Paulsen EE et al. (2018) LAG-3 in non-small-cell lung cancer: expression in primary tumors and metastatic lymph nodes is associated with improved survival. Clin Lung Cancer 19(3):249–59 e2

50. Jung EH, Jang HR, Kim SH, Suh KJ, Kim YJ, Lee JH et al (2021) Tumor LAG-3 and NY-ESO-1 expression predict durable clinical benefits of immune checkpoint inhibitors in advanced non-small cell lung cancer. Thorac Cancer 12(5):619–630

51. Klumper N, Ralser DJ, Bawden EG, Landsberg J, Zarbl R, Kristiansen G et al. (2020) LAG3 (LAG-3, CD223) DNA methylation correlates with LAG3 expression by tumor and immune cells, immune cell infiltration, and overall survival in clear cell renal cell carcinoma. J Immunother Cancer 8(1)

52. Dancsok AR, Setsu N, Gao D, Blay JY, Thomas D, Maki RG et al (2019) Expression of lymphocyte immunoregulatory biomarkers in bone and soft-tissue sarcomas. Mod Pathol 32(12):1772–1785

53. Mochizuki K, Kawana S, Yamada S, Muramatsu M, Sano H, Kobayashi S et al (2019) Various checkpoint molecules, and tumor-infiltrating lymphocytes in common pediatric solid tumors: possibilities for novel immunotherapy. Pediatr Hematol Oncol 36(1):17–27

54. Rotte A, Jin JY, Lemaire V (2018) Mechanistic overview of immune checkpoints to support the rational design of their combinations in cancer immunotherapy. Ann Oncol 29(1):71–83

55. Saleh R, Sasidharan Nair V, Toor SM, Taha RZ, Murshed K, Al-Dhaheri M et al. (2020) Differential gene expression of tumor-infiltrating CD8(+) T cells in advanced versus early-stage colorectal cancer and identification of a gene signature of poor prognosis. J Immunother Cancer 8(2)

56. Saleh R, Toor SM, Khalaf S, Elkord E (2019) Breast cancer cells and PD-1/PD-L1 blockade upregulate the expression of PD-1, CTLA-4, TIM-3 and LAG-3 immune checkpoints in CD4(+) T Cells. Vaccines (Basel) 7(4)

57. Saleh R, Toor SM, Elkord E (2020) Targeting TIM-3 in solid tumors: innovations in the preclinical and translational realm and therapeutic potential. Expert Opin Ther Targets 24(12):1251–1262

58. The instant disclosure provides antibodies that specifically bind to LAG-3 (e.g., human LAG-3) and antagonize LAG-3 function (2018) Also provided are pharmaceutical compositions comprising these antibodies, nucleic acids encoding these antibodies, expression vectors and host cells for making these antibodies, and methods of treating a subject using these antibodies. https://usptoreport/patent/grant/10.844.119

Immunotherapy in Genitourinary Malignancy: Evolution in Revolution or Revolution in Evolution

7

Kevin Lu, Kun-Yuan Chiu, and Chen-Li Cheng

7.1 Background

Over the past years, advances in cancer immunotherapy, with the success of immune checkpoint inhibitors (ICI), have transformed the therapeutic landscapes of cancer management [1, 2]. Immunotherapy, the 5th pillar of cancer care after surgery, radiotherapy, cytotoxic chemotherapy, and precision therapy (molecular targeted therapy), is revolutionizing the standard of care in certain patients and tumor types, especially previously hard-to-treat malignancies [3–5]. Genitourinary malignancies are one of those benefiting from these new therapeutic options. Given modest clinical benefits of interleukin (IL)-2 or IFN-α for metastatic renal

K. Lu (✉) · K.-Y. Chiu · C.-L. Cheng
Division of Urology, Taichung Veterans General Hospital, 1650, Taiwan Boulevard Sect. 4, Taichung, Taiwan 40705
e-mail: kevinlu0620@mail2000.com.tw

K. Lu
Department of Surgery and Clinical Competences Center, Department of Medical Education, Taichung Veterans General Hospital, 1650, Taiwan Boulevard Sect. 4, Taichung, Taiwan 40705

K. Lu · K.-Y. Chiu
School of Medicine, National Yang-Ming Chiao Tung University, Taipei, Taiwan

K. Lu
Department of Urology, Kaohsiung Medical University Chung-Ho Memorial Hospital, Kaohsiung, Taiwan

School of Medicine, Kaohsiung Medical University, Kaohsiung, Taiwan

K.-Y. Chiu
Department of Applied Chemistry, National Chi Nan University, Nantou, Taiwan

C.-L. Cheng
Institute of Medicine, Chung Shan Medical University, Taichung, Taiwan

cell carcinoma (RCC) and unsatisfactory durable response of intravesical Bacillus Calmette-Guerin (BCG) therapy for early-stage non-muscle-invasive bladder cancers (NMIBC) in the past decades, the enthusiasm for exploration of novel therapeutics has established the important role of ICIs in the therapeutic arsenals of genitourinary malignancies [6–15]. Recently, large phase II and III clinical trials demonstrate meaningful survival benefits and durable clinical response with the use of checkpoint inhibitors-based therapies in RCC, UC, and some prostate cancer (PC) [16–24]. Despite best efforts and advances in understanding immunobiology, the benefits are limited to a minority of unselected patients, with the potential of leading to robust stratification processes. Now come the next hurdles: figuring out which patients best respond to ICI and which patients won't respond to ICI? How best to approach ICI therapies to extend/maximize the treatment response as long as possible? How to overcome therapeutic resistance by specific concurrent immunomodulators or targeted therapy or chemotherapy [25–34]? The role of ICI in combination or sequencing with chemotherapy or other targeted therapies or other immunomodulating therapeutics or radiotherapy in the early disease, neoadjuvant, adjuvant, and metastatic setting is actively under exploration [35–40]. In the "omics" era, deeper understanding of biology of genitourinary malignancy and translational immunology, as well as discovery of predictive and prognostic biomarkers may open up future avenues of clinical innovation and development for genitourinary cancer therapies and possibly point to next therapeutic era on the horizons (Fig. 7.1) [4, 5, 41–51].

7.2   Immunotherapy in Urothelial Carcinoma

The roots of cancer immunotherapy date back to 1891, when William B. Coley injected Coley's toxin, streptococcal organisms, into patients and observed the regression of malignant tumors, particularly sarcoma, but with limited clinical efficacy. Later in the 1960s, Thomas and Burnet put forward the theory of cancer immune surveillance. It took about one-half of a century for this theory to be accepted [33, 52]. In 1976, Morale et al. began with intravesical BCG instillation to treat NMIBC. Subsequently, National Cancer Institute (NCI) conducted controlled trials, based on Morale's work in 1978. Lamm et al. report the first successful controlled trial in 1980. Subsequently, Brosman, Netto, Martinez-Pineiro, and many other study groups report BCG to be superior to chemotherapy in 1982 and later. Subsequently, intravesical BCG instillation serves as the recommended treatment of choice for NMIBC. Since then, it has been more than 40 years since BCG was first used as an immunotherapy to treat NMIBC [6, 9]. BCG was internalized by healthy and malignant urothelial cancer cells, which stimulate an inflammatory response, release cytokines and chemokine, and trigger recruitment of immune cells into bladder wall to eradicate cancer cells. For those unresponsive or refractory to BCG, there is a high risk of local progression and distant metastasis. The standard of care for BCG-relapse or refractory patients with high-grade NMIBC is radical cystectomy. Those deemed medically unfit for cystectomy are treated

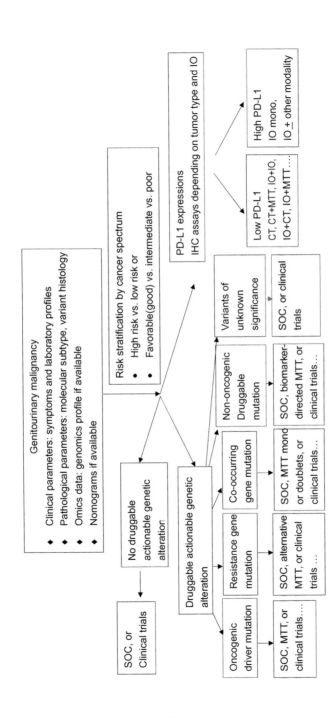

Fig. 7.1 Proposed algorithm of genitourinary malignancy treatment. IHC immunohistochemistry; IO immunotherapy; CT chemotherapy; MTT molecular targeted therapy; PD-L1 program cell death ligand-1; SOC standard of care

with salvage intravesical therapy. KEYNOTE-057 (NCT02625961), a multi-center, single-arm, phase 2 trial enrolled 148 patients with high-risk NMIBC receiving pembrolizumab 200 mg every three weeks, which was defined as persistent disease despite adequate BCG therapy, disease recurrence after an initial tumor-free state following adequate BCG therapy, or T1 disease following a single induction course of BCG; adequate BCG therapy was defined as administration of at least 5 of 6 doses of an initial induction course plus either at least 2 of 3 doses of maintenance therapy or 2 of 6 doses of a second induction course. Of 148 patients, 96 patients had BCG-unresponsive carcinoma in situ (CIS) with or without papillary tumors. Patients received pembrolizumab 200 mg every 3 weeks for 24 months until unacceptable toxicity, persistent or recurrent high-risk NMIBC, or progressive disease. The major efficacy outcome measures were complete response (as defined by negative results for cystoscopy (with transurethral resection of bladder tumor/biopsies as applicable), urine cytology, and computed tomography urography imaging), and duration of response. The complete response rate in the 96 patients with high-risk BCG-unresponsive NMIBC with CIS was 41% and median duration of response was 16.2 months. Forty-six percent of responding patients experienced a complete response lasting at least 12 months [53, 54]. On January 8, 2020, the Food and Drug Administration (FDA) approved pembrolizumab for the treatment of two distinct patient groups with BCG-unresponsive, high-risk, NMIBC with CIS with or without papillary tumors: those who are ineligible for surgery, and those who refuse surgery.

In patients with metastatic UC (mUC), median survival with standard cisplatin-based first-line chemotherapy, either Methotrexate/Vinblastine/Adriamycin/Cisplatin (MVAC) or Gemcitabine/Cisplatin (GC) is approximately 11–15 months. Despite high response rates (45–60%), cure rates remain low and responses are not durable. There are several independent prognostic factors to predict for shorter survival in patients with mUC who experience treatment failure with prior platinum-based therapy, such as low hemoglobin (<10 g/dl), Eastern Cooperative Oncology Group Performance Status (ECOG PS) ≥ 1, and the presence of liver metastasis. In patients with multiple adverse prognostic factors, the survival benefit of second-line treatment might be limited, and they may be spared therapies that can lead to unnecessary toxicity or be directed toward novel approaches [45, 55, 56]. In the past several years, five drugs-atezolizumab, pembrolizumab, avelumab, nivolumab, and durvalumab, were initially approved by FDA for use in patients with mUC and progression following platinum-based chemotherapy. However, two drugs have been voluntarily withdrawn. Durvalumab was granted accelerated approval in the United States in May 2017 based on promising tumor response rates and duration of response data observed in the phase 1/2 Study 1108 (NCT01693562) [57]. This study evaluated the safety and efficacy of durvalumab in advanced solid tumors, including previously treated bladder cancer. However, continued approval was dependent upon results from the phase 3 DANUBE trial (NCT02516241) in the first-line metastatic bladder cancer setting, which did not meet its primary endpoints in 2020 [58]. On February 22, 2021, AstraZeneca announced it has voluntarily withdrawn the indication for durvalumab in previously treated adult patients with locally advanced or metastatic bladder cancer in the United States, aligned with FDA guidance for evaluating indications with accelerated approvals that did not meet post-marketing requirements, as part of a broader industry-wide evaluation. However, this withdrawal does not affect indications outside the United States and does not impact other approved indications for durvalumab within or outside the United States. Another drug, atezolizumab, a monoclonal antibody designed to bind with the protein Program Death-Ligand 1 (PD-L1), was granted accelerated approval in 2016 for the treatment of prior platinum-treated mUC, based on the results of cohort 2 from the IMvigor210 study [59]. Continued approval for this indication was contingent upon the results of IMvigor211, the original post-marketing requirement for the prior platinum-treated mUC indication. This study did not meet its primary endpoint of overall survival (OS) in the PD-L1–high patient population (11.1 months vs. 10.6 months; Hazard ratio (HR) = 0.87; 95% Confidence Interval (CI): 0.63–1.21; $p = 0.41$) [60]. Subsequently, the FDA designated the IMvigor130 study as the post-marketing requirement, which will continue until the final analysis [61]. However, as the treatment landscape in second-line mUC has rapidly evolved with the emergence of new treatment options, Genentech, a member of the Roche Group, announced that the company voluntarily withdraw the US indication for atezolizumab in patients with prior platinum-treated mUC. However, there is one positive finding in second-line mUC setting. The multi-center, randomized, phase III KEYNOTE-045 trial (NCT02256436) compared pembrolizumab with chemotherapy (paclitaxel, docetaxel, or vinflunine) in 542 patients with advanced UC that recurred or progressed after platinum-based chemotherapy. OS was significantly longer in the pembrolizumab group compared to the chemotherapy group (10.1 months vs. 7.3 months; HR = 0.70; 95% CI: 0.57–0.85; $p = 0.00015$). However, progression-free survival (PFS) did not differ significantly between the two groups (2.1 months vs. 3.3 months; HR = 0.96; 95% CI: 0.79–1.16; $p = 0.31$). The objective response rate (ORR) was also higher with pembrolizumab (21% vs. 11%), and the median duration of response (DOR) was not reached with pembrolizumab versus 4.4 months with chemotherapy. In addition, pembrolizumab prolonged the time to deterioration in comparison to chemotherapy [62–64]. Among the remaining two drugs-avelumab and nivolumab, the multi-center, single-arm, phase II CheckMate 275 trial (NCT02387996) enrolled 270 patients with metastatic or unresectable locally advanced UC to evaluate the efficacy and safety of nivolumab. After a minimum follow-up of 34 months, ORR, median PFS, and median OS in all treated patients were 20.7%, 1.9 months, and 8.6 months, respectively [65, 66]. No phase III study is planned for nivolumab monotherapy in the second-line setting for mUC. A pooled analysis of two cohorts from the multi-center, phase Ib JAVELIN Solid Tumor trial (NCT01772004) assessed the efficacy and safety of avelumab in 249 patients with locally advanced or metastatic UC that had progressed after at least one previous platinum-based chemotherapy. After a median follow-up of 31.9 months, the confirmed ORR was 16.5% (95% CI: 12.1–21.8%; complete response in 4.1% and partial response in 12.4%). Median DOR was 20.5 months

(95% CI: 9.7 months to not estimable). Median PFS was 1.6 months (95% CI: 1.4–2.7 months) and the 12-month PFS rate was 16.8% (95% CI: 11.9–22.4%). Median OS was 7.0 months (95% CI: 5.9–8.5 months) and the 24-month OS rate was 20.1% (95% CI: 15.2%–25.4%) [67].

Though standard first-line therapy for mUC is cisplatin-based doublets or quadruplets, patient factors such as performance status, renal dysfunction, neuropathy, hearing impairment, or congestive heart failure make 30–50% of patients unsuitable for cisplatin-based therapies. ICI monotherapy or combination therapy might provide alternative options for cisplatin-ineligible or platinum-ineligible patients. In the first-line setting, there are two drugs—pembrolizumab and atezolizumab evaluated in phase II trials. In IMvigor 210, cohort 1 (NCT02108652), after median follow-up of 17 months, median OS was 16.3 months and ORR was 24% with 9% CR [68, 69]. The KEYNOTE-052 phase II trial (NCT02335424) evaluated 374 platinum-ineligible patients. The median OS was 11.3 months and ORR was 29% with 7% CR [69–72]. Based on those data, atezolizumab and pembrolizumab were both granted accelerated approval by FDA in 2017 for cisplatin-ineligible patients. Subsequently, phase III trials of immunotherapy and chemotherapy combination versus immunotherapy alone versus chemotherapy are underway. In 2018, FDA issued an alert that preliminary data analysis showed a decrease in survival for bladder cancer patients with low PD-L1 receiving mono-immunotherapy with pembrolizumab in KEYNOTE-361 (NCT02853305) or atezolizumab in IMvigor 130 (NCT02807636) versus chemotherapy as first-line therapy [73]. Further mature results are awaited. This year, in April 2021, the FDA's Oncologic Drugs Advisory Committee (ODAC) voted 5 to 3 to uphold the accelerated approval of pembrolizumab in the first-line treatment of patients with cisplatin-ineligible and carboplatin-ineligible locally advanced or metastatic urothelial carcinoma though KEYNOTE-361 trial did not meet its prespecified dual primary endpoints of overall survival or progression-free survival, versus standard-of-care chemotherapy. Unmet needs remain for appropriate patients newly diagnosed with certain types of advanced UC who are not eligible for platinum-containing chemotherapy. On August 31, 2021, the FDA granted a full approval to pembrolizumab for the treatment of patients with locally advanced or metastatic urothelial carcinoma who are not eligible for any platinum-containing chemotherapy.

Currently, the ESMO, EAU, and NCCN guidelines recommend cisplatin-backbone combination chemotherapy for cisplatin-eligible patients with advanced or metastatic UC. Now, we faced the clinical vignette whether we would monitor until progression and start second-line therapy at that point, or switch to an alternative strategy—maintenance therapy in patients achieving at least stable disease after first-line platinum-based chemotherapy. In the multi-center, phase III JAVELIN Bladder 100 trial (NCT02603432), 700 patients with locally advanced or metastatic UC who did not have disease progression during first-line chemotherapy (4–6 cycles of Gemcitabine plus Cisplatin or Carboplatin) were randomized to receive best supportive care with or without maintenance avelumab. The addition of maintenance avelumab to best supportive care significantly prolonged OS versus best supportive care alone both in the overall population and in the PD-L1 positive population (21.4 months vs. 14.3 months; HR = 0.69; 95% CI: 0.56–0.86; p = 0.001; not reached vs. 17.1 months, 95% CI: 0.40–0.79; p < 0.001, respectively) [74]. On June 30, 2020, FDA-approved avelumab for maintenance treatment of patients with locally advanced or mUC that has not progressed with first-line platinum-containing chemotherapy (Fig. 7.2).

Even though immune checkpoint inhibitors have superseded chemotherapy in platinum-refractory UC, there are many roadblocks ahead. The efficacy is often unpredictable and limited (20–40% with checkpoint blockers). The heterogeneity of tumors might impede efficacy. Primary or adaptive resistance developed over time after treatment. Difficulty to identify clinically significant biomarkers, lack of reliable biomarkers, and uncertainty of path forward for combination therapy remain challenging. Financial toxicity is always an issue needed to be resolved.

7.3 Immunotherapy in Renal Cell Carcinoma

Historically, RCCs are resistant to chemotherapy and radiotherapy [75]. The clinical observation of spontaneous regression of metastases after surgical removal or irradiation of primary RCC—"abscopal effect" and infiltration of immune cell into primary tumor offer insights into RCC is an immunogenic cancer and a potential target for immunotherapy. In 1976, the discovery of IL-2, naturally occurring cytokine with the ability to expand and differentiate T-cell population with anti-tumor activity, revolutionized the field of cancer immunotherapy. In 1992, US Food and Drug Administration (FDA) approved high-dose IL-2 for the treatment of locally advanced or metastatic RCC (mRCC), based on the results of seven phase II clinical trials with ORR of 14%, including 5% complete response (CR) [13, 76, 77]. However, the survival benefits of HD-IL2 were restricted in selected fit patients. Side effects with HD-IL2 are extremely common and severe. Toxicity and restricted accessibility of HD-IL2, which needs to be administered in specialized centers trained to manage its adverse events, make it a poor standard therapy. Subsequent approval of antiangiogenic molecularly targeted therapies that inhibit the vascular endothelial growth factors (VEGF) and proliferation signaling pathway-mammalian target of rapamycin (mTOR) swiftly changed the treatment landscape of advanced RCC. Although antiangiogenic tyrosine kinase inhibitors targeted VEGF (VEGF-TKI) have improved the response rate (RR) and OS for mRCC, and have been the cornerstone of treatment in mRCC, the majority of patients eventually experience disease progression over time upon these therapies. Thus, novel treatment approaches, including immunotherapeutic strategies, are warranted to further improve survival in mRCC.

Nivolumab, a programmed death 1 (PD-1) checkpoint inhibitor, was approved by the FDA in 2015 for advanced RCC patients who had been treated previously with anti-angiogenic therapies, based on the result of the randomized, phase III CheckMate-025 trial (NCT01668784). Nivolumab significantly improved OS compared with everolimus (mTOR inhibitor) (median OS 25.0 vs. 19.6 months; HR

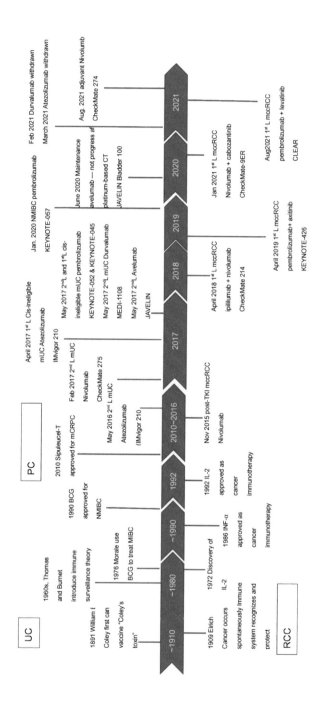

Fig. 7.2 Timeline of FDA-approved immunotherapy in genitourinary malignancy. BCG Bacillus Calmette-Guerin; NMIBC non-muscle invasive bladder cancer; TKI tyrosine kinase inhibitor; ccRCC clear cell Renal cell carcinoma; PC prostate cancer; UC urothelial carcinoma

= 0.73, 95% CI: 0.57–0.93, p = 0.002). The ORR was statistically superior with nivolumab in comparison with everolimus (25 vs. 5%; odds ratio (OR): 5.98, p < 0.001). Importantly, four patients (1%) achieved CR in nivolumab arm. Nivolumab had a much better overall safety profile compared with everolimus (Grade ≥ 3 adverse events 19% vs. 37%) [78–80]. Later, ICI has been quickly incorporated into treatment strategies for mRCC and moved to frontline therapies.

Multiple strategies, including combinations of nivolumab and ipilimumab or combinations of ICI and VEGF-TKI or ICI and hypoxia-inducible factor-2α (HIF-2α) inhibitors-belzutifan in management of patients with previously untreated mRCC have been applied, based on risk stratification of Memorial Sloan Kettering Cancer Center (MSKCC) or International Metastatic Renal Cell Carcinoma Database Consortium (IMDC) prognostic factor criteria. In phase III Check-Mate 214 trial (NCT02231749), 1096 patients with previously untreated clear cell mRCC were randomized to receive sunitinib or immunotherapy combination (ipilimumab and nivolumab) every three weeks for four cycles, followed by maintenance monotherapy nivolumab every two weeks. At initial interim analysis, there were significant OS benefits in the intermediate and poor-risk group in nivolumab-ipilimumab combination compared with sunitinib arm (HR: 0.63, 99.8% CI: 0.44–0.89). The PFS was numerically greater in nivolumab-ipilimumab combination but did not reach statistical significance (11.6 months vs. 8.4 months; HR: 0.82, 99.1% CI: 0.64–1.05). The ORR was significantly improved in nivolumab-ipilimumab combination compared with sunitinib arm (42% vs. 27%, p < 0.001), with CR of 9% in nivolumab-ipilimumab combination versus 1% in sunitinib. An updated 30-month follow-up, 11% CR was observed in nivolumab-ipilimumab combination. However, toxicity of nivolumab-ipilimumab combination was notable, with 22% discontinuing therapy because of toxicity [81–84]. Despite that toxicity, quality of life data demonstrated significant difference in favor of nivolumab-ipilimumab combination. Given the risk of serious immune-related adverse events by administering nivolumab-ipilimumab combination, shared-decision making with close surveillance is essential.

Because of angiogenesis-driven nature of mRCC, combing ICI with agents targeting angiogenesis axis was assessed in the several trials. The combination of anti-VEGF monoclonal antibody bevacizumab with anti-PD-L1 atezolizumab compared with sunitinib was conducted in the first-line setting in IMmotion 151 trial (NCT02420821). Preliminary report revealed PFS was superior in the combination arm versus sunitinib arm (11.2 months vs. 7.7 months; HR: 0.74; 95% CI: 0.57–0.96, p = 0.02), but median OS was not reached in either arm in intention-to-treat (ITT) (HR: 0.81; 95% CI: 0.63–1.03, p = 0.09). Mature OS is awaiting [85, 86]. The phase III JAVELIN Renal 101 trial (NCT02684006) compared avelumab plus axitinib (VEGHR inhibitors) with sunitinib (VEGRR inhibitors) in 886 patients with treatment-naïve mRCC. In PD-L1-positive cohort, median PFS was 13.8 months in avelumab and axitinib combination compared with 7.2 months in sunitinib (HR:0.61; 95% CI: 0.47–0.79, p < 0.001); death from any cause was 13.7% in avelumab and axitinib combination compared with 15.2% in sunitinib (stratified HR: 0.82; 95% CI = 0.53–1.28, p = 0.38). In

ITT cohort, median PFS was 13.8 months in avelumab and axitinib combination compared with 8.4 months in sunitinib (HR:0.69; 95% CI: 0.56–0.84, $p <$ 0.001). The OS data are still immature [87–90]. Three exciting positive trials were recently published for combination of immunotherapy with molecularly targeted therapy and switching current treatment paradigms. The phase III KEYNOTE-426 trial (NCT02853331) compared pembrolizumab plus axitinib with sunitinib in 861 patients with previously untreated advanced clear cell renal cell carcinoma (ccRCC). The dual primary endpoints were OS and PFS in the ITT population. Overall, 31%, 56% and 13% of patients were classified as IMDC favorable risk, intermediate risk, and poor risk, respectively. At an updated follow-up of 43 months, a statistically significant improvement in OS was observed in pembrolizumab plus axitinib combination in comparison with sunitinib (46 months vs. 40 months; HR: 0.73; 95% CI: 0.66–0.88, $p < 0.001$). The median PFS and DOR was significantly longer in pembrolizumab plus axitinib combination (16 months vs. 11 months; HR: 0.68; 95% CI: 0.58–0.80, $p < 0.001$; 24 months vs. 15 months, respectively). The ORR was better in pembrolizumab plus axitinib combination compared with sunitinib arm (60% vs. 40%, $p < 0.001$), with CR of 10% in pembrolizumab plus axitinib combination versus 4% in sunitinib. However, 10.7% of patients had to discontinue therapy in pembrolizumab plus axitinib combination because of toxicity. Further, there are higher than expected Grade 3 or more hepatic transaminitis in pembrolizumab plus axitinib combination [91, 92]. The reason for higher hepatic transaminitis is, to date, unknown. The exact causes for higher risk hepatic transaminitis in pembrolizumab plus axitinib combination were needed to be clarified for optimizing usage of pembrolizumab plus axitinib combination in clinical context. Despite these adverse events, in 2019, the FDA-approved pembrolizumab plus axitinib for the first-line treatment of patients with advanced RCC in terms of survival benefit (OS and PFS) of KEYNOTE-426 (NCT02853331). The phase III CheckMate 9ER trial (NCT03141177) randomized 651 patients with previously untreated advanced RCC to nivolumab plus cabozantinib (a tyrosine kinase inhibitors inhibit MET (hepatocyte growth factor receptor protein) and VEGFR, RET, GAS6 receptor (AXL), KIT, and Fms-like tyrosine kinase-3) or sunitinib. Overall, 22%, 58%, and 20% of patients were classified as IMDC favorable risk, intermediate risk, and poor risk, respectively. At median follow-up of 18.1 months, a statistically significant improvement in OS was observed in nivolumab plus cabozantinib combination in comparison with sunitinib (not reach vs. not reach; HR: 0.60; 95% CI: 0.40–0.89, $p = 0.001$). The median PFS and DOR were significantly greater in nivolumab plus cabozantinib combination (17 months vs. 8 months; HR: 0.51; 95% CI: 0.41–0.64, $p < 0.001$; 20 months vs. 12 months, respectively). The ORR was much better in nivolumab plus cabozantinib combination compared with sunitinib arm (56% vs. 27%, $p < 0.001$), with CR of 8% in nivolumab plus cabozantinib combination versus 5% in sunitinib. Efficacy benefits were consistent across all risk groups in nivolumab plus cabozantinib combination. Although adverse events of any cause of Grade 3 or higher occurred in nivolumab plus cabozantinib combination with 19.7% patients discontinuing at least one of the trial drugs owing to toxicity, patients reported better health-related quality of life in nivolumab plus cabozantinib combination arm than in sunitinib arm [93, 94]. Based on these positive findings of CheckMate-9ER (NCT03141177), the FDA approved the combination of nivolumab and cabozantinib as first-line treatment for patients with advanced RCC on January 22, 2021. In the third positive trial, phase III CLEAR trial (Study 307/KEYNOTE-581; NCT02811861), 1069 patients with treatment-naïve advanced RCC were randomized to lenvatinib (multi-kinase inhibitors inhibit vascular endothelial growth factor receptors VEGFR1, 2, and 3, as well as fibroblast growth factor receptors (FGFR) 1, 2, 3 and 4, platelet-derived growth factor receptor (PDGFR) alpha, c-Kit, and the RET proto-oncogene) plus pembrolizumab, lenvatinib plus everolimus or sunitinib. Overall, 33%, 56%, and 11% of patients were classified as IMDC favorable risk, intermediate risk, and poor risk, respectively. After a median follow-up of 26.6 months, median PFS was longer in lenvatinib plus pembrolizumab than in sunitinib (23.9 months vs. 9.2 months; HR: 0.39; 95% CI: 0.32–0.49; $P < 0.001$) and was longer in lenvatinib plus everolimus than in sunitinib (14.7 months vs. 9.2 months; HR:0.65; 95% CI, 0.53 0.80; $P < 0.001$). Overall survival was longer in lenvatinib plus pembrolizumab than with sunitinib (HR: 0.66; 95% CI: 0.49–0.88; $P = 0.005$) but was not longer in lenvatinib plus everolimus in with sunitinib (HR: 1.15; 95% CI: 0.88–1.50; $P = 0.30$). The ORR was much better in lenvatinib plus pembrolizumab combination and lenvatinib plus everolimus compared with sunitinib arm (71% vs. 54% vs. 36%), with CR of 16% in lenvatinib plus pembrolizumab combination, 10% in lenvatinib plus everolimus versus 4% in sunitinib. Grade 3 or higher adverse events emerged or worsened during treatment in 82.4% of the patients in lenvatinib plus pembrolizumab, with discontinuation of lenvatinib, pembrolizumab, or both drugs in 37.2% of patients (lenvatinib, 25.6%; pembrolizumab, 28.7%; both drugs, 13.4%); dose reduction of lenvatinib in 68.8% of patients; and interruption of lenvatinib, pembrolizumab, or both drugs in 78.4% of patients [95, 96]. Subsequently, on August 10, 2021, FDA-approved the combination of lenvatinib plus pembrolizumab for first-line treatment of adult patients with advanced renal cell carcinoma, based on the positive results of CLEAR (Study 307/KEYNOTE-581; NCT02811861) (Table 7.1).

Given promising results from ongoing and completed trials, many issues remain. How do we build on nivolumab and ipilimumab from CheckMate 214, pembrolizumab plus axitinib from KEYNOTE-426, nivolumab plus cabozantinib from CheckMate 9ER, and lenvatinib plus pembrolizumab from CLEAR? As clinical practice changes, how will we measure our ongoing phase 3 trials against older control—sunitinib, past standard of care? Are all anti-PD-1 and anti-PD-L1 inhibitors equivalent? How do we increase CRs—reaching flat tail of survival curve? Do we stop therapy from CRs or near-CRs and When? How about duration of therapy? When should we stop in responders?

Could we treat beyond progression? What is optimal schedule of treatment (q2wk vs. q3wk vs. q4wk)? Is flexible-dose based on body surface area or body-weight better than fixed flat dose? How do we manage the toxicity with these combinations to increase adherence or patient compliance? What biomarkers are

Table 7.1 Comparison among trials of upfront therapy for patients with metastatic clear cell renal cell carcinoma

Characteristics	CheckMate 214		KEYNOTE-426		CheckMate-9ER		CLEAR	
	Ipilimumab + Nivolumab N = 550	Sunitinib N = 546	Pembrolizumab + axitinib N = 432	Sunitinib N = 429	Nivolumab + cabozantinib N = 323	Sunitinib N = 328	Pembrolizumab + lenvatinib N = 355	Sunitinib N = 357
Control arm	Sunitinib		Sunitinib		Sunitinib		Sunitinib	
Primary endpoint	PFS, OS, and ORR		PFS and OS		PFS		PFS	
Prior nephrectomy	82%		83%		69%		74%	
IMDC risk								
Favorable	23		31		23		33	
Intermediate	61		56		58		56	
Poor	17		13		19		11	
ORR (CR)%	42 (9)	27 (1)	60 (10)	40 (4)	56 (8)	27 (5)	71 (16)	36 (4)
PFS	11.6	8.4	16	11	17	8	23.9	9.2
HR: 95% CI	HR: 0.82 99.1% CI: 0.64–1.05		HR: 0.68 95% CI: 0.58–0.80, $p < 0.001$		HR: 0.51 95% CI: 0.41–0.64, $p < 0.001$		HR: 0.39 95% CI: 0.32–0.49, $P < 0.001$	
OS	NR	38.4	46	40	NR	NR	NR	NR
HR; 95% CI	HR:0.63 99.8% CI: 0.44–0.89		HR: 0.73 95% CI: 0.66–0.88, $p < 0.001$		HR: 0.60 95% CI: 0.40–0.89, $p = 0.001$		HR: 0.66 95% CI: 0.49–0.88, $P = 0.005$	

CI confidence interval; *CR* complete response; *HR* hazard ratio; *IMDC* International Metastatic Renal Cell Carcinoma Database Consortium; *NR* not reach; *ORR* objective response rate; *OS* overall survival; *PFS* progression-free survival there which will improve our outcomes and select patients for immunotherapy combination and immunotherapy and VEGF-TKI combinations?

7.4   Immunotherapy in Prostate Cancer

Since the discovery of androgen deprivation therapy (ADT) by Dr. Charles Huggins in 1941, ADT has been the cornerstone of treatment for advanced prostate cancer. Despite initial success of ADT, most men eventually stop responses and progress to castration resistance within 1–3 years. The advent of intensified upfront therapy either with docetaxel or novel androgen receptor-axis-targeted therapies has heralded a new era after several decades of a "stagnant landscape". Critical questions regarding how best to customize their delivery and/or incorporate novel therapeutics to maximize clinical benefits are yet unanswered. Further, there are unmet clinical needs for therapeutic strategies, which can transform metastatic castration-resistant prostate cancer (mCRPC) from lethal disease to chronic disease, even cure. Could immunotherapy tsunami change the journey of mCRPC?

Traditionally, prostate cancer has been considered to be an immune desert. Unlike other solid tumors, prostate cancer has never shown a strong immune infiltrate within the tumor. Also, prostate tumors present metabolically hostile tumor microenvironment with increased glycolysis, which suppresses T-cell function. Meanwhile, prostate cancer has a low tumor mutation burden, which results in low neoantigen expression compared with other tumor types. As a result, immunotherapy for prostate cancer might be less effective. Despite low somatic alteration burden, prostate cancer expresses higher number of DNA damage and repair gene defects. Mutations in DNA damage and repair genes especially in members of the homologous recombination repair pathway both somatic and germline might make prostate cancer immune sensitive. Further, it has been observed that ADT sensitizes prostate cancer to patient's cell-mediated immune response. Immunotherapy might offer therapeutic potential for selected metastatic prostate cancer patients [22, 97–102].

In a double-blind, placebo-controlled, multi-center, phase III IMPACT trial (NCT00065442), 512 patients with asymptomatic or minimally symptomatic mCRPC were randomized to receive placebo or sipuleucel-T (intravenously every two weeks for three cycles). At a median follow-up of 36.5 months, sipuleucel-T prolongs median survival by 4.1 months compared with results in those treated with placebo (25.8 months vs. 21.7 months; HR = 0.759; $p = 0.017$) [102–106]. In 2010, FDA-approved sipuleucel-T, a dendritic cell vaccine activated using prostate acid phosphatase and granulocyte colony-stimulating factor, for asymptomatic or minimally symptomatic mCRPC. However, due to high costs, logistics of infusion, and unavailability of sipuleucel-T outside the US, its use remains limited. Recently, checkpoint inhibitors have been investigated in several clinical trials. In two phase III trials—CA184-043 and CA184-095 in mCRPC, despite long-lasting

CR after ipilimumab in some mCRPC patients, ipilimumab showed no significant survival improvement compared with placebo [107, 108]. Currently, there are no ICI approved for PC except pembrolizumab granted accelerated approval for adult and pediatric patients with unresectable or metastatic, microsatellite instability-high (MSI-H) or mismatch repair deficient (dMMR) solid tumors that have progressed following prior treatment and who have no satisfactory alternative treatment options by FDA on May 23, 2017.

For prostate cancer, based on current evidence of sipuleucel-T and checkpoint inhibitors, prostate cancer is an immunologically recognized disease, not totally "immune cold" or "immune desert." The questions on whether mCRPC patients could benefit from other immunologic treatments such as vaccines, antibody-drug conjugates, bispecific antibodies, or CAR T cells are still not clear. Also, there are no available biomarkers or genomics clues to predict "right" population to benefit immunotherapy. Is it better to treat in earlier disease, rather than late disease with large tumor burden? Should we be interrogating tumor/peripheral blood/or bone marrow throughout the disease dynamic in order to profile immune changes throughout the tumor microenvironment? Is there any way to predict who will respond to an immune therapy in patients without mutations relevant to agnostic indications of ICI? More research is needed to explore potential genetic and/or biomarkers that facilitate to identify which patients will respond to ICI.

7.5   Immunotherapy in Testicular Cancer

Testicular germ-cell tumors (GCT) are one of curable cancers with highest cure rate in solid tumors with advanced-stage disease. Given high 5-year survival rate in testicular GCTs, approximately 10–40% of patients fail first-line cisplatin-based regimens and require second-line salvage therapies, either with standard-dose salvage chemotherapy or high-dose chemotherapy with/without stem cell transplant. However, subsequent relapses are incurable and tough to tackle, with grave prognoses. Currently, there are no FDA-approved agents for testicular GCTs [109]. In retrospective studies, testicular GCTs expressed PD-L1. High PD-L1 expression in testicular GCTs was correlated with poor prognosis, compared with low PD-L1 expression, with inferior OS and PFS. These findings give insight to potential benefit of ICI for patients with testicular GCTs, though testes are immunologically privileged sites. In some series of patients' cohorts, ICI seems to be efficient in carefully selected patients with platinum-refractory GCT. However, predictive markers associated with tumor response are not yet known. In a phase II trial of pembrolizumab in patients with platinum-refractory germ-cell tumors— a Hoosier Cancer Research Network Study GU14-206, pembrolizumab is well tolerated but does not demonstrate clinically meaningful single-agent activity in refractory GCT [14]. Until more mature definitive data became available, ICI in testicular GCTs remains an experimental approach and larger prospective clinical trials are warranted.

7.6   Immunotherapy in Penile Cancer

Despite substantial advances in systemic therapy in the past decade, the prognosis of advanced penile squamous cell carcinoma (PSCC) remains disproportionally poor. Effective salvage therapies for penile squamous cell carcinoma after failure of platinum-based therapy were limited. Due to dismal character of metastatic PSCC, there is urgent need for alternative treatment options with more favorable outcomes and a lower toxicity profile. Genomic analyses, such as next-generation sequencing, have provided transformative knowledge on the genomic roadmaps and tumor microenvironment of PSCC. Approximately one-quarter of patients with metastatic PSCC have clinically actionable genomic alterations in mammalian targets of rapamycin, DNA repair, and receptor tyrosine kinase pathways. And PSCC is genetically similar to other HPV-driven cancers. In addition, 40–60% of PSCC tumors show strong PD-L1 expression, and the frequency of mutational signatures is suggestive of low immunotherapy resistance, pointing to the potential of immunotherapy as new modality for PSCC. There are multiple ongoing trials to explore the role of immunotherapy in this context, including combined and sequential targeted therapies and HPV-targeted therapies. Owing to the rarity of this condition, it is of great importance to pursue further research to unveil this disease [10, 11, 50, 109–112].

7.7   Perioperative Immunotherapy as Window of Opportunity to Cure—Paradigm-Changing

Adjuvant therapy is a treatment strategy to improve cure rate and prolonged survival after surgical removal of tumor. Despite numerous worthwhile attempts, no adjuvant immunotherapy has shown a consistent clinical benefit with broad use in clinical practice. This year, two positive signals of adjuvant immunotherapy trials might change current treatment paradigms. The randomized, double-blind, placebo-controlled phase III CheckMate 274 trial (NCT02632409), 709 patients who were within 120 days of radical resection of UC of the bladder or upper urinary tract (renal pelvis or ureter) at high risk of recurrence were randomized to receive nivolumab 240 mg or placebo by intravenous infusion every 2 weeks until recurrence or until unacceptable toxicity for a maximum treatment duration of 1 year. The primary efficacy endpoint was disease-free survival (DFS) in the ITT population and in patients with tumors expressing PD-L1 $\geq 1\%$. At a prespecified interim analysis, a statistically significant improvement in DFS was demonstrated in patients on the nivolumab arm versus placebo for both primary endpoints. In the ITT analysis, the median DFS was 20.8 months in nivolumab arm compared with placebo arm with 10.8 months (HR 0.70; 98.22% CI: 0.55–0.90, $p < 0.001$). For patients with tumors expressing PD-L1 $\geq 1\%$, median DFS was not reached in nivolumab arm versus 8.4 months in placebo arm (HR 0.55; 98.72% CI: 0.35–0.85, $p < 0.001$). In an exploratory analysis of patients with PD-L1-negative tumors (58%), the unstratified DFS HR estimate was 0.83 (95% CI: 0.64–1.08). OS data is immature with 33% of deaths in the overall randomized population. In the upper tract urothelial carcinoma subpopulation, 37 deaths occurred (20 in the nivolumab arm, 17 in the placebo arm) [113]. On August 19, 2021, FDA-approved nivolumab for the adjuvant treatment of patients with UC who are at high risk of recurrence after undergoing radical resection. This is the first FDA approval for adjuvant treatment of patients with high-risk UC.

In phase 3 KEYNOTE-564 trial (NCT03142334), 994 patients with histologically confirmed clear cell RCC who had undergone nephrectomy up to 12 weeks before randomization were randomized to receive either pembrolizumab at a dose of 200 mg every 3 weeks for approximately 1 year or placebo for the same schedule. The primary endpoint of the trial was investigator-assessed DFS, and a key secondary endpoint was OS. Another important endpoint of the trial was safety. At the prespecified interim analysis, pembrolizumab arm was associated with significantly longer DFS than placebo (DFS at 24 months, 77.3% vs. 68.1%; HR for recurrence or Death $= 0.68$; 95% CI: 0.53–0.87, $P = 0.002$). The estimated percentage of patients who remained alive at 24 months was 96.6% in the pembrolizumab arm and 93.5% in the placebo group (HR for death $= 0.54$; 95% CI: 0.30–0.96). Regarding safety, all-cause adverse effects were reported in 96.3% in adjuvant pembrolizumab arm versus 91.1% in placebo arm. All-cause toxicities resulted in treatment discontinuation in 20.7% in pembrolizumab arm (alanine aminotransferase increased (1.6%), adrenal insufficiency (1%), colitis (1%), listed adverse events $\geq 1\%$) and 2.0% of those in placebo arm [114, 115]. On Aug 10, 2021, FDA has granted priority review to a new supplemental biologics license application (sBLA) for pembrolizumab as an adjuvant treatment in patients with RCC who are at intermediate-high or high risk of recurrence after nephrectomy or following nephrectomy and resection of metastatic lesions. The regulatory agency is expected to reach a decision on the sBLA by December 10, 2021, under the Prescription Drug User Fee Act.

Given these positive signals of better DFS and manageable side effects from two adjuvant immunotherapy trials in genitourinary cancer, some issues are needed to be addressed. In clinical settings, patients were risk stratified by clinical and pathological adverse features. Could circulating tumor DNA profile be integrated into clinical decision tree to refine risk stratification to identify patients truly at high risk for recurrence or disease progression and avoid overtreatment for low-risk patients, and to monitor disease status or therapeutic response? Could adjuvant therapy delay subsequent therapy? Could adjuvant therapy negatively impact efficacy of subsequent therapies? As no clinically meaningful change of quality of life and acceptable safety profile is present, could DFS be a clinical meaningful surrogate—quality and/or quantity time under the context of potent subsequent therapies? Further studies to explore clinical benefits of adjuvant immunotherapy are awaiting.

7.8   Conclusion

"Tumor Biology is King, Selection is Queen, Technical maneuvers are the Prince and Princess." Occasionally the prince and princess try to usurp the throne, sometimes with temporary apparent victories, usually to no long-term avail; they almost always fail to overcome the powerful forces of the King and Queen ~Dr. Blake Cady" [116].

Exciting scientific advances in the understanding of the biology of genitourinary malignancy growth have opened up future avenues of clinical development for genitourinary malignancy therapies and possibly point to the next treatment era on the horizon. However, ideal strategy for cancer care is to provide not just more time, but more quality time: there remain unmet needs for novel therapies that exploit molecular or genetic pathways to extend survival without compromising health-related quality of life for patients with advanced genitourinary malignancies. Treatment paradigm has been moving from "one fits for all" to "deliver the right treatment to the right patient at the right time". There are many great questions answered in upcoming and ongoing proof-of-concept clinical trials, and it will be interesting to see how the field evolves and how patient outcomes improve with further research. "Precision and personalized therapy" will be the future of cancer therapy.

Contributors KL was involved in the conception, design, or planning of the study. KL, KYC, and CLC provided resource and administrative support. KL collected the data and analyzed data. KL accessed and verified the underlying data. KL and KYC wrote the first draft of the manuscript with contributions from all authors. KL, KYC, and CLC critically reviewed or revised the manuscript for important intellectual content. All authors reviewed the interim drafts and final version of the manuscript and agree with its content and submission. KL had final responsibility for the decision to submit this manuscript for publication.

Declaration of Interests All authors declare no competing interests.

Conflict of Interest The authors confirm they have no conflicts of interests to declare.

Funding Sources The authors declare no funding.

References

1. Robert C, Soria JC, Eggermont AM (2013) Drug of the year: programmed death-1 receptor/programmed death-1 ligand-1 receptor monoclonal antibodies. Eur J Cancer 49(14):2968–2971
2. Eggermont AM, Kroemer G, Zitvogel L (2013) Immunotherapy and the concept of a clinical cure. Eur J Cancer 49(14):2965–2967
3. González-Del-Alba A, Arranz J, Bellmunt J, Maroto JP, Fernández-Calvo O, Valderrama BP et al (2020) Latest progress in molecular biology and treatment in genitourinary tumours. Clin Transl Oncol 22(12):2175–2195

4. Gevaert T, Montironi R, Lopez-Beltran A, Van Leenders G, Allory Y, De Ridder D et al (2018) Genito-urinary genomics and emerging biomarkers for immunomodulatory cancer treatment. Semin Cancer Biol 52(Pt 2):216–227
5. Tsiatas M, Mountzios G, Curigliano G (2016) Future perspectives in cancer immunotherapy. Ann Transl Med 4(14):273
6. Singh AK, Netea MG, Bishai WR (2021) BCG turns 100: its nontraditional uses against viruses, cancer, and immunologic diseases. J Clin Invest 131(11)
7. Shih KW, Chen WC, Chang CH, Tai TE, Wu JC, Huang AC et al (2021) Non-muscular invasive bladder cancer: re-envisioning therapeutic journey from traditional to regenerative interventions. Aging Dis 12(3):868–885
8. Sfakianos JP, Salome B, Daza J, Farkas A, Bhardwaj N, Horowitz A (2021) Bacillus Calmette-Guerin (BCG): Its fight against pathogens and cancer. Urol Oncol 39(2):121–129
9. Lobo N, Brooks NA, Zlotta AR, Cirillo JD, Boorjian S, Black PC et al (2021) 100 years of Bacillus Calmette-Guérin immunotherapy: from cattle to COVID-19. Nat Rev Urol 1–12
10. Peyraud F, Allenet C, Gross-Goupil M, Domblides C, Lefort F, Daste A et al (2020) Current management and future perspectives of penile cancer: an updated review. Cancer Treat Rev 90:102087
11. Ahmed ME, Falasiri S, Hajiran A, Chahoud J, Spiess PE (2020) The immune microenvironment in penile cancer and rationale for immunotherapy. J Clin Med 9(10)
12. Zarrabi K, Paroya A, Wu S (2019) Emerging therapeutic agents for genitourinary cancers. J Hematol Oncol 12(1):89
13. Rosenberg SA (2014) IL-2: the first effective immunotherapy for human cancer. J Immunol 192(12):5451–5458
14. Adra N, Einhorn LH, Althouse SK, Ammakkanavar NR, Musapatika D, Albany C et al (2018) Phase II trial of pembrolizumab in patients with platinum refractory germ-cell tumors: a Hoosier cancer research network study GU14-206. Ann Oncol 29(1):209–214
15. Semaan A, Haddad FG, Eid R, Kourie HR, Nemr E (2019) Immunotherapy: last bullet in platinum refractory germ cell testicular cancer. Future Oncol 15(5):533–541
16. Lavacchi D, Pellegrini E, Palmieri VE, Doni L, Mela MM, Di Maida F et al (2020) Immune checkpoint inhibitors in the treatment of renal cancer: current state and future perspective. Int J Mol Sci 21(13)
17. Rassy EE, Khoury Abboud RM, Ibrahim N, Assi T, Aoun F, Kattan J (2018) The current state of immune checkpoint inhibitors in the first-line treatment of renal cancer. Immunotherapy 10(12):1047–1052
18. Gill DM, Hahn AW, Hale P, Maughan BL (2018) Overview of current and future first-line systemic therapy for metastatic clear cell renal cell carcinoma. Curr Treat Options Oncol 19(1):6
19. Venkatachalam S, McFarland TR, Agarwal N, Swami U (2021) Immune checkpoint inhibitors in prostate cancer. Cancers (Basel) 13(9)
20. Lopez-Beltran A, Cimadamore A, Blanca A, Massari F, Vau N, Scarpelli M et al (2021) Immune checkpoint inhibitors for the treatment of bladder cancer. Cancers (Basel) 13(1)
21. Thana M, Wood L (2020) Immune checkpoint inhibitors in genitourinary malignancies. Curr Oncol 27(Suppl 2):S69–S77
22. Slovin SF (2020) Immunotherapy for prostate cancer: treatments for the "Lethal" phenotype. Urol Clin North Am 47(4):469–474
23. Heidegger I, Necchi A, Pircher A, Tsaur I, Marra G, Kasivisvanathan V et al (2020) A systematic review of the emerging role of immune checkpoint inhibitors in metastatic castration-resistant prostate cancer: will combination strategies improve efficacy? Eur Urol Oncol
24. Fay EK, Graff JN (2020) Immunotherapy in prostate cancer. Cancers (Basel) 12(7)
25. Ukleja J, Kusaka E, Miyamoto DT (2021) Immunotherapy combined with radiation therapy for genitourinary malignancies. Front Oncol 11:663852
26. Ingles Garces AH, Au L, Mason R, Thomas J, Larkin J (2019) Building on the anti-PD1/PD-L1 backbone: combination immunotherapy for cancer. Expert Opin Investig Drugs 28(8):695–708

27. Collazo-Lorduy A, Galsky MD (2016) Combining chemotherapy and immune checkpoint blockade. Curr Opin Urol 26(6):508–513
28. Kgatle MM, Boshomane TMG, Lawal IO, Mokoala KMG, Mokgoro NP, Lourens N et al (2021) Immune checkpoints, inhibitors and radionuclides in prostate cancer: promising combinatorial therapy approach. Int J Mol Sci 22(8)
29. Goff PH, Bhakuni R, Pulliam T, Lee JH, Hall ET, Nghiem P (2021) Intersection of two checkpoints: could inhibiting the DNA damage response checkpoint rescue immune checkpoint-refractory cancer? Cancers (Basel) 13(14)
30. Einsele H, Borghaei H, Orlowski RZ, Subklewe M, Roboz GJ, Zugmaier G et al (2020) The BiTE (bispecific T-cell engager) platform: development and future potential of a targeted immuno-oncology therapy across tumor types. Cancer 126(14):3192–3201
31. De Cicco P, Ercolano G, Ianaro A (2020) The new era of cancer immunotherapy: targeting myeloid-derived suppressor cells to overcome immune evasion. Front Immunol 11:1680
32. Crispen PL, Kusmartsev S (2020) Mechanisms of immune evasion in bladder cancer. Cancer Immunol Immunother 69(1):3–14
33. Chakravarty D, Huang L, Kahn M, Tewari AK (2020) Immunotherapy for metastatic prostate cancer: current and emerging treatment options. Urol Clin North Am 47(4):487–510
34. Benitez JC, Remon J, Besse B (2020) Current panorama and challenges for neoadjuvant cancer immunotherapy. Clin Cancer Res 26(19):5068–5077
35. Liu KG, Gupta S, Goel S (2017) Immunotherapy: incorporation in the evolving paradigm of renal cancer management and future prospects. Oncotarget 8(10):17313–17327
36. Sharabi AB, Lim M, DeWeese TL, Drake CG (2015) Radiation and checkpoint blockade immunotherapy: radiosensitisation and potential mechanisms of synergy. Lancet Oncol 16(13):e498-509
37. Carosella ED, Ploussard G, LeMaoult J, Desgrandchamps F (2015) A systematic review of immunotherapy in urologic cancer: evolving roles for targeting of CTLA-4, PD-1/PD-L1, and HLA-G. Eur Urol 68(2):267–279
38. Rey-Cárdenas M, Guerrero-Ramos F, Gómez de Liaño Lista A, Carretero-González A, Bote H, Herrera-Juárez M et al (2021) Recent advances in neoadjuvant immunotherapy for urothelial bladder cancer: what to expect in the near future. Cancer Treat Rev 93:102142
39. Kuusk T, Abu-Ghanem Y, Mumtaz F, Powles T, Bex A (2021) Perioperative therapy in renal cancer in the era of immune checkpoint inhibitor therapy. Curr Opin Urol 31(3):262–269
40. Gandhy SU, Madan RA, Aragon-Ching JB (2020) The immunotherapy revolution in genitourinary malignancies. Immunotherapy 12(11):819–831
41. Tsiatas M, Grivas P (2016) Immunobiology and immunotherapy in genitourinary malignancies. Ann Transl Med 4(14):270
42. Sarkis J, Assaf J, Alkassis M (2021) Biomarkers in renal cell carcinoma: towards a more selective immune checkpoint inhibition. Transl Oncol 14(6):101071
43. Maiorano BA, Schinzari G, Ciardiello D, Rodriquenz MG, Cisternino A, Tortora G et al (2021) Cancer vaccines for genitourinary tumors: recent progresses and future possibilities. Vaccines (Basel) 9(6)
44. Bogen JP, Grzeschik J, Jakobsen J, Bähre A, Hock B, Kolmar H (2021) Treating bladder cancer: engineering of current and next generation antibody-, fusion protein-, mRNA-, cell- and viral-based therapeutics. Front Oncol 11:672262
45. Tripathi A, Grivas P (2020) The utility of next generation sequencing in advanced urothelial carcinoma. Eur Urol Focus 6(1):41–44
46. Schmidt AL, Siefker-Radtke A, McConkey D, McGregor B (2020) Renal cell and urothelial carcinoma: biomarkers for new treatments. Am Soc Clin Oncol Educ Book 40:1–11
47. Roviello G, Catalano M, Nobili S, Santi R, Mini E, Nesi G (2020) Focus on biochemical and clinical predictors of response to immune checkpoint inhibitors in metastatic urothelial carcinoma: where do we stand? Int J Mol Sci 21(21)
48. Mo Q, Li R, Adeegbe DO, Peng G, Chan KS (2020) Integrative multi-omics analysis of muscle-invasive bladder cancer identifies prognostic biomarkers for frontline chemotherapy and immunotherapy. Commun Biol 3(1):784

49. Beaumont KG, Beaumont MA, Sebra R (2020) Application of single-cell sequencing to immunotherapy. Urol Clin North Am 47(4):475–485
50. Aydin AM, Chahoud J, Adashek JJ, Azizi M, Magliocco A, Ross JS et al (2020) Understanding genomics and the immune environment of penile cancer to improve therapy. Nat Rev Urol 17(10):555–570
51. Andolfi C, Bloodworth JC, Papachristos A, Sweis RF (2020) The urinary microbiome and bladder cancer: susceptibility and immune responsiveness. Bladder Cancer 6(3):225–235
52. Mehta K, Patel K, Parikh RA (2017) Immunotherapy in genitourinary malignancies. J Hematol Oncol 10(1):95
53. Balar AV, Kamat AM, Kulkarni GS, Uchio EM, Boormans JL, Roumiguié M et al (2021) Pembrolizumab monotherapy for the treatment of high-risk non-muscle-invasive bladder cancer unresponsive to BCG (KEYNOTE-057): an open-label, single-arm, multicentre, phase 2 study. Lancet Oncol 22(7):919–930
54. Lenfant L, Aminsharifi A, Seisen T, Rouprêt M (2020) Current status and future directions of the use of novel immunotherapeutic agents in bladder cancer. Curr Opin Urol 30(3):428–440
55. Witjes JA, Bruins HM, Cathomas R, Compérat EM, Cowan NC, Gakis G et al (2021) European Association of urology guidelines on muscle-invasive and metastatic bladder cancer: summary of the 2020 guidelines. Eur Urol 79(1):82–104
56. Gakis G (2020) Management of muscle-invasive bladder cancer in the 2020s: challenges and perspectives. Eur Urol Focus 6(4):632–638
57. O'Donnell PH, Arkenau HT, Sridhar SS, Ong M, Drakaki A, Spira AI et al (2020) Patient-reported outcomes and inflammatory biomarkers in patients with locally advanced/metastatic urothelial carcinoma treated with durvalumab in phase 1/2 dose-escalation study 1108. Cancer 126(2):432–443
58. Powles T, van der Heijden MS, Castellano D, Galsky MD, Loriot Y, Petrylak DP et al (2020) Durvalumab alone and durvalumab plus tremelimumab versus chemotherapy in previously untreated patients with unresectable, locally advanced or metastatic urothelial carcinoma (DANUBE): a randomised, open-label, multicentre, phase 3 trial. Lancet Oncol 21(12):1574–1588
59. Necchi A, Joseph RW, Loriot Y, Hoffman-Censits J, Perez-Gracia JL, Petrylak DP et al (2017) Atezolizumab in platinum-treated locally advanced or metastatic urothelial carcinoma: post-progression outcomes from the phase II IMvigor210 study. Ann Oncol 28(12):3044–3050
60. Powles T, Durán I, van der Heijden MS, Loriot Y, Vogelzang NJ, De Giorgi U et al (2018) Atezolizumab versus chemotherapy in patients with platinum-treated locally advanced or metastatic urothelial carcinoma (IMvigor211): a multicentre, open-label, phase 3 randomised controlled trial. Lancet 391(10122):748–757
61. Galsky MD, Arija JÁA, Bamias A, Davis ID, De Santis M, Kikuchi E et al (2020) Atezolizumab with or without chemotherapy in metastatic urothelial cancer (IMvigor130): a multicentre, randomised, placebo-controlled phase 3 trial. Lancet 395(10236):1547–1557
62. Fradet Y, Bellmunt J, Vaughn DJ, Lee JL, Fong L, Vogelzang NJ et al (2019) Randomized phase III KEYNOTE-045 trial of pembrolizumab versus paclitaxel, docetaxel, or vinflunine in recurrent advanced urothelial cancer: results of >2 years of follow-up. Ann Oncol 30(6):970–976
63. Vaughn DJ, Bellmunt J, Fradet Y, Lee JL, Fong L, Vogelzang NJ et al (2018) Health-related quality-of-life analysis from KEYNOTE-045: a phase III study of pembrolizumab versus chemotherapy for previously treated advanced urothelial cancer. J Clin Oncol 36(16):1579–1587
64. Bellmunt J, de Wit R, Vaughn DJ, Fradet Y, Lee JL, Fong L et al (2017) Pembrolizumab as second-line therapy for advanced urothelial carcinoma. N Engl J Med 376(11):1015–1026
65. Ohyama C, Kojima T, Kondo T, Naya Y, Inoue T, Tomita Y et al (2019) Nivolumab in patients with unresectable locally advanced or metastatic urothelial carcinoma: CheckMate 275 2-year global and Japanese patient population analyses. Int J Clin Oncol 24(9):1089–1098

66. Sharma P, Retz M, Siefker-Radtke A, Baron A, Necchi A, Bedke J et al (2017) Nivolumab in metastatic urothelial carcinoma after platinum therapy (CheckMate 275): a multicentre, single-arm, phase 2 trial. Lancet Oncol 18(3):312–322
67. Vaishampayan U, Schöffski P, Ravaud A, Borel C, Peguero J, Chaves J et al (2019) Avelumab monotherapy as first-line or second-line treatment in patients with metastatic renal cell carcinoma: phase Ib results from the JAVELIN Solid Tumor trial. J Immunother Cancer 7(1):275
68. Balar AV, Galsky MD, Rosenberg JE, Powles T, Petrylak DP, Bellmunt J et al (2017) Atezolizumab as first-line treatment in cisplatin-ineligible patients with locally advanced and metastatic urothelial carcinoma: a single-arm, multicentre, phase 2 trial. Lancet 389(10064):67–76
69. Suzman DL, Agrawal S, Ning YM, Maher VE, Fernandes LL, Karuri S et al (2019) FDA approval summary: atezolizumab or pembrolizumab for the treatment of patients with advanced urothelial carcinoma ineligible for cisplatin-containing chemotherapy. Oncologist 24(4):563–569
70. Vuky J, Balar AV, Castellano D, O'Donnell PH, Grivas P, Bellmunt J et al (2020) Long-term outcomes in KEYNOTE-052: phase II study investigating first-line pembrolizumab in cisplatin-ineligible patients with locally advanced or metastatic urothelial cancer. J Clin Oncol 38(23):2658–2666
71. Grivas P, Plimack ER, Balar AV, Castellano D, O'Donnell PH, Bellmunt J et al (2020) Pembrolizumab as first-line therapy in cisplatin-ineligible advanced urothelial cancer (KEYNOTE-052): outcomes in older patients by age and performance status. Eur Urol Oncol 3(3):351–359
72. Balar AV, Castellano D, O'Donnell PH, Grivas P, Vuky J, Powles T et al (2017) First-line pembrolizumab in cisplatin-ineligible patients with locally advanced and unresectable or metastatic urothelial cancer (KEYNOTE-052): a multicentre, single-arm, phase 2 study. Lancet Oncol 18(11):1483–1492
73. Powles T, Csőszi T, Özgüroğlu M, Matsubara N, Géczi L, Cheng SY et al (2021) Pembrolizumab alone or combined with chemotherapy versus chemotherapy as first-line therapy for advanced urothelial carcinoma (KEYNOTE-361): a randomised, open-label, phase 3 trial. Lancet Oncol 22(7):931–945
74. Powles T, Park SH, Voog E, Caserta C, Valderrama BP, Gurney H et al (2020) Avelumab maintenance therapy for advanced or metastatic urothelial carcinoma. N Engl J Med 383(13):1218–1230
75. Hizal M, Sendur MAN, Bilgin B, Akinci MB, Sener Dede D, Yalcin B (2020) A historical turning point for the treatment of advanced renal cell carcinoma: inhibition of immune checkpoint. Curr Med Res Opin 36(4):625–635
76. Waldmann TA (2018) Cytokines in cancer immunotherapy. Cold Spring Harb Perspect Biol 10(12)
77. Roviello G, Zanotti L, Correale P, Gobbi A, Wigfield S, Guglielmi A et al (2017) Is still there a role for IL-2 for solid tumors other than melanoma or renal cancer? Immunotherapy 9(1):25–32
78. Motzer RJ, Escudier B, McDermott DF, George S, Hammers HJ, Srinivas S et al (2015) Nivolumab versus everolimus in advanced renal-cell carcinoma. N Engl J Med 373(19):1803–1813
79. Escudier B, Sharma P, McDermott DF, George S, Hammers HJ, Srinivas S et al (2017) CheckMate 025 randomized phase 3 study: outcomes by key baseline factors and prior therapy for nivolumab versus everolimus in advanced renal cell carcinoma. Eur Urol 72(6):962–971
80. Motzer RJ, Escudier B, George S, Hammers HJ, Srinivas S, Tykodi SS et al (2020) Nivolumab versus everolimus in patients with advanced renal cell carcinoma: updated results with long-term follow-up of the randomized, open-label, phase 3 CheckMate 025 trial. Cancer 126(18):4156–4167
81. Motzer RJ, Tannir NM, McDermott DF, Arén Frontera O, Melichar B, Choueiri TK et al (2018) Nivolumab plus Ipilimumab versus sunitinib in advanced renal-cell carcinoma. N Engl J Med 378(14):1277–1290

82. Cella D, Grünwald V, Escudier B, Hammers HJ, George S, Nathan P et al (2019) Patient-reported outcomes of patients with advanced renal cell carcinoma treated with nivolumab plus ipilimumab versus sunitinib (CheckMate 214): a randomised, phase 3 trial. Lancet Oncol 20(2):297–310

83. Albiges L, Tannir NM, Burotto M, McDermott D, Plimack ER, Barthélémy P et al (2020) Nivolumab plus ipilimumab versus sunitinib for first-line treatment of advanced renal cell carcinoma: extended 4-year follow-up of the phase III CheckMate 214 trial. ESMO Open 5(6):e001079

84. Escudier B, Motzer RJ, Tannir NM, Porta C, Tomita Y, Maurer MA et al (2020) Efficacy of nivolumab plus ipilimumab according to number of IMDC risk factors in CheckMate 214. Eur Urol 77(4):449–453

85. Flippot R, Escudier B, Albiges L (2018) Immune checkpoint inhibitors: toward new paradigms in renal cell carcinoma. Drugs 78(14):1443–1457

86. Lalani AA, McGregor BA, Albiges L, Choueiri TK, Motzer R, Powles T et al (2019) Systemic treatment of metastatic clear cell renal cell carcinoma in 2018: current paradigms, use of immunotherapy, and future directions. Eur Urol 75(1):100–110

87. Motzer RJ, Penkov K, Haanen J, Rini B, Albiges L, Campbell MT et al (2019) Avelumab plus axitinib versus sunitinib for advanced renal-cell carcinoma. N Engl J Med 380(12):1103–1115

88. Choueiri TK, Motzer RJ, Rini BI, Haanen J, Campbell MT, Venugopal B et al (2020) Updated efficacy results from the JAVELIN Renal 101 trial: first-line avelumab plus axitinib versus sunitinib in patients with advanced renal cell carcinoma. Ann Oncol 31(8):1030–1039

89. Motzer RJ, Robbins PB, Powles T, Albiges L, Haanen JB, Larkin J et al (2020) Avelumab plus axitinib versus sunitinib in advanced renal cell carcinoma: biomarker analysis of the phase 3 JAVELIN Renal 101 trial. Nat Med 26(11):1733–1741

90. Uemura M, Tomita Y, Miyake H, Hatakeyama S, Kanayama HO, Numakura K et al (2020) Avelumab plus axitinib vs sunitinib for advanced renal cell carcinoma: Japanese subgroup analysis from JAVELIN Renal 101. Cancer Sci 111(3):907–923

91. Rini BI, Plimack ER, Stus V, Gafanov R, Hawkins R, Nosov D et al (2019) Pembrolizumab plus axitinib versus sunitinib for advanced renal-cell carcinoma. N Engl J Med 380(12):1116–1127

92. Powles T, Plimack ER, Soulières D, Waddell T, Stus V, Gafanov R et al (2020) Pembrolizumab plus axitinib versus sunitinib monotherapy as first-line treatment of advanced renal cell carcinoma (KEYNOTE-426): extended follow-up from a randomised, open-label, phase 3 trial. Lancet Oncol 21(12):1563–1573

93. Choueiri TK, Powles T, Burotto M, Escudier B, Bourlon MT, Zurawski B et al (2021) Nivolumab plus cabozantinib versus sunitinib for advanced renal-cell carcinoma. N Engl J Med 384(9):829–841

94. Yu EM, Linville L, Rosenthal M, Aragon-Ching JB (2021) A contemporary review of immune checkpoint inhibitors in advanced clear cell renal cell carcinoma. Vaccines (Basel) 9(8)

95. Motzer R, Alekseev B, Rha SY, Porta C, Eto M, Powles T et al (2021) Lenvatinib plus pembrolizumab or everolimus for advanced renal cell carcinoma. N Engl J Med 384(14):1289–1300

96. Lee CH, Shah AY, Rasco D, Rao A, Taylor MH, Di Simone C et al (2021) Lenvatinib plus pembrolizumab in patients with either treatment-naive or previously treated metastatic renal cell carcinoma (study 111/KEYNOTE-146): a phase 1b/2 study. Lancet Oncol 22(7):946–958

97. Shore ND, Drake CG, Lin DW, Ryan CJ, Stratton KL, Dunshee C et al (2020) Optimizing the management of castration-resistant prostate cancer patients: a practical guide for clinicians. Prostate 80(14):1159–1176

98. Vitkin N, Nersesian S, Siemens DR, Koti M (2019) The tumor immune contexture of prostate cancer. Front Immunol 10:603

99. Venturini NJ, Drake CG (2019) Immunotherapy for prostate cancer. Cold Spring Harb Perspect Med 9(5)

100. Cha HR, Lee JH, Ponnazhagan S (2020) Revisiting immunotherapy: a focus on prostate cancer. Cancer Res 80(8):1615–1623
101. Nair SS, Weil R, Dovey Z, Davis A, Tewari AK (2020) The tumor microenvironment and immunotherapy in prostate and bladder cancer. Urol Clin North Am 47(4s):e17–e54
102. Sardana R, Mishra SK, Williamson SR, Mohanty A, Mohanty SK (2020) Immune checkpoints and their inhibitors: reappraisal of a novel diagnostic and therapeutic dimension in the urologic malignancies. Semin Oncol 47(6):367–379
103. Flanigan RC, Polcari AJ, Shore ND, Price TH, Sims RB, Maher JC et al (2013) An analysis of leukapheresis and central venous catheter use in the randomized, placebo controlled, phase 3 IMPACT trial of Sipuleucel-T for metastatic castrate resistant prostate cancer. J Urol 189(2):521–526
104. Schellhammer PF, Chodak G, Whitmore JB, Sims R, Frohlich MW, Kantoff PW (2013) Lower baseline prostate-specific antigen is associated with a greater overall survival benefit from sipuleucel-T in the immunotherapy for prostate adenocarcinoma treatment (IMPACT) trial. Urology 81(6):1297–1302
105. Crawford ED, Petrylak DP, Higano CS, Kibel AS, Kantoff PW, Small EJ et al (2015) Optimal timing of sipuleucel-T treatment in metastatic castration-resistant prostate cancer. Can J Urol 22(6):8048–8055
106. George DJ, Nabhan C, DeVries T, Whitmore JB, Gomella LG (2015) Survival outcomes of sipuleucel-T phase III studies: impact of control-arm cross-over to salvage immunotherapy. Cancer Immunol Res 3(9):1063–1069
107. Kwon ED, Drake CG, Scher HI, Fizazi K, Bossi A, van den Eertwegh AJ et al (2014) Ipilimumab versus placebo after radiotherapy in patients with metastatic castration-resistant prostate cancer that had progressed after docetaxel chemotherapy (CA184-043): a multicentre, randomised, double-blind, phase 3 trial. Lancet Oncol 15(7):700–712
108. Fizazi K, Drake CG, Beer TM, Kwon ED, Scher HI, Gerritsen WR et al (2020) Final analysis of the ipilimumab versus placebo following radiotherapy phase III trial in postdocetaxel metastatic castration-resistant prostate cancer identifies an excess of long-term survivors. Eur Urol 78(6):822–830
109. McGregor BA, Sonpavde GP (2020) Rare genitourinary malignancies: current status and future directions of immunotherapy. Eur Urol Focus 6(1):14–16
110. Aydin AM, Cheriyan S, Spiess PE (2019) Treating advanced penile cancer: where do we stand in 2019? Curr Opin Support Palliat Care 13(3):249–254
111. Resch I, Abufaraj M, Hübner NA, Shariat SF (2020) An update on systemic therapy for penile cancer. Curr Opin Urol 30(2):229–233
112. de Vries HM, Ottenhof SR, Horenblas S, van der Heijden MS, Jordanova ES (2019) Defining the tumor microenvironment of penile cancer by means of the cancer immunogram. Eur Urol Focus 5(5):718–721
113. Bajorin DF, Witjes JA, Gschwend JE, Schenker M, Valderrama BP, Tomita Y et al (2021) Adjuvant nivolumab versus placebo in muscle-invasive urothelial carcinoma. N Engl J Med 384(22):2102–2114
114. Choueiri TK, Tomczak P, Park SH, Venugopal B, Ferguson T, Chang YH et al (2021) Adjuvant pembrolizumab after nephrectomy in renal-cell carcinoma. N Engl J Med 385(8):683–694
115. McKay RR (2021) The promise of adjuvant immunotherapy in renal-cell carcinoma. N Engl J Med 385(8):756–758
116. Cady B (1997) Basic principles in surgical oncology. Arch Surg 132(4):338–346

Immune-Based Therapeutic Interventions for Acute Myeloid Leukemia

8

Fabiana Perna, Manuel R. Espinoza-Gutarra, Giuseppe Bombaci, Sherif S. Farag, and Jennifer E. Schwartz

8.1   Introduction

Acute Myeloid Leukemia (AML) is a myeloid malignancy characterized by the neoplastic transformation of primitive hematopoietic stem or progenitor cells. Abnormal cell proliferation and differentiation result in a high level of immature malignant cells and fewer differentiated red blood cells, platelets, and white blood cells. With an estimated 21,000 new cases of AML in the United States in 2020, AML represents the second most common type of leukemia overall and the most common acute leukemia presenting in the adult patient, with a median age at onset of approximately 65 years [1]. Treatment outcomes and survival have been historically dismal for patients afflicted with this condition, as evidenced by 5-year overall survival rates of less than 30% [2]. Despite much progress in the molecular understanding of AML pathogenesis, since the 1970s, AML-directed treatment protocols rely on Cytarabine- and Anthracycline-based combinations [3]. Allogeneic stem cell transplantation (allo-SCT) offers the best opportunity at cure for those patients with high-risk AML deemed "medically fit" to tolerate intensive therapies [4].

Allo-SCT is the most fundamental form of immunotherapy and provides historically the evidence that AML is an immuno-responsive disease. The ability of donor T and NK cells to recognize and eradicate AML is responsible for the graft-versus-leukemia (GvL) effect, which is a major mechanism for the curative effect of allo-SCT. This was identified early in the development of stem cell transplantation; initially through the clinical observation of its relation to Graft versus Host Disease (GvHD) [5] and then demonstrated through dramatic improvement

F. Perna (✉) · M. R. Espinoza-Gutarra · G. Bombaci · S. S. Farag · J. E. Schwartz
Department of Medicine, Division of Hematology/Oncology, Indiana University School of Medicine, Indianapolis, USA
e-mail: fabperna@iu.edu

© The Author(s), under exclusive license to Springer Nature Switzerland AG 2022   225
P. Hays (ed.), *Cancer Immunotherapies*, Cancer Treatment and Research 183,
https://doi.org/10.1007/978-3-030-96376-7_8

via the use of Donor Lymphocyte Infusions (DLI) [6]. However, there are significant limitations for allo-SCT. It has the best outcomes in first complete remission (CR1), and many patients with high-risk AML do not achieve a first remission [7]. Donor allo-immune responses also affect healthy tissues (GvHD), carrying significant risks of treatment-related morbidity and mortality. GvHD and GvL often co-occur and, a major barrier to exploiting the full immunotherapeutic benefit of donor immune cells against patient leukemia is the immunosuppression required to treat GvHD. Understanding the antigenic basis of immune responses in GvL compared to GvHD that might go beyond the incompatibility at minor and major histocompatibility antigens will help develop more beneficial targeted therapies.

We will discuss alternative and synergistic therapeutic rationales that have been developed to harness the anti-leukemic immune responses against patient AML cells. We will analyze the use of immune checkpoint inhibitors such as Ipilimumab and Nivolumab in unleashing the endogenous immune system against AML cells, the advantages of using bispecific antibodies recruiting T cells to attack AML cells independently of the T cell receptor (TCR) specificity and the novel uses of Chimeric Antigen Receptor (CAR) T cell therapy redirecting T cell specificity and antibody-mediated therapies recognizing AML cell surface molecules. We will emphasize the disease-specific challenges in developing safe and effective immunotherapies. Finally, we will discuss the most relevant areas of preclinical and clinical research in the field.

8.2   Immune Checkpoint Inhibitors Reactivating Endogenous T Cell Responses

The release of negative regulators of immune activation (immune checkpoints) that limit antitumor responses has resulted in unprecedented rates of long-lasting tumor responses in patients with solid tumors. This can be mainly achieved by antibodies blocking the cytotoxic T lymphocyte antigen-4 (CTLA-4) or the programmed death-1 (PD-1) pathway, either alone or in combination [8, 9]. The main premise for inducing an immune response is the pre-existence of antitumor T cells that were limited by specific immune checkpoints.

In AML models, binding of PD-1 to PD-L1 results in suppression of proliferation and immune response of T cells [10] and PD-1 positive T cells have been detected in bone marrow aspirates of patients [11, 12] however, CTLA-4 and PD-1/PD-L1 inhibitor monotherapy in AML patients have resulted in mixed response rates.

Berger et al., reported that a single administration of CT-011, a humanized IgG1 monoclonal antibody targeting PD-1 led to clinical benefit in patients with advanced hematologic malignancies including AML. The objectives of this phase I study were to assess the dose-limiting toxicities, to determine the maximum tolerated dose, and to study the pharmacokinetics of CT-011 administered once to patients with advanced hematologic malignancies. Seventeen patients were treated with escalating doses of CT-011 ranging from 0.2 to 6 mg/kg. The study showed the antibody to be safe and well tolerated in this patient population. No single maximum tolerated dose was defined. Clinical benefit was observed in 33% of the patients with one complete remission (CR). Sustained elevation in the percentage of peripheral blood $CD4^+$ lymphocytes was observed up to 21 days following CT-011 treatment [13].

Davids et al., conducted a phase 1/1b multicenter study to determine the safety and efficacy of Ipilimumab (a monoclonal antibody targeting CTLA-4; anti-CTLA-4 inhibitor) in patients with relapsed AML after allo-SCT. Patients received induction therapy with Ipilimumab at a dose of 3 or 10 mg/kg of body weight every 3 weeks for a total of four doses, with additional doses every 12 weeks for up to 60 weeks in patients who had a clinical benefit. A total of 28 patients were enrolled. Among 22 patients who received a high dose of 10 mg/kg, 5 (23%) had a complete response, 2 (9%) had a partial response, and 6 (27%) had decreased tumor burden. Four patients had a durable response for more than 1 year. Grade 3–4 immune-related adverse events were noted in 6 (21%) of patients and development of GVHD to the point that it precluded further Ipilimumab treatment occurred in 4(14%) of patients [14].

Reville et al., designed a phase II clinical trial to study the efficacy and safety of a PD-1 inhibitor Nivolumab as single-agent maintenance in high-risk AML in remission not being considered for allo-SCT. Eligible patients had AML in remission (defined as CR, CR with incomplete hematologic recovery [CRi], or partial remission [PR]). Patients were defined as having high-risk disease by any of the following: in 1st CR with secondary AML; high-risk cytogenetics at diagnosis; fms-related tyrosine kinase three internal tandem duplication mutated at diagnosis; the presence of measurable residual disease assessed by flow cytometry at time of enrollment; or 2nd CR or greater regardless of disease characteristics at the time of initial diagnosis. Patients received induction and at least one cycle of consolidation chemotherapy and achieved a CR within 12 months of protocol enrollment. Patients received Nivolumab at a dose of 3 mg/kg intravenously every 2 weeks. Cycles are repeated every 28 days in the absence of disease progression or unacceptable toxicity. After cycle 6, patients received Nivolumab every 4 weeks. After cycle 12, patients received Nivolumab every 3 months until disease relapse. The primary outcome was recurrence-free survival (RFS) rate at six months. Secondary outcomes were to evaluate measurable residual disease (MRD) by flow cytometry as a predictor of response. 15 patients were enrolled and median RFS was 8.48 months and only two out of nine patients that were MRD positive on enrollment converted to MRD negative. 5 patients experienced non-immune-related adverse events (AE) and six patients experienced immune AEs including two patients with Grade 3 pneumonitis and one patient with Grade 4 ALT elevation [15].

Given that, use of immunotherapy in combination with other agents has also been explored. The rationale behind this strategy also includes potential increase in neoantigen generation upon chemotherapy and hypomethylating drugs. Ravandi et al., aimed to assess the addition of Nivolumab to frontline therapy with idarubicin and cytarabine in patients with newly diagnosed AML or high-risk

myelodysplastic syndrome (MDS). The primary endpoint was event free survival. Nivolumab was started at 3 mg/kg dosing on day 24 and continued every
2 weeks for up to 1 year in responders; at a median follow-up of 17.25 months the
median RFS for responders was 18.54 months with 43% of patients proceeding to
allo-SCT where Grade 3 to 4 GvHD was observed in 26% of patients [16].

Daver et al., evaluated the combination of Nivolumab and Azacytidine in a
phase II trial of 70 patients with relapsed/refractory (R/R) AML. The overall
response rate (ORR) was 33%, including 15 (22%) CR/CRi, 1 PR, and seven
patients with hematologic improvement maintained >6 months. Six patients (9%)
had stable disease >6 months. The ORR was 58% and 22%, in hypomethylating
agent (HMA)-naive ($n = 25$) and HMA-pretreated ($n = 45$) patients, respectively.
Grade 3 to 4 immune-related adverse events occurred in 8 (11%) patients [17].

An immune checkpoint inhibitor targeting TIM-3 (T cell immunoglobulin and
mucin domain 3) showed reduced tumor progression and improved antitumor T
cell responses in preclinical models [18]. Researchers have found that TIM-3 is
expressed in leukemic stem cells (LSC) and progenitors [19, 20], and its expression is associated with disease progression [21]. Sabatolimab (MBG453) is a high
affinity, humanized, IgG4 antibody targeting TIM-3 which is currently being studied in a clinical trial. Borate et al., reported the first results of such multicenter,
open-label, phase 1b dose-escalation study to evaluate MBG453 combined with
decitabine or azacitidine in patients with high-risk MDS or AML (NCT03066648).
Patients with high-risk MDS and newly diagnosed (ND) or R/R AML were
eligible if they were HMA "i" and not candidates for intensive chemotherapy.
Primary objectives were safety/tolerability. For de novo AML patients treated
with decitabine, the ORR was 41% and for R/R AML patients 24%; 4 CR and
1 CRi were reported in the de novo cohort and all patients that responded in
the R/R cohort had CRi. For patients treated with azacytidine, ORR was 27%
for the de novo AML cohort and only one patient achieved CR. Both combinations were well tolerated. With MBG453+Dec and MBG453+Aza, respectively
most common Grade 3/4 AEs were thrombocytopenia (41%; 52%), febrile neutropenia (46%; 21%), neutropenia (42%; 38%), and anemia (25%; 28%). For
MBG453+Dec, only 4 pts (3 AML, 1 MDS) experienced ≥1 potentially immune-
related (IR) Gr ≥ 3 AE (ALT increase, arthritis, hepatitis, hypothyroidism, rash).
For MBG453+Aza, no patient experienced treatment-related Grade ≥3 potential
IR AEs. No treatment-related Grade 4 IR AEs or deaths were seen with either
combination [22].

Magrolimab is a monoclonal antibody against CD47 and a macrophage checkpoint inhibitor designed to interfere with recognition of CD47 by the SIRPα
receptor on macrophages, thus blocking the "do not eat me" signal used by cancer
cells to avoid being ingested by macrophages [23]. Sallman et al., reported the
first results of the phase 1b trial for treatment-"i" AML patients unfit for intensive
chemotherapy. 68 patients (39 patients with high-risk MDS and 29 AML patients)
were treated with magrolimab and azacitidine. ORR in evaluable MDS patients
was 91% and 64% in evaluable AML patients. Clinical activity was also observed
in the p53 mutant subset associated with a worse prognosis and high rates of

chemoresistance and relapse [24]. In this study the ORR of 12 TP53 mutant AML patients was 75% (42% CR, 33% CRi, 17% SD, 8% PD) and, out of the entire cohort, 91% of MDS and 100% of AML responding patients maintained their response at 6 months. Adverse events included anemia (38%), fatigue (21%), neutropenia (19%), thrombocytopenia (18%) and infusion reaction (16%); only one patient discontinued treatment due to adverse events [25].

In conclusion, the addition of immune checkpoint inhibitors to AML treatment remains under investigation. A major roadblock is the fact that AML patients present a low mutation burden (13–16 mutations per megabase in de novo AML) compared to solid tumors [26]. Due to this, immune recognition of potential neoantigens, arising from mutated genes, is less probable and thus less likely to be responsive to immune checkpoint blockade monotherapy [27–31]. Yet, HLA expression and proteasomal antigen processing are often downregulated in AML as a mechanism of tumor immune escape [32] thus, preventing the AML antigen presentation capacity to cytotoxic T cells.

8.3   Bispecific Antibodies Recruiting T Cells Independently of T Cell Receptor (TCR) Specificity

Bispecific antibodies are recombinant proteins that recruit T cells, through CD3 engagement, and target tumor cells through binding to a tumor-associated antigen. The engagement of CD3 (part of the T cell receptor) induces both proliferation of CD4 and CD8 T cells and cytotoxic activity by CD8 and in part CD4 cells against the target, to eliminate cancer cells [33]. Binding between effector cells and tumor cells facilitates the formation of cytolytic synapses inducing the apoptosis of the tumor cells and under specific circumstances the release of cytokines to amplify the immunological response by involving other immune cells or induce T cells proliferation [34].

In this scenario, the class of bispecific antibodies (bsAbs), also known as dual-targeting molecules, includes antibodies or single-chain variable fragments (scFv) derived from the Fab fragment of the IgG immunoglobulin, composed of the VH and VL domains attached with a linker, to physically bridge two or more cells [35]. Different formats of bispecific antibodies have been developed such as bispecific T cell engagers (BiTEs), tandem antibodies combining two scFVs for each target and thus maintaining the avidity of a bivalent antibody, DuoBody antibodies, affinity-tailored adaptors for T cells, dual affinity retargeting (DART) antibodies and tetravalent bispecific antibodies (Fig. 8.1). Dual affinity retargeting antibodies (DARTs) consist of variable domains of two antigen-binding specificities linked to two independent polypeptide chains. Each variable domain is formed by associating one VL segment on one chain with a VH segment on a second chain, covalently linked via disulfide bridge.

A major challenge in translating the success of bispecific antibody constructs in B-cell CD19$^+$ malignancies to AML has been the identification of suitable target antigens.

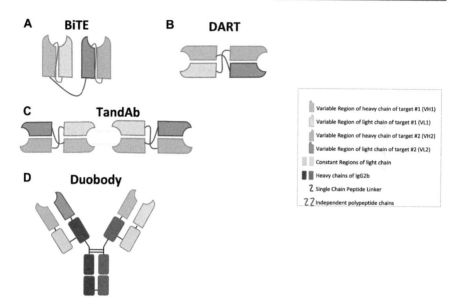

Fig. 8.1 Bispecific antibodies. **a** Bispecific T Cell Engager (BiTE): two single chain variable fragments (scFv) derived from Fab fragment of the IgG immunoglobulin, composed of VH and VL domains per antigen attached with a single polypeptide linker. **b** Dual affinity retargeting (DART): variable Regions of two antigen-binding specificities linked by two independent polypeptide chains. Each variable domain is formed by associating one VL segment on one chain and a VH segment on another chain, linked via disulfide bridge. **c** Tandem antibodies: tetravalent antibodies consisting of two VLs/VHs pairs from two distinct Fv, meaning they do not carry Fc domains. **d** Duobody: full-length human IgG bispecific antibody (bsAb) recognizing two targets, generated by controlled Fab-arm exchange

AMG 330 is a BiTE targeting CD3 and CD33 which was found to have activity in preclinical models, taking advantage of the frequent expression of CD33 in AML [36]. Ravandi et al., evaluated the efficacy of AMG 330 in a phase I trial for R/R AML (NCT#02520427). Eligible patients were ≥18 years old with >5% blasts in bone marrow and ≥1 line/s of prior therapy. 7 patients out of 55 enrolled achieved CR including four with incomplete hematologic recovery and a median duration of response of 58.5 days. 100% of patients reported AEs with 67% of patients experiencing cytokine release syndrome (CRS) and 13% having Grade 3 or greater CRS [37]. In vitro, long-term cultures of primary AML blasts in the presence of AMG330 induced T cell activation, PD-L1 overexpression on T cells, release of IFN-γ, TNF-α, IL-2, IL-10, and IL-6 [38, 39] and reduced bone marrow immune-suppressive $CD14^+$ HLA-DR^{low} monocytic like myeloid-derived suppressor cells [40]. These findings provided the rationale for combining AMG330 with pembrolizumab in an ongoing clinical trial. Recently, a new approach based on the combination of T cell engagers with immune checkpoint blockade in a single molecule, using a bifunctional checkpoint inhibitory T cell-engaging (CiTE) antibody that combines T cell redirection to CD33 on AML cells with locally restricted immune checkpoint blockade has been proposed, which could lead to increased clinical activity by combining the effects of checkpoint inhibitors and bispecific antibodies in a single molecule [41]. A challenge of AMG330 is its short half-life requiring continuous intravenous infusion. Other CD33xCD3 bispecific antibodies in clinical trials include AMG 673 (NCT03224819), AMV564 (NCT03144245), GEM333 (NCT03516760) and JNJ-67571244 (NCT03915379) in adult patients with R/R AML.

Flotetuzumab is a bispecific CD3 and CD123 dual affinity retargeting (DART) protein that engages T cells with CD123 expressing neoplastic cells [42]. CD123 has been identified both in blast cells [43] as well as Leukemic Stem Cells (LSC) [44]; it is expressed in around 45% of leukemic cells and has been associated with worse prognosis in AML, with patients expressing higher rates of CD123 showing significantly lower rates of response to induction when compared to patients with a lower expression of CD123 [45]. It confers a survival advantage over normal hematopoietic stem cells that do not express it via the formation of heterodimers with the cytokine receptor common beta chain CD131, whose activation-induced tyrosine phosphorylation that results in the activation of the JAK-STAT signaling pathway [46]. In an early phase trial, 38 patients with R/R AML deemed to be primary induction failure (PIF) or early relapse (ER) were treated with flotetuzumab at 500 ng/kg/day. The ORR rate was 42.1%, with seven patients achieving CR, with 68.8% proceeding to receive SCT. PIF patients showed an ORR of 45.8%, with five patients achieving CR. ORR for ER patients was 35.7% with two patients achieving CR. Median OS was 4.5 months. CRS occurred in all patients, including 13.3% at Grade 3 or above. T cell infiltration and the presence of CD123+ cells in the bone marrow are associated with response, on the other hand, nonresponders had elevated PD-L1 levels, suggesting the use of PD-L1 inhibitors as a potential way to overcome resistance [47]. Current trials are ongoing in R/R AML or Intermediate/High-risk MDS [48].

Uy et al., reported the results of a multicenter, open-label, phase 1/2 study of flotetuzumab in 88 adults with relapsed/refractory AML. 42 in a dose-finding segment and 46 at the recommended phase 2 dose (RP2D) of 500 ng/kg per day. Among 30 patients with PIF or ER treated at the RP2D, the CR/CRh (CR with partial hematologic recovery) rate was 26.7%, with a CR/CRh/CRi (CR with incomplete count recovery) of 30.0%. In patients achieving CR or CRh median OS was 10.2 months with 12-month survival rate of 50%. Notably, flotetuzumab induced CR even in TP53 mutated patients enrolled that represent a subset of AML patients with especially poor response to chemotherapy and consistently dismal outcomes. Patients with TP53 abnormalities who achieved a CR experienced encouraging OS (median 10.3 months; range, 3.3–21.3 months) [49, 50]. 24 PIF and 14 ER AML patients were treated at the RP2D (median age 63 years); the large majority (94.7) had non-favorable risk by ELN 2017 criteria. The overall CR/CRh/CRi rate was 42.1%, with 68.8% of patients subsequently able to undergo allo-SCT. Median OS was 4.5 months, with a median follow-up time of 10.8 months, in the entire population, and 7.7 months in responding patients. CRS (cytokine release syndrome) was the most common side effect, with all patients experiencing mild-to-moderate (Grade ≤ 2) CRS. The incidence of CRS progressively decreased during dosing at RP2D, allowing outpatient treatment in most cases [49].

Vibecotamab (XmAb14045) is a bispecific monoclonal antibody, with a full-length immunoglobulin molecule requiring intermittent infusions, targeting both CD123 and CD3 and stimulating targeted T cell-mediated killing of CD123-expressing cells. Recently, the development of a dual-targeting triple body CD33-CD16-CD123 (SPM-2) agent, with binding sites for target antigens CD33 and CD123, and for CD16 to engage NK as cytolytic effectors were reported, with promising clinical activity. Blasts from all 29 patients, including patients with genomic alterations associated with an unfavorable genetic subtype, were lysed at nanomolar concentrations of SPM-2. Maximum susceptibility was observed for cells with a combined density of CD33 and CD123 above 10,000 copies/cell [51]. In vivo studies with Vibecotamab in animal models showed good activity against CD123 cells [52]. A phase I study evaluated 104 elderly patients with a median age of 63 years who were heavily pretreated. ORR was 14% with two patients obtaining CR and 3 CRi [53]. Other CD123xCD3 bispecific antibodies in early phase clinical trials in patients with R/R AML include SAR440334 (NCT03594955), APVO436 (NCT03647800), an optimized ADAPTIR bispecific antibody, and JNJ-63709178 (NCT02715011).

APVO436 is an ADAPTIR (modular protein technology) molecule, consisting of BsAbs containing two sets of binding domains linked to immunoglobulin Fc domains, targeting CD3 and CD123 designed to reduce inflammatory cytokine release after T cell activation, as it does not activate T cells unless the molecule is bound to both target and effector cells [54]. In vitro and in vivo activity was demonstrated in CD123 expressing patient samples and mouse AML models [55]. Phase I trial evaluating its safety and tolerability in R/R AML or MDS patients is currently ongoing [56].

MCLA-117 is a modified full-length human bispecific IgG and is the only CLEC12A x CD3 bispecific antibody currently in a clinical trial in AML patients (NCT03038230). The target antigen, C-type lectin domain family 12 member A (CLEC12A aka CLL-1) is expressed in the majority of AML cases, including on LCSs, but has not been detected on healthy HSCs, making it an attractive target [57–59]; however, trials are yet to be reported.

AMG 427 is a CD3 x FLT3 BiTE and is being evaluated in a phase I clinical trial in adults with R/R AML (NCT03541369).

In addition to T cell-based immunotherapeutic approaches, NK cytotoxicity is also engaged through CD16 and CD33 Bispecific Killer Cell engagers (BiKE), showing vitro activity in killing AML [60] and MDS samples as well suppressing MDSCs [61]. A limitation to this approach is the short in vivo survival of NK cells, which led to the development of Trifunctional Killer Cell Engagers (TriKE) which add IL-15 to the molecule to promote NK-cell persistence. GTB-3550 is a TriKE currently in early phase trials which have shown promising response rates and no observed CRS in nine patients reported so far [62].

8.4 Chimeric Antigen Receptor (CAR) T Cells Redirecting T Cell Specificity

Chimeric Antigen Receptor (CAR)-T cell therapy has had a dramatic effect on the management of CD19$^+$ pediatric Acute Lymphoblastic Leukemia (ALL) with relapsed/refractory patients obtaining nearly 100% complete remission [63] as well as significant impact in adult ALL with still impressive CR rates of 70%, leading to regulatory approval [64]. CARs are synthetic receptors that retarget and reprogram T cells [65]. Unlike the physiological T cell receptor, which engages HLA-peptide complexes, CARs bind to native cell surface molecules that do not require antigen processing or HLA expression for tumor recognition [66]. CAR T cells can therefore recognize cell surface target antigens on any HLA background or in tumor cells that have downregulated HLA expression or proteasomal antigen processing, two mechanisms promoting tumor immune escape [67]. Multiple CAR T products are in trial for ALL and non-Hodgkin lymphoma (CD19$^+$ diseases) globally now with a variety of vectors and approaches. As such, CAR T cells are the first gene/cell therapies for cancer that have attained regulatory approval (Fig. 8.2). CAR T cells are bespoke products that first need native T cells to be collected from patient through leukapheresis, then T cells are engineered to express the CAR DNA transduced and integrated through a viral vector and are then expanded and cryopreserved ex vivo. Finally, they are tested prior to infusion to ensure sterility and adequate T cell dosage [68]. Patients might or might not receive bridging therapy to attain disease control during the CAR T manufacturing process; however, they must all receive lymphodepleting regimens to facilitate CAR T clonal expansion in vivo [69].

The application of CAR therapy for AML is challenged by the lack of suitable antigens, clonal heterogeneity, and increased recognition of bone marrow immunosuppressive factors, which has hindered therapeutic success. Identifying a target antigen that is exclusively expressed on AML cells including leukemia stem cells (LSCs) is one of the foremost challenges in this emerging field [70] (Fig. 8.3).

The first clinical experience of CAR T-cell therapy in AML was reported by Ritchie et al. who utilized a second-generation CAR—which includes a co-stimulatory domain in addition to the CD3-zeta domain [71]—directed against the Lewis Y antigen. Following fludarabine-containing preconditioning, four patients received up to 1.3×10^9 total T cells, of which 14–38% expressed the CAR. One patient achieved a cytogenetic remission whereas another with active leukemia had a reduction in peripheral blood (PB) blasts and a third showed a protracted remission. They demonstrated trafficking to the BM in those patients with the greatest clinical benefit. Furthermore, in a patient with leukemia cutis, CAR T cells infiltrated proven sites of disease. They also demonstrated that infused CAR T cells persisted for up to 10 months, but ultimately all patients experienced progression of disease despite CAR T persistence. No Grade 3 or 4 adverse events were observed [72].

A Chinese phase I clinical trial studied the feasibility of anti-CD33 CAR in the treatment of R/R AML. Only one patient was treated. Suggestions of a beneficial

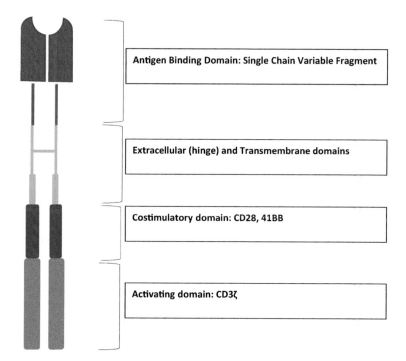

Fig. 8.2 Structure of a second-generation Chimeric Antigen Receptor. In red, the single chain variable fragment (scFv) providing antigen specificity and binding. In green on top, the extracellular domain (also known as hinge) and the transmembrane domain which crosses the cellular membrane. In purple, the costimulatory domain which is required for CAR T cell persistence. In dark green, the activating CD3zeta domain which is responsible of the cytotoxic activity

effect were present, but severe side effects such as fever, CRS, and pancytopenia were reported [73]. Tambaro et al., more recently reported the results of single-center, single-arm, phase I clinical trial (NCT03126864) investigating the feasibility and safety of autologous T cells, modified to express a CD33-targeted CAR with 4-1BB and CD3ζ endo-domains and co-expressed with truncated human epidermal growth factor receptor (HER1t) in patients with R/R AML. The goals of this phase I clinical trial were to assess the feasibility and safety of adoptive transfer of the autologous CD33-CAR T cells and identify the recommended phase II dose. Ten adults with R/R AML were enrolled; patients had received a median of 5 (range 3–8) prior treatment regimens; three underwent prior allo-SCT. Only 2/11 patients were able to be infused due to production failure or rapid progression of disease [74]. Recently, an approach to treat AML by targeting CD33 combines CD33-targeted CAR T cells with the transplantation of hematopoietic stem cells that have been engineered to knock out CD33 in human hematopoietic stem/progenitor CD34+ cells (HSPC). This was achieved by using the CRISPR/Cas9 technology which allows to edit genes, turning selected genes on and off quickly within cells [75]. Of note, CD33-KO HSPCs maintain their ability

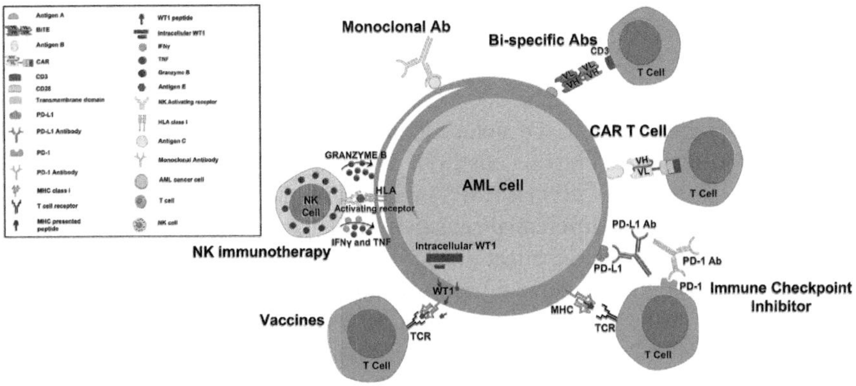

Fig. 8.3 Immune-based therapeutic interventions for AML. In clockwise fashion, we see: bispecific antibodies, targeting CD3 on T cells and an AML-associated cell surface antigen on AML cells; CAR T cells targeting an AML-associated cell surface antigen; immune checkpoint inhibitors blocking PD-L1 on the AML cell and PD-1 on the T cell through antibodies, thus facilitating the formation of a MHC/TCR complex; vaccines promoting the expansion of T cells reacting against AML cells upon prior exposure to an antigen such as WT1 protein; NK immunotherapy based on NK activating receptors, which interact with HLA class I (classical) on the AML cell surface, leading to direct NK cytotoxicity via granzymes, IFNγ and TNF. Finally, monoclonal antibodies directly binding to antigens on the AML cell surface

to engraft and repopulate normal hematopoiesis. A clinical trial evaluating safety and toxicity of such approach is currently under design for R/R AML patients [76]. CD33 has been also tested as target for preclinical validation of CAR engineered CIK cells [77]. Cytokine-induced killer (CIK) cells represent a terminally differentiated $CD3^+ CD56^+$ T cell population with both T cell and NK-cell-like phenotype and functionality. CD33 CAR-CIK cells showed significant anti-leukemic activity in vitro and, when used in patient-derived AML xenograft models, were capable of contrasting AML development both as early treatment and in mice with established disease. These data may support further development and implementation of CD33 CAR-CIK cells into early clinical trials.

Preclinical models have explored alternative targets such as CD123 [78], CLL-1 [79], CD70 [80] and FLT-3 [81], among others [82]. A case report with CD123 CARs was part of a reduced-intensity conditioning regimen in haploidentical SCT for R/R AML, leading to CR [83, 84]. Phase I clinical trials with CD123 CARs are currently ongoing (NCT03766126, NCT02159495).

Autologous NKG2D CAR T cells proved ineffective in a phase I trial for patients with R/R AML and MM [85]. Cui et al., investigated the clinical therapeutic efficacy and safety of CD38-targeted CAR T (CAR T-38) cells in 6 AML patients who experienced relapse post-allo-HSCT (NCT04351022). Four weeks after the initial infusion of CAR T-38 cells, four of six (66.7%) patients achieved CR or CRi; the median CR or CRi time was 191 (range 117–261) days. The cumulative relapse rate at 6 months was 50%. The median OS and leukemia-free survival (LFS) times were 7.9 and 6.4 months, respectively. One case relapsed

117 days after the first CAR T-38 cell infusion, with remission achieved after the second CAR T-38 cell infusion. All six patients experienced clinically manageable side effects [86].

In a human xenograft mouse model, CLL-1 CAR-Ts mediated anti-leukemic activity prolonging survival. Of note, normal progenitor cells were able to form colonies of differentiated blood cells [87]. This means that the differentiation and proliferation potentials of normal HSPCs remained intact following CLL-1 CAR T treatment, suggesting a differential sensitivity of leukemic versus normal myeloid compartment to the cytotoxic effect of CLL-1 CAR T cells [88]. Bu et al., reported the results from a phase I clinical trial evaluating the toxicity and efficacy of the CLL-1 CAR T in the treatment of pediatric R/R AML. 3 pediatric AML patients were enrolled and, CAR T cells were given a dose at $0.2 - 1 \times 10^6$/kg with a single dose. All patients achieved CR after CAR T treatment and 2/3 patients achieved MRD negativity and were successfully bridged to allo-SCT. CRS occurred in all three patients (2 patients experienced Grade 1 CRS and one patient Grade 2 CRS); no neurotoxicity occurred. All patients suffered pancytopenia, granulocytopenia, and monocytopenia [89]. Liu et al., evaluated the toxicity and efficacy of CLL1-CD33 CAR T cells in the treatment of R/R AML and explored the possibility of using a reduced-intensity conditioning of HSCT after such CAR therapy. 9 R/R AML patients were enrolled (NCT03795779). Eight dual CARs were manufactured from autologous cells, and a ninth was from an HLA-matched sibling donor. All patients who received conditioning of fludarabine and cyclophosphamide and CAR T cells were given a dose-escalation at $1 \sim 3 \times 10^6$/kg with a single or split dose. Within 4 weeks post-CAR T cell infusion, 7 of 9 patients were MRD negative by flow cytometry, 2 of 9 had no response, one of which was $CD33^+/CLL1^-$, indicating the importance of CLL1 target in the CAR T treatment. For the seven patients who reached MRD negativity, six patients moved to a subsequent HSCT, except one patient, who withdrew from the study due to a personal issue. One of six received a standard myeloablative conditioning; 5 of 6 received a reduced intensity regimen. Five patients successfully engrafted with a persistent full chimerism, except one who died of sepsis on day +6 before engraftment. CRS occurred in eight patients (3 Grade I, 3 Grade II, and 2 Grade III). Neurotoxicity occurred in four patients (1 Grade I and 3 Grade III). All patients suffered Grade IV pancytopenia, five had a mildly increased liver enzyme, four had a coagulation disorder, four with diarrhea, one with skin rash, 1 with renal insufficiency, three with sepsis, two with fungi infection, and three with pneumonia [90].

A splice variant of the adhesive receptor CD44 has been used as CAR target in AML. CD44 is a glycoprotein involved in cell–cell and cell–matrix interactions and is widely expressed both in hematologic and solid neoplasms. The isoform contains the variant domain 6 of CD44 gene (CD44v6) and contributes to the cancer stem cell phenotype. CD44v6-targeted CAR T cells in AML and MM are being explored in a phase I/II clinical trial, to evaluate their safety and early efficacy (NCT04097301).

CD8+ and CD4+ T cells expressing an FLT3-specific CAR have been developed, demonstrating potent reactivity against wild-type or mutated FLT3 AML cells [91].

Of interest, the FLT3-inhibitor crenolanib increased surface expression of FLT3 thus eliciting CAR T cells recognition and activity. However, FLT3-CAR T-cells also recognized normal HSCs in vitro and in vivo, being profoundly myelotoxic.

Currently, there are about 50 active trials registered to use CAR T technology for the treatment of AML from clinicaltrials.gov.

8.5 Vaccines

The results of vaccine-based studies support the principle that antigen-specific immune responses may be elicited in AML patients. Vaccines are an inoculation of antigens with or without an adjuvant, in an attempt to induce host Cytotoxic T Cells (CTLs) to mount an immune response against cancer cells based on exposure and recognition of neoplastic epitopes that ideally are not expressed on normal cells [92]. Among the antigens used for this purpose, Wilms' tumor 1 (WT1) has probably accumulated the most relevant evidence in the clinical setting; other target antigens, such as Proteinase-3 (PR3), preferentially expressed antigen of melanoma (PRAME) and receptor for hyaluronic acid-mediated motility (RHAMM) are under clinical investigation for the treatment of AML.

WT-1 is a protein encoded by the WT-1 gene, a tumor suppressor gene associated with the development of Wilms' Tumor, from which it was named. Mutations in exon 7 and 9 of WT1 have been recurrently identified in AML and associated with poorer prognosis and chemotherapy resistance. It is highly expressed in AML and is known to be leukemogenic [93]. WT-1 peptide vaccine was noted to induce expansion of WT-1 specific T cells [94] which have been used in the treatment and prevention of relapsed AML and MDS. Clinical studies have recruited a relatively low number of patients and different epitopes have all been noted to be safe with only 8% (7/88) of patients experiencing any Grade III-IV toxicity [95]. Clinical activity has been modest with some patients maintaining long-term CR [96] and one patient achieving CR after relapsed disease [97]. Maslak et al., conducted a phase 2 study investigating a multivalent WT1 peptide vaccine (Galinpepimut-S) in adults with AML in CR1. Patients received six vaccinations administered over 10 weeks with the potential to receive six additional monthly doses if they remained in CR1. 22 AML patients were treated. Fifteen patients (68%) relapsed, and 10 (46%) died. Median disease-free survival from CR1 was 16.9 months. The vaccine was well tolerated, with the most common toxicities being Grade 1/2 injection site reactions (46%), fatigue (32%), and skin induration (32%) [98]. Post HSCT use of this strategy decreased relapse rates and improved OS compared to historical controls in a single-arm study [99].

Proteinase 3 (P3) is a serine proteinase present in the primary granules of neutrophils. Preliminary studies demonstrated that PR3-specific CD8 T cell responses can be elicited in patients with AML, offering the rational to exploit this antigen in leukemia vaccination strategies [100]. Recently, PR1, an HLA-A2-restricted peptide derived from both proteinase 3 and neutrophil elastase, has been tested in a phase I/II peptide vaccination trial. Qazilbash et al., evaluated safety, immunogenicity, and clinical activity of PR1 vaccination in a phase I/II trial. 66 HLA-A2$^+$ patients with AML (42), chronic myeloid leukemia (CML: 13), or MDS (11) received three to six PR1 peptide vaccinations, administered subcutaneously every 3 weeks at dose levels of 0.25, 0.5 or 1.0 mg. Primary end points were safety and immune response, assessed by doubling of PR1/HLA-A2 tetramer-specific CTL, and the secondary end point was clinical response. Of the 53 evaluable patients with active disease, 12 (24%) had objective clinical responses (complete: 8; partial: 1 and hematologic improvement: 3). Immune responses were noted in 35 of 66 (53%) patients [101].

Van de Loosdrecht reported the results of phase II study (NCT03697707) showing the capability of the allogeneic leukemia-derived dendritic cell vaccine, DCP-001 to convert MRD positive patients to negative. Eligible patients were AML patients ineligible for HSCT and in CR1 but still MRD positive. Patients received a primary vaccination regimen of four times 25.10^6 or 50.10^6 cells per vaccination, biweekly, followed by two booster vaccinations (10.10^6 cells/vaccination) at weeks 14 and 18 after start of treatment. Primary endpoints of this trial were the safety and tolerability of the two vaccination schedules and the effect of vaccination on the MRD status. 10 patients were enrolled and dosed within the study, completing the first dose cohort of 25.10^6 cells/vaccination. Four patients could be evaluated for MRD; two patients became MRD negative at the first timepoint after the initial vaccinations (week 14) and remained negative until end of active FU (week 32), two other patients remained in CR, but with MRD positivity. All vaccinations were well tolerated and adverse events to the vaccine were limited to local injection site reactions such as redness, swelling, and warmth (maximum Grade 2) [102].

Ho et al., conducted a phase I clinical trial in which high-risk AML or MDS patients were immunized with irradiated, autologous, GM-CSF-secreting tumor cells early after allo-SCT. 9 of 10 subjects who completed vaccination achieved durable CR, with a median follow-up of 26 months. Six long-term responders showed marked decreases in the levels of soluble NKG2D ligands, and three demonstrated normalization of cytotoxic lymphocyte NKG2D expression as a function of treatment. Local and systemic reactions were qualitatively similar to those previously observed in non-transplanted, immunized solid-tumor patients [103].

Based on the initial demonstration that polyclonal memory CD8+T cell responses to PRAME-specific peptides can be detectable in patients with different hematologic malignancies, including AML [104], other studies addressed the capacity to elicit anti-leukemic response. By using an in vitro model with peptide-loaded allogeneic dendritic cells, PRAME-reactive and functional CTLs could be isolated from healthy individuals and maintained in culture, but not from AML patients, even when samples were collected in patients that had recently achieved a CR [105]. Although not conclusive and possibly dependent on the type of strategy used to elicit anti-leukemic response, these results suggest that important differences may exist for AML patients and further studies are warranted to establish the potential of PRAME as effective target antigen for AML vaccination.

A phase I clinical trial included 10 patients with hematologic malignancies, who were vaccinated with a highly immunogenic CD8+T cell epitope peptide derived from RHAMM. In seven of 10 patients, an increase of fully functional RHAMM-specific CD8+effector T cells was observed, and some clinical responses were reported in 1/3 AML patients [106].

The mixed results of WT-1 and other antigens-directed treatments have further highlighted the need of identifying novel immunogenic tumor-specific antigens (TSAs) [107].

8.6   Adoptive NK-Cell Therapies

Natural killer (NK) cells are lymphocytes belonging to the innate immune system. Their cytotoxic activity is tightly regulated by the balanced expression of a wide and complicated pattern of activating and inhibitory receptors, among which inhibitory receptors called killer-immunoglobulin-like receptors (KIRs) have been extensively characterized [108]. Based on their crucial role, several strategies of NK-cell-based immunotherapy are currently under active preclinical and clinical investigation and early data from clinical studies have been reported [109]. A wide variety of sources of therapeutic NK cells have been proposed both at preclinical and early clinical level, including haploidentical NK cells, umbilical cord blood NK cells, stem cell derived NK cells, NK-cell lines, adaptive NK cells, cytokine-induced memory-like NK cells, and CAR NK cells [110].

Infusion of haploidentical NK cells has a major role in the eradication of residual AML cells after haploidentical SCT [111], boosting graft-versus-leukemia (GVL) effects without exacerbating graft versus host disease (GVHD). In the setting of haploidentical transplantation, the KIR-KIR-L mismatch in the graft-versus-leukemia direction, which defines the donor's NK alloreactivity, has been shown to significantly impact the post-transplant clinical outcome by virtue of its effect in regulating donor NK-cell cytotoxicity against AML cells [112].

Outside the transplant setting, Miller and colleagues [113] demonstrated that adoptive immunotherapy with haploidentical NK cells in AML patients was feasible, safe, and effective in AML patients, likely exerting their effect through killer immunoglobulin receptors (KIRs) and nonclassical HLA loci. The trial evaluated 19 patients with poor-risk AML—defined as primary refractory disease, relapsed disease after chemotherapy, secondary AML arising from MDS or relapsed disease at least 3 months after allo-SCT—out of which five achieved morphologic CR after infusion of haploidentical activated NK cells along with subcutaneous IL-2 with prior lymphodepleting treatment. In a retrospective analysis, the authors also showed that a better response was associated with KIR-L mismatch between donor and recipient. Several reports demonstrated the feasibility and early clinical efficacy of selecting and infusing highly purified, T cell-depleted, KIR-mismatched NK cells to consolidating remission in high-risk patients with AML [114]. Of note, the percentage of donor-derived alloreactive NK-cell clones before infusion signif-icantly correlated with relapse occurrence, indicating that a functional dose of 2×10^5/kg alloreactive NK cells may be predictive of response [115].

To potentiate effector cell function of infused NK cells several approaches, both in vivo and in vitro, of NK-cell activation have been tested. To this end, in vivo interleukin IL-2 administration has been widely used, resulting in expansion and cytotoxicity enhancement of infused NK cells [115]. However, IL-2 is well-known to increase the number and the suppressive function of regulatory T cells (Tregs) [110], which in turn antagonize NK cells by reducing their expansion capacity and effector function [116]. Accordingly, the number of circulating Tregs after NK-cell infusion critically influences the capacity of infused NK cells to expand and to kill AML cells. Of note, depletion of host Tregs by an IL-2 diphtheria toxin fusion protein was associated with increased NK-cell expansion and higher response rates in adults with relapsed AML, resulting in better overall and disease-free survival [117]. To overcome some of the limitations associated with IL-2 administration—such as considerable side effects including capillary leak syndrome, cytopenias, hypotension, renal failure, fever/rigors, liver toxicity, and neurotoxicity [118]—other cytokines are currently under investigation for in vivo activation of NK cells. IL-15 stimulates both NK cells and T cells. IL-15 in complex with its IL-15R-α receptor induces a potent proliferative signal that expands NK-cell and CD8 T cell populations in murine models [119]. Recently, the results of two first-in-human phase 1 and phase 2 trials of recombinant human IL-15 (rhIL-15) given with haplo-NK-cell therapy after lymphodepletion to treat R/R AML have been reported. 42 patients received either intravenous (iv) or subcutaneous (sc) rhIL-15. Robust in vivo NK-cell expansion at day 14 was observed in a significant fraction of patients and correlated with the achievement of some complete responses. Of note, rhIL-15 induced higher fractions of in vivo NK-cell expansion and remission compared with previous trials with IL-2 −35% versus 10%, respectively but it was associated with previously unreported CRS after subcutaneous but not intravenous dosing [120].

Activation of NK cells to be used as a means of adoptive immunotherapy may be also obtained by priming NK cells before infusion. Ex vivo activation may have the advantage to avoid some of the detrimental and toxic effects associated with in vivo cytokines administration. To this aim, overnight activation with IL-2 was used to prime haploidentical NK cells before infusion in a cohort of R/R AML and high-risk-MDS patients [121]. NK cells were enriched from non-mobilized donor PB mononuclear cells (PBMC) after T- and B-cell depletion. The number of responses, including CR, was promising, allowing in some cases to proceed to allo-SCT.

Biological studies suggest a correlation between clinical response and the fre-quency of donor NK cells at day 7/14 as well as reduced activation of CD8 T cells and lower levels of inflammatory cytokines after infusion [121].

Human NK cells primed with the CTV-1 leukemia cell line lysate CNDO-109 have been shown to exhibit enhanced cytotoxicity against NK-cell-resistant cell lines [122]. On this basis, a phase I trial with CNDO-109-activated donor-derived haploidentical NK cells (CNDO-109-NK cells) was conducted in AML patients who had obtained CR at high-risk for relapse as part of a consolidation strategy. Before CNDO-109-NK-cell administration, patients were treated with lymphodepleting fludarabine/cyclophosphamide. CNDO-109-NK cells were well tolerated and durable CRs were observed in some cases, which may suggest a clinical activity of the whole strategy warranting further studies [123].

In the development of NK-cell-based adoptive immunotherapy, one major limitation has been the collection of a sufficient number of NK cells to reach a therapeutic effect. For this reason, several attempts to expand NK cells have been made both at preclinical and clinical levels. Most studies have used cytokines, such as IL-2 and/or IL-15, as a means of ex vivo NK-cell expansion [124].

Recently, other approaches have been tested combining cytokine-based priming with additional stimuli. One method to expand NK cells ex vivo using K562 feeder cells expressing membrane-bound IL-21 has been recently reported. This approach expands highly functional NK cells up to 35,000-fold in 3 weeks [125], providing the preclinical rationale for a phase 1 study to determine safety, feasibility, and maximum tolerated dose [126]. Thirteen patients with myeloid malignancies in remission, including AML, at high-risk of disease occurrence, were treated before and after haploidentical SCT to prevent relapse. Lymphodepleting chemotherapy was administered before infusion. No infusion reactions or dose-limiting toxicities occurred. Of interest, NK-cell reconstitution was superior to that observed in a similar group of patients not receiving NK cells. Clinically, infusion of NK cells was associated with improved low relapse and incidence of viral infections. To achieve high number of functional NK cells, NK-cell product for adoptive immunotherapy was also generated ex vivo from CD34+ hematopoietic stem and progenitor cells (HSPC) which had been obtained from partially HLA-matched umbilical cord blood units [127]. Moving from the hypothesis that generating more homogeneous and well-defined allogeneic NK-cell products may be associated with better clinical results, the authors had previously developed a culture system where CD34+ HSPC isolated from allogeneic umbilical cord blood (UCB) were expanded and differentiated into NK cells in the presence of IL15 and IL2, resulting in a clinically relevant dose of highly purified NK cells [128].

These allogeneic HSPC-NK cells showed cytolytic activity against AML cells in vitro and in vivo mouse models [129]. Ten AML patients in CR received escalating doses of HSPC-NK cells after lymphodepleting chemotherapy without additional in vivo cytokine administration. No infusion-related adverse events were noted; no CRS or GVHD was reported either. Of interest, despite the absence of in vivo cytokine boosting, HSPC-NK cells were clearly detectable in a significant fraction of patients. Donor chimerism was documented also in the BM of infused patients. From a clinical point of view, it is noteworthy that some patients harboring MRD in bone marrow became MRD negative. These findings suggest that HSPC-NK-cell adoptive transfer is a promising and, importantly, potential "off-the-shelf" strategy for NK-cell-based adoptive immunotherapy in AML.

8.7   Monoclonal Antibodies

Antibodies are naturally produced by B lymphocytes as a reaction to antigen exposure and fulfill several roles in the immune system across various isotypes. They were first approved for the treatment of solid and hematologic malignancies in the 1990s [130] and are now widely used in cancer treatment. Monoclonal Antibodies (MAbs) exert their anti-cancer properties through antibody-dependent cell-mediated cytotoxicity (ADCC) and phagocytosis which depend on native NK cells and monocytes/macrophages against neoplastic cells via FcγR recognition of the Fc domain in IgG molecules [131]. Mutations of FcγR genes may influence tumor clinical response to antibody therapy through expression of lower affinity receptors in certain subsets [132]. IgG is the most commonly used backbone for therapeutic MAbs [133], exerting its function through the patient's immune system [134] as well altering the downstream cellular transduction pathways of malignant cells [135]. IgG can also induce complement-mediated cytotoxicity through fixation of complement and activation of its classical pathway [136] and this accounts for some of their in vivo efficacy [137]. Some MAbs can alter downstream cellular signal transduction within cancer cells leading to decreased cell growth rates and chemosensitization [138]. Other MAbs are unable to produce anti-neoplastic effects on their own but can still have useful applications by delivering a cytotoxic payload which can include chemotherapy, toxins, radioisotopes, and cytokines [130].

The first ADC Gemtuzumab Ozogamicin (GO) or Mylotarg targeted CD33 which is a transmembrane cell adhesion and interaction protein that appears to be limited to myeloid cells and is expressed in about 90% of patients with AML [139]; it was initially approved by the FDA in 2000 in the R/R setting. However, it was withdrawn from the market after a confirmatory phase III trial failed to show benefit [140]. It was then reapproved in September 2017 based on the ALFA-0701 trial, a randomized phase III open-label trial which compared standard chemotherapy with or without GO in 271 patients which reported superior EFS (Event Free Survival) in the GO arm 17 versus 10 months; however, there was no improvement in OS [141]. Hills et al. published a meta-analysis including 3325 patients which showed benefit in patients with favorable or intermediate-risk cytogenetics including reduced risk or relapse and improvement in 5 year OS; however, this meta-analysis did not include the updated survival data from the original ALFA-0701 study, where the survival advantage lost its statistical significance [142]. Additionally, the development of hepatic veno-occlusive disease (VOD) limits its use to CEBPA-mutated AML [143]. There are currently several other CD33-linked ADC in investigation for the treatment of AML [144].

CD123 has been targeted using a recombinant fusion protein that consists of human IL-3 fused to a truncated diphtheria toxin (DT) payload; it binds to CD123 and is then internalized to release the DT into the leukemic cell, producing inhibition of protein synthesis and apoptosis of targeted cells [145]. In December 2018, it received FDA approval for the treatment of Blastic Plasmacytoid Dendritic Cell

Neoplasm (BPDCN) in adults and children over 2 years old based on an open-label, multicohort study on which 47 patients were randomized to either 7 or 12 ug/kg on days 1–5 of each 21-day cycle. An ORR of 90% was achieved and 45% of patients went on to undergo SCT; with survival rates at 18 and 24 months of 59% and 52%, respectively; of note, there were 15 previously treated patients who had an ORR of 67% and a median OS of 8.5 months. Side effects included elevation of liver enzymes (60%), peripheral edema (51%), thrombocytopenia (49%), and capillary leak syndrome in up to 19% of the patients [146]. Retrospective data evaluating Tagraxofusp vs Hyper-CVAD or CHOP-based regimens showed no difference in CR and OS; with patients who were able to receive allogeneic SCT having significantly longer survival (HR: 0.16) [147]. A current trial is evaluating the role of Tagraxofusp maintenance post SCT in BPDCN patients [148]. Another CD123 targeting molecule is IMGN632, a humanized anti-CD123 antibody linked to indolinobenzodiazepine pseudodimer, which induced cell death through alkylation of the DNA of the CD123 expressing cell without crosslinking it [149]. A phase 1/2 dose-escalation study of 74 R/R patients (67 AML and 7 BPDCN); with 70% of AML patients being adverse risk by ELN 2017 criteria and showing an overall response rate of 20%. This increased to 32% when evaluating non-secondary AML; ORR for BPDCN was 43%. Side effects include diarrhea, febrile neutropenia, [149], and nausea [150]. IMGN632 is currently being evaluated in R/R CD123+ hematologic malignancies [151] as well as its use in conjunction with hypomethylating agents in CD123+ AML [152].

The humanized anti-CD47 antibody Magrolimab was shown to have in vitro and in vivo activity against AML through promotion of phagocytosis [153], which was shown to be increased in preclinical models by the addition of azacitidine [154]. Magrolimab is currently being studied in combination with azacytidine in patients with treatment-naive AML who are deemed unfit for intensive chemotherapy, ORR was 65% and the CR rate was 44%. Of note, in the TP53mut subset, which this cohort was enriched for, ORR was 71% and CR rate was 48%; cytopenias, particularly on-target anemia, occurred at acceptable rates [155].

CD70 a member of the Tumor Necrosis Factor Receptor (TNFR) superfamily which is expressed on activated immune cells and binds to CD27 to induce immune suppression in several pathologic states [156]. CD70 expression was demonstrated in CD34+ AML cells and its activation caused disease progression, and its expression was also associated with worse prognosis [157]. This led to the development of cusatuzumab, a human CD70 monoclonal antibody that was initially shown to have activity against LSCs in vitro and in mouse models, as well as in combination with azacitidine [158]. Based on these results two early phase clinical trials are currently underway evaluating cusatuzumab clinical efficacy in AML in association with Hypomethylating Agents (HMA). CULMINATE (NCT04023526) is evaluating the combination of cusatuzumab at 20 mg/kg and azacytidine in de novo or secondary AML unfit for intensive therapy [159]. A pre-planned interim analysis in 52 patients showed a CR rate of 27% and a composite CR (CRc) rate of 40% and results are currently being evaluated by the study sponsor [160]; a second phase Ib study ELEVATE (NCT04150887) is currently evaluating cusatuzumab with azacitidine and venetoclax in a similar patient population [161].

Other potential targets that are preferentially expressed on leukemic blasts and LSC include CD44 [162], CD96 [163], CLL-1 [164], LILRB4 [165] which are currently under investigation [166].

8.8   Active Areas of Research

The anti-leukemic response in allo-HSCT mediated by a graft-versus-leukemia effect shows that donor T cells and NK cells can recognize and eliminate leukemia cells. The positive correlation of lymphocyte recovery after chemotherapy and response rates and rare cases of spontaneous remission in AML patients upon activation of the immune system through pathogen infections support the immunogenicity of leukemia [167]. Despite this evidence, the best targets for immunotherapy of AML remain unknown. Targeting leukemia-associated antigens that are up-regulated normal cell surface gene products such as CD33 or CD123 raises safety concerns often referred to as "on-target off-tumor" toxicity, given the expression of this class of targets in normal tissues and organs including normal counterparts (i.e., hematopoietic stem and progenitor cells) or even non-hematopoietic tissues [168]. On the other hand, leukemia-specific antigens (i.e., neoantigens) are generally derived from leukemia-specific mutations in protein-coding genes (e.g., WT1 and NPM1). However, AML bears few genetic mutations and such neoantigens are often expressed intracellularly, and their potential as immunotherapeutic targets depends on HLA expression or proteasomal antigen processing, two mechanisms often downregulated in AML as a mechanism of tumor immune escape.

Little is known about the genetic mutational quality rather than quantity and target antigen expression. Multiple mechanisms driven by frequently recurring AML mutations, including epigenetic mechanisms, such as DNA methylation or, RNA splicing, may qualitatively and quantitatively alter the leukemia cell surface proteome, regulating the surface abundance of immunotherapy targets. Neoantigen production does not only result from DNA mutations and altered RNA splicing may lead to neoepitopes production [169–173]. Recent studies showed that perturbing RNA splicing in combination with immune checkpoint blockade may promote control of established tumors [169]. Splicing modulation generates many novel mRNAs derived from large-scale events, including inclusion of intronic sequence into mature mRNA, juxtaposition of exons not normally spliced together, and exons with abnormal 50 or 30 ends. Each can result in the production of peptides or proteins containing novel sequences potentially contributing to disease immunogenicity.

We previously developed a target discovery strategy for CAR therapy of AML by integrating large sets of proteomic and transcriptomic data from malignant and normal tissues. As we did not identify single targets with an ideal expression profile, we identified combinatorial pairings that could potentially enhance selective targeting of AML cells [174]. As such, an active area of research investigations in this field involves combinatorial targeting that may address the issue of AML clonal heterogeneity and prevent antigen escape.

As immunosuppressive bone marrow factors are recognized in AML, many groups are trying to overcome such additional challenges to develop effective immunotherapy for AML. T cell dysfunction has been documented at initial diagnosis [175] as well as during progression [176] and restored when CR is obtained [177]. The mechanisms behind this include regulatory T cells (Tregs) [178], myeloid-derived suppressor cells (MDSC) [179], cytokines [180] and immune checkpoints such as CTLA-4 [181], PD-1 [182], TIM-3 [183] and T cell immunoglobulin and immunoreceptor tyrosine-based inhibitory motif (ITIM) domain (TIGIT) [184]. Tregs are a specialized subpopulation of T cells that act to suppress immune responses by inhibiting T cell proliferation and cytokine production. The number of Tregs in AML patients is significantly increased, contributing to the immune escape of AML [185]. MDSCs facilitate tumor growth and development by suppressing T cell function; patients with AML exhibit increased presence of MDSCs in their peripheral blood, and this is associated with a shorter OS [179]. TIGIT is a T cell coinhibitory receptor; its expression on CD8+ T cells is elevated in AML patients. It contributes to functional T cell impairment while associating with poor clinical outcome in AML [184]. Cytokines exert profound effects on AML. While pro-inflammatory mediators such as IL-1β, TNF-α and IL-6 tend to generally increase AML aggressiveness, anti-inflammatory mediators such as TGF-β and IL-10 appear to impede AML progression. Dysregulation of the complex interactions between pro- and anti-inflammatory cytokines in AML may create a pro-tumorigenic microenvironment with effects on leukemic cell proliferation, survival, and drug resistance [186]. Thus, the interplay between leukemic blasts and the bone marrow microenvironment affects response to therapy including immunotherapy [187]. Hematopoietic stem cells on the endosteal surface of the bone marrow also interact with a variety of cellular and extracellular components, such as osteoblasts, macrophages, collagen, and laminin fibers. These cells can condition and reshape the microenvironment, facilitating leukemia cell proliferation and survival. Previous studies showed that co-culture of AML blasts with bone marrow stromal cells stimulated blasts' survival and inhibited chemotherapy-induced apoptosis, highlighting the critical role of the microenvironment with implications for treatment strategies for AML patients [188, 189].

Chemotherapeutic agents induce immunogenic cell death that leads to the release of neoantigens and damage-associated molecular patterns (DAMPS) such as ATP into the extracellular space. Extracellular ATP can bind to purinergic P2RX7 receptors on dendritic cells (DCs), thus triggering activation of the inflammasome and release of IL-1β in the lymph node. Hyperactive DCs induce strong antigen-specific responses, which in combination with IL-1β binding to IL-1R1 on T cells may generate cytotoxic responses such as IFNγ and granzyme B release by CD8$^+$ T cells and NK cells and kill malignant cells. Inflammasomes may potentiate DCs to promote long-lived memory T cells.

On the other side, activation of the innate immune system through various ligands and signaling pathways is also an important driver of MDS and AML. The DAMPs, or alarmins, which activate the inflammasome pathway via the TLR4/NLR signaling cascade cause the lytic cell death of normal hematopoietic stem and progenitor cells (HSPCs), ineffective hematopoiesis, and β-catenin-induced proliferation of cancer cells, leading to the development of MDS/AML phenotype [190]. Many studies suggest the crucial role of the immune system, inflammation, and inflammasomes in the pathogenesis of myeloid malignancies. The interleukin 1 receptor accessory protein (IL1RAP) is overactivated in MDS/AML HSPCs and is enriched in high-risk diseases with worse prognosis [191]. IL1RAP can act as a coactivator for FLT3 signaling thus playing a stimulatory role in malignant myeloid expansion [192]. Overexpression of genes involved in innate immune pathways is reported in over 50% of MDS patients [193]. Of note, Höckendorf et al., suggested a tumor suppressor role of the inflammasome in AML [194]. Thus, players involved in the inflammasome pathway may soon represent novel pharmacological targets against MDS/AML. For example, the IL-1β neutralizing antibody Canakinumab, the soluble decoy IL-1 receptor Rilonacept, and the recombinant IL-1 receptor antagonist Anakinra, currently approved for autoimmune diseases are already in clinical trials for AML patients (NCT04810611 and NCT04239157). Similarly, anti-IL1RAP/CD3 bispecific antibodies are under investigation (NCT02842320). Thus, inflammasomes and immune response pathways have opened avenues for exciting new drug targets for AML.

References

1. Dores GM et al (2012) Acute leukemia incidence and patient survival among children and adults in the United States, 2001–2007. Blood 119(1):34–43
2. Shah A et al (2013) Survival and cure of acute myeloid leukaemia in England, 1971–2006: a population-based study. Br J Haematol 162(4):509–516
3. Yates JW et al (1973) Cytosine arabinoside (NSC-63878) and daunorubicin (NSC-83142) therapy in acute nonlymphocytic leukemia. Cancer Chemother Rep 57(4):485–488
4. Santos GW et al (1983) Marrow transplantation for acute nonlymphocytic leukemia after treatment with busulfan and cyclophosphamide. N Engl J Med 309(22):1347–1353
5. Weiden PL et al (1979) Antileukemic effect of graft-versus-host disease in human recipients of allogeneic-marrow grafts. N Engl J Med 300(19):1068–1073
6. Kolb H et al (1990) Donor leukocyte transfusions for treatment of recurrent chronic myelogenous leukemia in marrow transplant patients. Blood 76(12):2462–2465
7. Schiller GJ (2013) High-risk acute myelogenous leukemia: treatment today … and tomorrow. Hematology Am Soc Hematol Educ Program 2013:201–208
8. Hodi FS et al (2010) Improved survival with Ipilimumab in patients with metastatic melanoma. N Engl J Med 363(8):711–723
9. Wolchok JD (2015) PD-1 blockers. Cell 162(5):937
10. Zhang L, Gajewski TF, Kline J (2009) PD-1/PD-L1 interactions inhibit antitumor immune responses in a murine acute myeloid leukemia model. Blood 114(8):1545–1552
11. Daver N et al (2016) Defining the immune checkpoint landscape in patients (pts) with Acute Myeloid Leukemia (AML). Blood 128(22):2900–2900
12. Chen C et al (2020) Expression patterns of immune checkpoints in acute myeloid leukemia. J Hematol Oncol 13(1):28

13. Berger R et al (2008) Phase I safety and pharmacokinetic study of CT-011, a humanized antibody interacting with PD-1, in patients with advanced hematologic malignancies. Clin Cancer Res 14(10):3044–3051
14. Davids MS et al (2016) Ipilimumab for patients with relapse after allogeneic transplantation. N Engl J Med 375(2):143–153
15. Reville PK et al (2021) Nivolumab maintenance in high-risk acute myeloid leukemia patients: a single-arm, open-label, phase II study. Blood Cancer J 11(3):60
16. Ravandi F et al (2019) Idarubicin, cytarabine, and nivolumab in patients with newly diagnosed acute myeloid leukaemia or high-risk myelodysplastic syndrome: a single-arm, phase 2 study. Lancet Haematol 6(9):e480–e488
17. Daver N et al (2019) Efficacy, safety, and biomarkers of response to azacitidine and Nivolumab in relapsed/refractory acute myeloid leukemia: a nonrandomized, open-label, phase II study. Cancer Discov 9(3):370
18. Acharya N, Sabatos-Peyton C, Anderson AC (2020) Tim-3 finds its place in the cancer immunotherapy landscape. J Immunother Cancer 8(1)
19. Jan M et al (2011) Prospective separation of normal and leukemic stem cells based on differential expression of TIM3, a human acute myeloid leukemia stem cell marker. Proc Natl Acad Sci U S A 108(12):5009–5014
20. Haubner S et al (2019) Coexpression profile of leukemic stem cell markers for combinatorial targeted therapy in AML. Leukemia 33(1):64–74
21. Asayama T et al (2017) Functional expression of Tim-3 on blasts and clinical impact of its ligand galectin-9 in myelodysplastic syndromes. Oncotarget 8(51):88904–88917
22. Borate U et al (2020) Anti-TIM-3 antibody MBG453 in combination with hypomethylating agents in patients with high-risk myelodysplastic syndrome and acute myeloid leukemia: a phase 1 study. In: Abstract presented at: the 25th European Hematology Association Congress
23. Chao MP et al (2019) Therapeutic targeting of the macrophage immune checkpoint CD47 in myeloid malignancies. Front Oncol 9:1380
24. Molica M et al (2021) TP53 mutations in acute myeloid leukemia: still a daunting challenge? Front Oncol 10(3368)
25. Sallman DA et al (2020) Tolerability and efficacy of the first-in-class anti-CD47 antibody magrolimab combined with azacitidine in MDS and AML patients: phase Ib results. J Clin Oncol 38(15_suppl):7507–7507
26. Papaemmanuil E et al (2016) Genomic classification and prognosis in acute myeloid leukemia. N Engl J Med 374(23):2209–2221
27. Rizvi NA et al (2015) Cancer immunology. Mutational landscape determines sensitivity to PD-1 blockade in non-small cell lung cancer. Science 348(6230):124–128
28. McGranahan N et al (2016) Clonal neoantigens elicit T cell immunoreactivity and sensitivity to immune checkpoint blockade. Science 351(6280):1463–1469
29. Alexandrov LB et al (2013) Signatures of mutational processes in human cancer. Nature 500(7463):415–421
30. Van Allen EM et al (2015) Genomic correlates of response to CTLA-4 blockade in metastatic melanoma. Science 350(6257):207–211
31. Lawrence MS et al (2013) Mutational heterogeneity in cancer and the search for new cancer-associated genes. Nature 499(7457):214–218
32. Vago L et al (2009) Loss of mismatched HLA in leukemia after stem-cell transplantation. N Engl J Med 361(5):478–488
33. Huehls AM, Coupet TA, Sentman CL (2015) Bispecific T-cell engagers for cancer immunotherapy. Immunol Cell Biol 93(3):290–296
34. Ross SL et al (2017) Bispecific T cell engager (BiTE(R)) antibody constructs can mediate bystander tumor cell killing. PLoS One. 12(8):e0183390
35. Guy DG, Uy GL (2018) Bispecific antibodies for the treatment of acute myeloid leukemia. Curr Hematol Malig Rep 13(6):417–425

36. Laszlo GS et al (2014) Cellular determinants for preclinical activity of a novel CD33/CD3 bispecific T-cell engager (BiTE) antibody, AMG 330, against human AML. Blood 123(4):554–561
37. Ravandi F et al (2020) Updated results from phase I dose-escalation study of AMG 330, a bispecific T-cell engager molecule, in patients with relapsed/refractory acute myeloid leukemia (R/R AML). J Clin Oncol 38(15_suppl):7508–7508
38. Krupka C et al (2014) CD33 target validation and sustained depletion of AML blasts in long-term cultures by the bispecific T-cell-engaging antibody AMG 330. Blood 123(3):356–365
39. Friedrich M et al (2014) Preclinical characterization of AMG 330, a CD3/CD33-bispecific T-cell-engaging antibody with potential for treatment of acute myelogenous leukemia. Mol Cancer Ther 13(6):1549–1557
40. Jitschin R et al (2018) CD33/CD3-bispecific T-cell engaging (BiTE(R)) antibody construct targets monocytic AML myeloid-derived suppressor cells. J Immunother Cancer 6(1):116
41. Herrmann M et al (2018) Bifunctional PD-1 x alphaCD3 x alphaCD33 fusion protein reverses adaptive immune escape in acute myeloid leukemia. Blood 132(23):2484–2494
42. Chichili GR et al (2015) A CD3xCD123 bispecific DART for redirecting host T cells to myelogenous leukemia: preclinical activity and safety in nonhuman primates. Sci Transl Med 7(289):289ra82
43. Cruz NM et al (2018) Selection and characterization of antibody clones are critical for accurate flow cytometry-based monitoring of CD123 in acute myeloid leukemia. Leuk Lymphoma 59(4):978–982
44. Jordan CT et al (2000) The IL-3 receptor alpha chain is a unique marker for human acute myelogenous leukemia stem cells. Leukemia 14(10):1777–1784
45. Testa U et al (2002) Elevated expression of IL-3Ralpha in acute myelogenous leukemia is associated with enhanced blast proliferation, increased cellularity, and poor prognosis. Blood 100(8):2980–2988
46. Muñoz L et al (2001) IL-3 receptor alpha chain (CD123) is widely expressed in hematologic malignancies. Haematologica 86(12):1261–1269
47. Godwin JE et al (2019) Flotetuzumab (FLZ), an investigational CD123 x CD3 bispecific Dart® protein-induced clustering of CD3+ T cells and CD123+ AML cells in bone marrow biopsies is associated with response to treatment in primary refractory AML patients. Blood 134(Supplement_1):1410–1410
48. Flotetuzumab in Primary Induction Failure (PIF) or Early Relapse (ER) Acute Myeloid Leukemia (AML)
49. Vadakekolathu J et al (2020) TP53 abnormalities correlate with immune infiltration and associate with response to flotetuzumab immunotherapy in AML. Blood Adv 4(20):5011–5024
50. Aldoss I et al (2020) Flotetuzumab as salvage therapy for primary induction failure and early relapse acute myeloid leukemia. Blood 136(Supplement 1):16–18
51. Braciak TA et al (2018) Dual-targeting triplebody 33–16–123 (SPM-2) mediates effective redirected lysis of primary blasts from patients with a broad range of AML subtypes in combination with natural killer cells. Oncoimmunology 7(9):e1472195
52. Chu SY et al (2014) Immunotherapy with long-lived anti-CD123 × anti-CD3 bispecific antibodies stimulates potent T cell-mediated killing of human AML cell lines and of CD123+ cells in monkeys: a potential therapy for acute myelogenous leukemia. Blood 124(21):2316–2316
53. Ravandi F et al (2020) Complete responses in relapsed/refractory acute myeloid leukemia (AML) patients on a weekly dosing schedule of Vibecotamab (XmAb14045), a CD123 x CD3 T cell-engaging bispecific antibody; initial results of a phase 1 study. Blood 136(Supplement 1):4–5
54. Hernandez-Hoyos G et al (2016) MOR209/ES414, a novel bispecific antibody targeting PSMA for the treatment of metastatic castration-resistant prostate cancer. Mol Cancer Ther 15(9):2155–2165

55. Comeau MR et al (2018) Abstract 1786: APVO436, a bispecific anti-CD123 x anti-CD3 ADAPTIR™ molecule for redirected T-cell cytotoxicity, induces potent T-cell activation, proliferation and cytotoxicity with limited cytokine release. Can Res 78(13 Supplement):1786
56. Study of APVO436 in Patients With AML or MDS
57. Morsink LM, Walter RB, Ossenkoppele GJ (2019) Prognostic and therapeutic role of CLEC12A in acute myeloid leukemia. Blood Rev 34:26–33
58. Wang YY et al (2017) Low CLL-1 expression Is a novel adverse predictor in 123 patients with De Novo CD34(+) acute myeloid leukemia. Stem Cells Dev 26(20):1460–1467
59. van Rhenen A et al (2007) Aberrant marker expression patterns on the CD34+CD38- stem cell compartment in acute myeloid leukemia allows to distinguish the malignant from the normal stem cell compartment both at diagnosis and in remission. Leukemia 21(8):1700–1707
60. Singer H et al (2010) Effective elimination of acute myeloid leukemic cells by recombinant bispecific antibody derivatives directed against CD33 and CD16. J Immunother 33(6):599–608
61. Gleason MK et al (2014) CD16xCD33 bispecific killer cell engager (BiKE) activates NK cells against primary MDS and MDSC CD33+ targets. Blood 123(19):3016–3026
62. GT Biopharma I (2021) GT biopharma announces interim GTB-3550 TRIKE™ monotherapy clinical trial results AT 2021 Raymond James human health innovation conference. Available from: https://www.gtbiopharma.com/news-media/press-releases/detail/225/gt-bio pharma-announces-interim-gtb-3550-trike
63. Majzner RG, Mackall CL (2019) Clinical lessons learned from the first leg of the CAR T cell journey. Nat Med 25(9):1341–1355
64. Shah BD et al (2021) KTE-X19 anti-CD19 CAR T-cell therapy in adult relapsed/refractory acute lymphoblastic leukemia: ZUMA-3 phase 1 results. Blood 138(1):11–22
65. Jensen MC, Riddell SR (2015) Designing chimeric antigen receptors to effectively and safely target tumors. Curr Opin Immunol 33:9–15
66. Sadelain M, Riviere I, Riddell S (2017) Therapeutic T cell engineering. Nature 545(7655):423–431
67. Zhou G, Levitsky H (2012) Towards curative cancer immunotherapy: overcoming posttherapy tumor escape. Clin Dev Immunol 2012:124187
68. Levine BL et al (2017) Global manufacturing of CAR T cell therapy. Mol Ther Methods Clin Dev 4:92–101
69. Klebanoff CA et al (2005) Sinks, suppressors and antigen presenters: how lymphodepletion enhances T cell-mediated tumor immunotherapy. Trends Immunol 26(2):111–117
70. Epperly R, Gottschalk S, Velasquez MP (2020) A bump in the road: how the hostile AML microenvironment affects CAR T cell therapy. Front Oncol 10:262
71. Hartmann J et al (2017) Clinical development of CAR T cells-challenges and opportunities in translating innovative treatment concepts. EMBO Mol Med 9(9):1183–1197
72. Ritchie DS et al (2013) Persistence and efficacy of second generation CAR T cell against the LeY antigen in acute myeloid leukemia. Mol Ther 21(11):2122–2129
73. Wang QS et al (2015) Treatment of CD33-directed chimeric antigen receptor-modified T cells in one patient with relapsed and refractory acute myeloid leukemia. Mol Ther 23(1):184–191
74. Tambaro FP et al (2021) Autologous CD33-CAR-T cells for treatment of relapsed/refractory acute myelogenous leukemia. Leukemia
75. Redman M et al (2016) What is CRISPR/Cas9? Arch Dis Child Educ Pract Ed 101(4):213–215
76. Borot F et al (2019) Gene-edited stem cells enable CD33-directed immune therapy for myeloid malignancies. Proc Natl Acad Sci U S A 116(24):11978–11987
77. Rotiroti MC et al (2020) Targeting CD33 in chemoresistant AML patient-derived xenografts by CAR-CIK cells modified with an improved SB transposon system. Mol Ther 28(9):1974–1986
78. Loff S et al (2020) Rapidly switchable universal CAR-T cells for treatment of CD123-positive Leukemia. Mol Ther Oncolytics 17:408–420

79. Wang J et al (2018) CAR-T cells targeting CLL-1 as an approach to treat acute myeloid leukemia. J Hematol Oncol 11(1):7
80. Sauer T et al (2021) CD70-specific CAR T cells have potent activity against acute myeloid leukemia without HSC toxicity. Blood 138(4):318–330
81. Shrestha E et al (2020) Preclinical development of anti-FLT3 CAR-T therapy for the treatment of acute myeloid leukemia. Blood 136(Supplement 1):4–5
82. Mardiana S, Gill S (2020) CAR T cells for acute myeloid Leukemia: state of the art and future directions. Front Oncol 10(697)
83. Study evaluating safety and efficacy of CAR-T cells targeting CD123 in patients with acute myelocytic Leukemia
84. Yao S et al (2019) Donor-derived CD123-targeted CAR T cell serves as a RIC regimen for haploidentical transplantation in a patient with FUS-ERG+ AML. Front Oncol 9:1358
85. Baumeister SH et al (2019) Phase I trial of autologous CAR T cells targeting NKG2D ligands in patients with AML/MDS and multiple myeloma. Cancer Immunol Res 7(1):100–112
86. Cui Q et al (2021) CD38-directed CAR-T cell therapy: a novel immunotherapy strategy for relapsed acute myeloid leukemia after allogeneic hematopoietic stem cell transplantation. J Hematol Oncol 14(1):82
87. Youn BS, Mantel C, Broxmeyer HE (2000) Chemokines, chemokine receptors and hematopoiesis. Immunol Rev 177:150–174
88. Tashiro H et al (2017) Treatment of acute myeloid leukemia with T cells expressing chimeric antigen receptors directed to C-type lectin-like molecule 1. Mol Ther 25(9):2202–2213
89. Bu C et al (2020) Phase I clinical trial of autologous CLL1 CAR-T therapy for pediatric patients with relapsed and refractory acute myeloid leukemia. Blood 136(Supplement 1):13–13
90. Liu F et al (2018) First-in-human CLL1-CD33 compound CAR T cell therapy induces complete remission in patients with refractory acute myeloid leukemia: update on phase 1 clinical trial. Blood 132:901
91. Jetani H et al (2018) CAR T-cells targeting FLT3 have potent activity against FLT3(-)ITD(+) AML and act synergistically with the FLT3-inhibitor crenolanib. Leukemia 32(5):1168–1179
92. Thomas S, Prendergast GC (2016) Cancer vaccines: a brief overview. Methods Mol Biol 1403:755–761
93. Sugiyama H (2005) Cancer immunotherapy targeting Wilms' tumor gene WT1 product. Expert Rev Vaccines 4(4):503–512
94. Mailänder V et al (2004) Complete remission in a patient with recurrent acute myeloid leukemia induced by vaccination with WT1 peptide in the absence of hematologic or renal toxicity. Leukemia 18(1):165–166
95. Di Stasi A et al (2015) Review of the results of WT1 peptide vaccination strategies for myelodysplastic syndromes and acute myeloid leukemia from nine different studies. Front Immunol 6(36)
96. Tsuboi A et al (2012) Long-term WT1 peptide vaccination for patients with acute myeloid leukemia with minimal residual disease. Leukemia 26(6):1410–1413
97. Keilholz U et al (2009) A clinical and immunologic phase 2 trial of Wilms tumor gene product 1 (WT1) peptide vaccination in patients with AML and MDS. Blood 113(26):6541–6548
98. Maslak PG et al (2018) Phase 2 trial of a multivalent WT1 peptide vaccine (galinpepimut-S) in acute myeloid leukemia. Blood Adv 2(3):224–234
99. Anguille S et al (2017) Dendritic cell vaccination as postremission treatment to prevent or delay relapse in acute myeloid leukemia. Blood 130(15):1713–1721
100. Scheibenbogen C et al (2002) CD8 T-cell responses to Wilms tumor gene product WT1 and proteinase 3 in patients with acute myeloid leukemia. Blood 100(6):2132–2137
101. Qazilbash MH et al (2017) PR1 peptide vaccine induces specific immunity with clinical responses in myeloid malignancies. Leukemia 31(3):697–704
102. van de Loosdrecht A et al (2020) Conversion from MRD positive to negative status in AML patients in CR1 after treatment with an allogeneic leukemia-derived dendritic cell vaccine. Blood 136(Supplement 1):13–14

103. Ho VT et al (2009) Biologic activity of irradiated, autologous, GM-CSF-secreting leukemia cell vaccines early after allogeneic stem cell transplantation. Proc Natl Acad Sci USA 106(37):15825–15830

104. Rezvani K et al (2009) Ex vivo characterization of polyclonal memory CD8+ T-cell responses to PRAME-specific peptides in patients with acute lymphoblastic leukemia and acute and chronic myeloid leukemia. Blood 113(10):2245–2255

105. van den Ancker W et al (2013) Priming of PRAME- and WT1-specific CD8(+) T cells in healthy donors but not in AML patients in complete remission: implications for immunotherapy. Oncoimmunology 2(4):e23971

106. Schmitt M et al (2008) RHAMM-R3 peptide vaccination in patients with acute myeloid leukemia, myelodysplastic syndrome, and multiple myeloma elicits immunologic and clinical responses. Blood 111(3):1357–1365

107. Ehx G et al (2021) Atypical acute myeloid leukemia-specific transcripts generate shared and immunogenic MHC class-I-associated epitopes. Immunity 54(4):737-752.e10

108. Lanier LL (2005) NK cell recognition. Annu Rev Immunol 23:225–274

109. Lichtenegger FS et al (2017) Recent developments in immunotherapy of acute myeloid leukemia. J Hematol Oncol 10(1):142

110. Myers JA, Miller JS (2021) Exploring the NK cell platform for cancer immunotherapy. Nat Rev Clin Oncol 18(2):85–100

111. Ruggeri L et al (2002) Effectiveness of donor natural killer cell alloreactivity in mismatched hematopoietic transplants. Science 295(5562):2097–2100

112. Ruggeri L et al (2007) Donor natural killer cell allorecognition of missing self in haploidentical hematopoietic transplantation for acute myeloid leukemia: challenging its predictive value. Blood 110(1):433–440

113. Miller JS et al (2005) Successful adoptive transfer and in vivo expansion of human haploidentical NK cells in patients with cancer. Blood 105(8):3051–3057

114. Rubnitz JE et al (2010) NKAML: a pilot study to determine the safety and feasibility of haploidentical natural killer cell transplantation in childhood acute myeloid leukemia. J Clin Oncol 28(6):955–959

115. Curti A et al (2016) Larger size of donor alloreactive NK cell repertoire correlates with better response to NK cell immunotherapy in elderly acute myeloid leukemia patients. Clin Cancer Res 22(8):1914–1921

116. Gasteiger G et al (2013) IL-2-dependent tuning of NK cell sensitivity for target cells is controlled by regulatory T cells. J Exp Med 210(6):1167–1178

117. Bachanova V et al (2014) Clearance of acute myeloid leukemia by haploidentical natural killer cells is improved using IL-2 diphtheria toxin fusion protein. Blood 123(25):3855–3863

118. Marabondo S, Kaufman HL (2017) High-dose IL-2 (IL-2) for the treatment of melanoma: safety considerations and future directions. Expert Opin Drug Saf 16(12):1347–1357

119. Cooper MA et al (2002) In vivo evidence for a dependence on IL-15 for survival of natural killer cells. Blood 100(10):3633–3638

120. Cooley S et al (2019) First-in-human trial of rhIL-15 and haploidentical natural killer cell therapy for advanced acute myeloid leukemia. Blood Adv 3(13):1970–1980

121. Björklund AT et al (2018) Complete remission with reduction of high-risk clones following haploidentical NK-cell therapy against MDS and AML. Clin Cancer Res 24(8):1834–1844

122. North J et al (2007) Tumor-primed human natural killer cells lyse NK-resistant tumor targets: evidence of a two-stage process in resting NK cell activation. J Immunol 178(1):85–94

123. Fehniger TA et al (2018) A phase 1 trial of CNDO-109-activated natural killer cells in patients with high-risk acute myeloid leukemia. Biol Blood Marrow Transplant 24(8):1581–1589

124. Fujisaki H et al (2009) Expansion of highly cytotoxic human natural killer cells for cancer cell therapy. Cancer Res 69(9):4010–4017

125. Denman CJ et al (2012) Membrane-bound IL-21 promotes sustained ex vivo proliferation of human natural killer cells. PLoS One 7(1):e30264

126. Ciurea SO et al (2017) Phase 1 clinical trial using mbIL21 ex vivo-expanded donor-derived NK cells after haploidentical transplantation. Blood 130(16):1857–1868

127. Dolstra H et al (2017) Successful transfer of umbilical cord blood CD34(+) hematopoietic stem and progenitor-derived NK cells in older acute myeloid leukemia patients. Clin Cancer Res 23(15):4107–4118

128. Spanholtz, J., et al., *Clinical-Grade generation of active NK cells from cord blood hematopoietic progenitor cells for immunotherapy using a closed-system culture process.* PLoS One, 2011. **6**(6): p. e20740.

129. Cany J et al (2013) Natural killer cells generated from cord blood hematopoietic progenitor cells efficiently target bone marrow-residing human leukemia cells in NOD/SCID/IL2Rg(null) mice. PLoS One 8(6):e64384

130. Adams GP, Weiner LM (2005) Monoclonal antibody therapy of cancer. Nat Biotechnol 23(9):1147–1157

131. Raghavan M, Bjorkman PJ (1996) Fc receptors and their interactions with immunoglobulins. Annu Rev Cell Dev Biol 12:181–220

132. Gavin PG et al (2017) Association of polymorphisms in FCGR2A and FCGR3A with degree of trastuzumab benefit in the adjuvant treatment of ERBB2/HER2–positive breast cancer. JAMA Oncol 3(3):335

133. DeVita VT, Lawrence TS, Rosenberg SA (2008) DeVita, Hellman, and Rosenberg's cancer: principles and practice of oncology. Wolters Kluwer/Lippincott Williams & Wilkins, Philadelphia

134. Steplewski Z, Lubeck MD, Koprowski H (1983) Human macrophages armed with murine immunoglobulin G2a antibodies to tumors destroy human cancer cells. Science 221(4613):865–867

135. Trauth BC et al (1989) Monoclonal antibody-mediated tumor regression by induction of apoptosis. Science 245(4915):301–305

136. Walport MJ (2001) Complement. First of two parts. N Engl J Med 344(14):1058–1066

137. Boross P et al (2011) The in vivo mechanism of action of CD20 monoclonal antibodies depends on local tumor burden. Haematologica 96(12):1822–1830

138. Li S et al (2005) Structural basis for inhibition of the epidermal growth factor receptor by cetuximab. Cancer Cell 7(4):301–311

139. Fenwarth L et al (2020) Biomarkers of Gemtuzumab Ozogamicin Response for Acute Myeloid Leukemia Treatment. Int J Mol Sci 21(16)

140. Petersdorf SH et al (2013) A phase 3 study of gemtuzumab ozogamicin during induction and postconsolidation therapy in younger patients with acute myeloid leukemia. Blood 121(24):4854–4860

141. Lambert J et al (2019) Gemtuzumab ozogamicin for de novo acute myeloid leukemia: final efficacy and safety updates from the open-label, phase III ALFA-0701 trial. Haematologica 104(1):113–119

142. Hills RK et al (2014) Addition of gemtuzumab ozogamicin to induction chemotherapy in adult patients with acute myeloid leukaemia: a meta-analysis of individual patient data from randomised controlled trials. Lancet Oncol 15(9):986–996

143. Cortes JE et al (2020) Prevention, recognition, and management of adverse events associated with gemtuzumab ozogamicin use in acute myeloid leukemia. J Hematol Oncol 13(1)

144. Han YC et al (2021) Development of highly optimized antibody-drug conjugates against CD33 and CD123 for acute myeloid leukemia. Clin Cancer Res 27(2):622–631

145. Frankel AE et al (2000) Characterization of diphtheria fusion proteins targeted to the human IL-3 receptor. Protein Eng 13(8):575–581

146. Pemmaraju N et al (2019) Tagraxofusp in blastic plasmacytoid dendritic-cell neoplasm. N Engl J Med 380(17):1628–1637

147. Yun S et al (2020) Survival outcomes in blastic plasmacytoid dendritic cell neoplasm by first-line treatment and stem cell transplant. Blood Adv 4(14):3435–3442

148. Tagraxofusp in treating patients with blastic plasmacytoid dendritic cell neoplasm after stem cell transplant

149. Kovtun Y et al (2018) A CD123-targeting antibody-drug conjugate, IMGN632, designed to eradicate AML while sparing normal bone marrow cells. Blood Adv 2(8):848–858

150. Daver NG et al (2019) Clinical profile of IMGN632, a novel CD123-targeting antibody-drug conjugate (ADC). In: Patients with relapsed/refractory (R/R) acute myeloid leukemia (AML) or blastic plasmacytoid dendritic cell neoplasm (BPDCN). Blood 134(Supplement_1):734
151. Study of IMGN632 in patients with untreated BPDCN and relapsed/refractory BPDCN
152. IMGN632 as monotherapy or with venetoclax and/or azacitidine for patients with CD123-positive acute myeloid leukemia
153. Liu J et al (2015) Pre-clinical development of a humanized anti-CD47 antibody with anti-cancer therapeutic potential. PLOS ONE 10(9):e0137345
154. Feng D et al (2018) Combination treatment with 5F9 and azacitidine enhances phagocytic elimination of acute myeloid leukemia. Blood 132:2729
155. Sallman D et al (2020) The first-in-class anti-CD47 antibody magrolimab combined with azacitidine is well-tolerated and effective in AML patients: phase 1b results. Blood 136(Supplement 1):330
156. Nolte MA et al (2009) Timing and tuning of CD27-CD70 interactions: the impact of signal strength in setting the balance between adaptive responses and immunopathology. Immunol Rev 229(1):216–231
157. Riether C et al (2017) CD70/CD27 signaling promotes blast stemness and is a viable therapeutic target in acute myeloid leukemia. J Exp Med 214(2):359–380
158. Riether C et al (2020) Targeting CD70 with cusatuzumab eliminates acute myeloid leukemia stem cells in patients treated with hypomethylating agents. Nat Med 26(9):1459–1467
159. Trudel GC et al (2020) CULMINATE: a phase II study of cusatuzumab + azacitidine in patients with newly diagnosed AML, ineligible for intensive chemotherapy. J Clin Oncol 38(15_suppl):TPS7565
160. Globenewswire. argenx announces 2021 corporate priorities and highlights recent achievements across immunology pipeline. 2021 July 28, 2021]. Available from: https://www.globenewswire.com/fr/news-release/2021/01/08/2155322/0/en/argenx-Announces-2021-Corporate-Priorities-and-Highlights-Recent-Achievements-Across-Immunology-Pipeline.html
161. Cusatuzumab in combination with background therapy for the treatment of participants with acute myeloid leukemia
162. Jin L et al (2006) Targeting of CD44 eradicates human acute myeloid leukemic stem cells. Nat Med 12(10):1167–1174
163. Liu F et al (2020) CD96, a new immune checkpoint, correlates with immune profile and clinical outcome of glioma. Sci Rep 10(1):10768
164. Ma H et al (2019) Targeting CLL-1 for acute myeloid leukemia therapy. J Hematol Oncol 12(1):41
165. Anami Y et al (2020) LILRB4-targeting antibody-drug conjugates for the treatment of acute myeloid leukemia. Mol Cancer Ther
166. Saito Y et al (2010) Identification of therapeutic targets for quiescent, chemotherapy-resistant human leukemia stem cells. Sci Transl Med 2(17):17ra9
167. Mozafari R, Moeinian M, Asadollahi-Amin A (2017) Spontaneous complete remission in a patient with acute myeloid leukemia and severe sepsis. Case Rep Hematol 2017:9593750
168. Perna F (2021) Safety starts with selecting the targets. Mol Ther 29(2):424–425
169. Lu SX et al (2021) Pharmacologic modulation of RNA splicing enhances anti-tumor immunity. Cell 184(15):4032-4047.e31
170. Jayasinghe RG et al (2018) Systematic analysis of splice-site-creating mutations in cancer. Cell Rep 23(1):270-281.e3
171. Dong C et al (2021) Intron retention-induced neoantigen load correlates with unfavorable prognosis in multiple myeloma. Oncogene
172. Kahles A et al (2018) Comprehensive analysis of alternative splicing across tumors from 8705 patients. Cancer Cell 34(2):211-224.e6
173. (2017) Integrated genomic characterization of pancreatic ductal adenocarcinoma. Cancer Cell 32(2):185–203.e13
174. Perna F et al (2017) Integrating proteomics and transcriptomics for systematic combinatorial chimeric antigen receptor therapy of AML. Cancer Cell 32(4):506–519 e5

175. Schnorfeil FM et al (2015) T cells are functionally not impaired in AML: increased PD-1 expression is only seen at time of relapse and correlates with a shift towards the memory T cell compartment. J Hematol Oncol 8:93
176. Zhou Q et al (2011) Coexpression of Tim-3 and PD-1 identifies a CD8+ T-cell exhaustion phenotype in mice with disseminated acute myelogenous leukemia. Blood 117(17):4501–4510
177. Lichtenegger FS et al (2014) Impaired NK cells and increased T regulatory cell numbers during cytotoxic maintenance therapy in AML. Leuk Res 38(8):964–969
178. Shenghui Z et al (2011) Elevated frequencies of CD4$^+$ CD25$^+$ CD127lo regulatory T cells is associated to poor prognosis in patients with acute myeloid leukemia. Int J Cancer 129(6):1373–1381
179. Pyzer AR et al (2017) MUC1-mediated induction of myeloid-derived suppressor cells in patients with acute myeloid leukemia. Blood 129(13):1791–1801
180. Mansour I et al (2016) Indoleamine 2,3-dioxygenase and regulatory T cells in acute myeloid leukemia. Hematology 21(8):447–453
181. LaBelle JL et al (2002) Negative effect of CTLA-4 on induction of T-cell immunity in vivo to B7-1+, but not B7-2+, murine myelogenous leukemia. Blood 99(6):2146–2153
182. Chen X et al (2008) Clinical significance of B7-H1 (PD-L1) expression in human acute leukemia. Cancer Biol Ther 7(5):622–627
183. Kikushige Y et al (2010) TIM-3 is a promising target to selectively kill acute myeloid leukemia stem cells. Cell Stem Cell 7(6):708–717
184. Kong Y et al (2016) T-cell immunoglobulin and ITIM domain (TIGIT) associates with CD8+ T-cell exhaustion and poor clinical outcome in AML patients. Clin Cancer Res 22(12):3057–3066
185. Wan Y et al (2020) Hyperfunction of CD4 CD25 regulatory T cells in de novo acute myeloid leukemia. BMC Cancer 20(1):472
186. Binder S, Luciano M, Horejs-Hoeck J (2018) The cytokine network in acute myeloid leukemia (AML): a focus on pro- and anti-inflammatory mediators. Cytokine Growth Factor Rev 43:8–15
187. Rashidi A, Uy GL (2015) Targeting the microenvironment in acute myeloid leukemia. Curr Hematol Malig Rep 10(2):126–131
188. Garrido SM et al (2001) Acute myeloid leukemia cells are protected from spontaneous and drug-induced apoptosis by direct contact with a human bone marrow stromal cell line (HS-5). Exp Hematol 29(4):448–457
189. Bendall LJ et al (1994) Bone marrow adherent layers inhibit apoptosis of acute myeloid leukemia cells. Exp Hematol 22(13):1252–1260
190. Chakraborty S et al (2021) Therapeutic targeting of the inflammasome in myeloid malignancies. Blood Cancer J 11(9):152
191. Barreyro L et al (2012) Overexpression of IL-1 receptor accessory protein in stem and progenitor cells and outcome correlation in AML and MDS. Blood 120(6):1290–1298
192. Mitchell K et al (2018) IL1RAP potentiates multiple oncogenic signaling pathways in AML. J Exp Med 215(6):1709–1727
193. Barreyro L, Chlon TM, Starczynowski DT (2018) Chronic immune response dysregulation in MDS pathogenesis. Blood 132(15):1553–1560
194. Hockendorf U et al (2016) RIPK3 restricts myeloid leukemogenesis by promoting cell death and differentiation of leukemia initiating cells. Cancer Cell 30(1):75–91

Off-the-Shelf Chimeric Antigen Receptor Immune Cells from Human Pluripotent Stem Cells

9

Handi Cao and Ryohichi Sugimura

9.1 The History of CAR-T Therapy

The invention of CAR-T cells and adoptive cell therapy (ACT) is a recent breakthrough. The use of patients' immune cells to treat cancers dated back to 1902 when Blumenthal and E. von Leyden tried to treat their cancer patients with suspension derived from autologous tumor tissue culture. Some beneficial effects can be noted in individuals but without significant disease remission [1]. ACT mainly involves the isolation of the patient's tumor-specific immune cells, especially T cells, genetic modification, the proliferation of these cells in vitro, and infusion back to the patient circulation following a lymphoid-depleting conditioning regimen, such as fludarabine and cyclophosphamide, for cancer treatment [2]. Three forms of adoptive T cell transfer have been developed for cancer immunotherapy, including tumor-infiltrating lymphocytes (TILs), T cell receptor (TCR) T cells, and chimeric antigen receptor (CAR) T cells [3]. There are many approaches to modify immune cells in the laboratory while CAR-T is successfully used in clinical trials. The first use of genetically engineered T cells following the aforementioned ACT canonical workflow for cancer treatment was reported in 1989 [4]. In the mid-1990s, the term CAR-T was first described but the results from the preclinical and clinical study were not satisfactory [5]. Nevertheless, as more and more modifications and improvements were applied to CAR-T design, the promising therapeutic effect of CAR-T therapy has been demonstrated and the huge success of CAR-T therapy emerged. FDA approved the first CAR-T therapy called tisagenlecleucel in August 2017 for children with relapsed B cell acute lymphoblastic leukemia treatment [1].

H. Cao · R. Sugimura (✉)
School of Biomedical Sciences, Li Ka Shing Faculty of Medicine, The University of Hong Kong, Hong Kong, China
e-mail: Rios@hku.hk

© The Author(s), under exclusive license to Springer Nature Switzerland AG 2022
P. Hays (ed.), *Cancer Immunotherapies*, Cancer Treatment and Research 183,
https://doi.org/10.1007/978-3-030-96376-7_9

CARs as synthetic receptors are generally composed of a specific domain from a monoclonal antibody that can detect corresponding tumor antigen, a T cell activation domain usually derived from the CDζ chain, and a linker domain that bridges the two domains. CAR-T cells can direct tumor cells automatically under the guidance of the antigen detection domain, then the T cell activation domain elicits downstream signals to activate T cells to perform antitumor response [6]. Eshhar's group showed that these CAR-T therapy-related synthetic receptors endow T cells with MHC-independent target recognition compared with engineered TCRs therapy [4, 7]. Eshhar developed the first-generation CAR-T cells targeting 2, 4, 6-trinitrophenyl (TNP)-bearing cells. They removed TCR variable regions and replaced them with antibody variable regions based on a similar structure. These CAR-T cells were composed of VH and VL chains derived from TNP antibody, TCR constant domain, and transmembrane segment. Nevertheless, the results from the initial clinical trial using the first-generation CAR-T cells did not display satisfactory antitumor effects [4]. The first-generation CAR is most likely to fail to fully engage genetically modified T cells because activation is initiated by antigen-dependent signals through the chimeric CD3ζ chain, independent of costimulation through accessory molecules [8]. To enhance the efficacy of CAR-T cells, many modifications were performed, leading to the generation of the second-generation CAR-T cells [9]. Second-generation CARs are improved by the addition of costimulatory domains, such as CD28, OX40, or 4-1BB (also known as CD137), linked with CD3ζ. Although the first-generation CARs displayed disappointing anti-cancer efficacy in clinical trials, the second-generation CARs targeting CD19 with costimulatory domains emerged as a great success in 2011 [10, 11]. CD19 has become a nearly ideal target in CAR-T therapy for B cell malignancies. More and more clinical trials of CAR-T targeting BCMA and CD22 have been carried out and showed significant anti-cancer effects in multiple myeloma and acute lymphoblastic leukemia, respectively [3]. In 2017, FDA has already two autologous second-generation CAR-T cells products due to the promising therapeutic effect in patients with hematologic malignancies, tisagenlecleucel (Kymriah, Novartis) and axicabtagene (Yescarta, Kite Pharma) targeting CD19, for the treatment of relapsed or refractory B cell acute lymphoblastic leukemia (ALL) and relapsed or refractory diffuse large B cell lymphoma and primary mediastinal large B cell lymphoma [12–14]. Third-generation CARs convey two costimulatory domains together to further enhance the antitumor activity [15, 16]. Nowadays, the fourth-generation CARs as the newest version have emerged with additional functional domains, which can precisely control CAR-T cell activity or further effectively enhance CAR-T potency [17]. Diaconu et al. reported that the inclusion of inducible pro-apoptotic protein caspase-9 (iC9) safety switch into the vector encoding the CAR can terminate the effect of CAR-T cells in a humanized mouse model by using chemical inducer of dimerization, which can efficiently eliminate $85\% \sim 90\%$ CARs once cytokine release syndrome (CRS) or severe toxicities occur [18]. Phase I trial of fourth-generation anti-CD19 CAR-T cells with iCasp9 suicide switch (4SCAR19) has been carried out [19]. TRUCK T cells refer to CAR-T cells with a transgenic "payload" and belong to another type of fourth-generation CAR. These

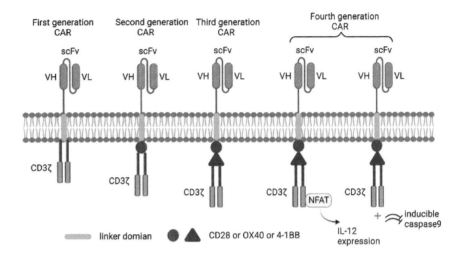

Fig. 9.1   Overview of CAR-T development history

TRUCK T cells can shape the tumor microenvironment by the inducible release of transgenic immune modifiers, such as IL-12, to eliminate antigen-negative cancer cells in the targeted lesion [20]. A dual CAR system has been developed that the first synthetic Notch receptor detected one antigen resulting in the second inducible CAR expression to recognize the other antigen [21]. The SUPERCAR system composed of a zipCAR and zipscFv is another novel CAR system. A zipCAR has a leucine zipper in place of antigen detection domain as the extracellular portion of the CAR. A zipscFv has antigen detection scFv fused to a cognate leucine zipper which can bind with leucine zipper located on the zipCAR. This design endows CAR-T with target antigen flexibility and fine tuneability [22]. After the approval of CAR-T therapy in 2017, increasing numbers of clinical trials have been registered and authorized to develop new products of CAR-T cell therapy. The effect of CAR-T conveyed with a single antigen seems restricted, caused by the limited capacity to discriminate tumor cells from healthy tissue. Researchers have started to study and evaluate the effect of combined sensing approaches by targeting two or more antigens (Fig. 9.1).

9.2   The Achievements and Existing Problems About CAR-T Therapy

CAR-T cells therapy has greatly revolutionized the landscape of hematologic malignancies treatment, especially for acute lymphoblastic leukemia (ALL) and diffuse large B cell lymphoma (DLBCL). In relapsed or refractory cancer patients who have no response to conventional therapy, complete responses (CRs) by CAR-T therapy are approximately 40~60% to aggressive lymphoma and 60 ~80% to ALL [23–25]. However, there are a significant proportion of patients who do not respond to this treatment regimen. The most important step for CAR-T therapy is to choose the unique antigen based on tumor characteristics. The unique antigen should only be expressed on tumor cells and not on other issues. Although the CD19 CAR-T product has been approved by FDA, it can target not only B malignant cells but also normal B cells. There continues to be a great need for further investigation into proper unique antigen discovery [26]. Severe toxicity, most notably CRS and neurotoxicity, is another hurdle for CAR-T therapy. The frequency of severe CRS and neurotoxicity generally range from 10 to 50%. Lisocabtagene maraleucel as the third product currently being explored in a clinical study for DLBCL treatment shows the exceptionally low frequency of severe adverse events with the same antitumor effect as axicabtagene. In this trial, only one patient showed Grade 3 CRS while the percentage of Grades 3 and 4 neurotoxicity was also low as 12% [25]. The syndrome of CRS includes fever, hemodynamic instability, hypoxia, and end-organ dysfunction, which is similar to systemic inflammatory response syndrome. FDA has approved IL-6 receptor blocker tocilizumab as an option for CRS treatment after CAR-T therapy. Delirium, aphasia, cerebral edema, and seizures are the syndrome of neurotoxicity. Levetiracetam as a type of anticonvulsants can be used for seizure prophylaxis and severe symptoms should be treated with corticosteroids.

CAR-T therapy has shown a promising therapeutic effect in hematologic malignancies while less successful in solid tumors [27]. The reasons why CAR-T therapy shows disappointing outcomes in solid tumors include the following factors. First, it is difficult for CAR-T cells to penetrate solid tumors owing to the massive physical barriers surrounding tumor tissues [28]. Second, the solid tumor forms an immune-suppressive microenvironment to hamper CAR-T antitumor activity by secreting inhibitory cytokines and recruiting immune-suppressive cells [29]. Lastly, tumor-specific antigens are highly heterogenous in a solid tumor, which is hostile to monoclonal antibody-guided therapy [30]. How to improve the antitumor effect of CAR-T cells therapy in solid tumor treatment is an urgent problem that needs to be resolved.

Off-the-shelf CAR-T cells will solve the issue of donor availability. Patient-derived autologous T cells have been the source of CAR-T. Autologous T cells have long persistence after adoptive transfer because they can evade host allogeneic immune response. However, autologous CAR-T cells therapy requires a bespoke manufacturing process for every patient after leukapheresis and display certain disadvantages. It takes approximately 3 weeks to produce enough CAR-T cells for autologous CAR-T cells therapy and the cost of CAR-T cells therapy is inevitably expensive [6]. Moreover, T cell quality is variable for cancer patients and is susceptible to be impaired by chemotherapeutic agents. Dysfunctional T cells isolated from the immunosuppressive tumor microenvironment in certain cancer patients lead to CAR-T cell therapy failure [31]. The application of 'off-the-shelf' allogeneic CAR-T cells has many potential advantages compared with autologous T cells if the inherent barriers caused by MHC mismatch can be resolved. Allogeneic CAR-T cells are usually derived from healthy donors who have a robust immune function, which can overcome immune defects of autologous T cells from cancer patients. Moreover, harnessing allogeneic CAR-T cells makes it possible to perform more rapid and less expensive treatment, which also simplifies the manufacturing process and standardizes CAR-T products [32]. In addition, parts of allogeneic CAR-T cells can be stored by cryopreservation when they have been manufactured; thus cancer patients can be simultaneously treated with the combination of CAR-T cells targeted different antigens. Peripheral blood mononuclear cells (PBMCs) from healthy donors are the main source of allogeneic CAR-T cells. In very rare cases, umbilical cord blood (UCB) can also be the source of allogeneic CAR-T cells. Indeed, T cells from UCB have a unique antigen-naïve condition associated with decreased incidence and severity of graft-versus-host disease (GVHD) [33]. Nowadays, more and more studies focus on self-renewable pluripotent stem cells such as induced pluripotent stem cells (iPSCs) or embryonic stem cells (ESCs) as the new source of allogeneic CAR-T cells [34]. These pluripotent stem cells can proliferate indefinitely and theoretically produce all other cells in the human body. Harnessing pluripotent stem cells to produce therapeutic cells has been of keen interest to regenerative medicine [35–37]. Application of iPSCs as the source can generate more homogeneous CAR-T cells because they are produced from one clonal engineered pluripotent cell line. Antibody-mediated graft rejection usually causes organ transplantation failure and the presence of donor-specific anti-HLA antibodies (DSAs) appears to impede the successful engraftment of donor cells [38]. For allogeneic CAR-T cells transfer, the levels of DSAs as the major barrier need to be assessed carefully [39]. The allogeneic approach leads to two major issues that need to be addressed promptly. First, it may cause life-threatening GVHD. GVHD is the main reason for morbidity in allogeneic CAR-T transplantation and $\alpha\beta$-T cells play the central role in the pathogenesis of both acute and chronic GVHD [40–43]. In GVHD, T cells express TNF family molecules and secret intracellular granule contents to damage target organs [44, 45]. HLA mismatches between donor and recipient elicit immune recognition, potentially causing graft rejection and GVHD. HLA-restricted TCR repertoire can recognize subtle structural differences of allogeneic HLA molecules, leading to T cell alloreactivity. The generation of allogeneic CAR-T cells by deletion of endogenous TCR is expected to reduce the chance of GVHD [46]. Second, these allogeneic CAR-T cells have a high chance to be eliminated by the host immune system, hampering the antitumor effect [34]. The antitumor effect of allogeneic CAR-T cells is determined by the initial expansion, length of persistence, and host immune rejection. According to the first-in-human report with CAR19-T cells manufactured using *piggyBac* transposon system, *piggyBac* CAR19-T cells induced CAR-T cell lymphoma in two of ten patients, while the same phenomenon has not been found with CAR19-T cells produced by viral vector [47]. This incidence indicates the needs of either lentiviral vectors for primary T cells, or safe-harbor loci (such as AAVS1 and human ROSA26) in pluripotent stem cell-derived T cells.

The reasons leading to CAR-T therapy failure include immune-suppressive tumor microenvironment, tumor antigen escape, CAR-T cell exhaustion, and persistence reduction. Individual conventional CAR-T cells can only recognize one

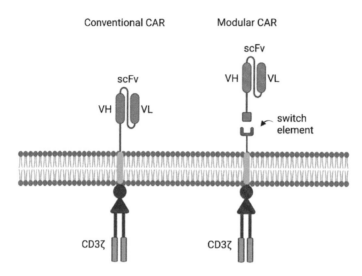

Fig. 9.2 The structure of conventional CAR and modular CAR

specific tumor antigen because of the fixed, single-antigen targeting capacity. Antigen loss of tumor tissue usually leads to therapy failure [48, 49]. The manufacturing of CAR-T cells targeting diverse tumor antigens is a promising approach to address this issue. Compared with the traditional CAR-T system, the modular or universal CAR-T technology utilizes a switch molecule to separate targeting and signaling elements. An adaptor or switch element in modular CAR-T cells replaces the antigen detection domain in conventional CAR-T cells. By choosing specific targets, the strategy would achieve better efficiency in the cold TME. This adaptor can be assembled with any specific tumor antigen and is required to bridge the immunological synapse [50] (Fig. 9.2).

According to the antigen expression of the patient's tumor, the modular CAR-T system can be flexibly adjusted with the corresponding tumor antigen, allowing for tailored therapy. Meanwhile, the modular or universal system can precisely control CAR-T activity by managing the adaptor function. The ability to titrate on adaptors enables halting of the administration of the adaptor, resulting in the blockade of CAR-T function without the effect on other T cells (Table 9.1).

9.3 Generation of CAR-immune cells from PSCs (examples, advances)

In 1998, the human ESCs were established by the James Thomson group for the first time [51]. In 2006, Shinya Yamanaka discovered that mouse somatic cells are capable to be reprogrammed to ESCs-like status by transducing four pivotal transcription factors (Klf4, Oct4, Sox2, and c-Myc), these cells are termed as iPSCs

Table 9.1 CAR-T clinical trial for solid tumors

Solid tumor	Target antigen	Target cell	CAR	Clinical trials
Glioblastoma	IL13Rα2	T cell	IL13Rα2 scFv-4-1-BB-CD3ζ	NCT02208362
Glioblastoma	EGFRvIII	T cell	EGFRvIII scFv-CD8 Hinge&TM-4-1BB-CD3ζ	NCT02209376
Neuroblastoma	L1-CAM	T cell	L1-CAM scFv-CD3ζ	NCT00006480
Neuroblastoma	GD2	T cell	GD2 scFv-CD3ζ	NCT00085930
Carcinomas	CD133	T cell	CD133 scFv-CD8a Hingle&TM-4-1BB-CD3ζ	NCT02541370
Colon cancer	CEA	T cell	CEA scFv-CD8 Hinge-CD28-CD3ζ	NCT01373047
Colon cancer	HER2	T cell	HER2 scFv-CD8 Hinge&TM-CD28-4-1BB-CD3ζ	NCT00924287
Pancreatic cancer	Mesothelin	T cell	Mesothelin scFv-4-1BB-CD3ζ	NCT01897415
Renal cell carcinoma	CAIX	T cell	CAIX scFv-CD16γ TM&Signal domain	Phase I/II
Prostate cancer	PSMA	T cell	PSMA scFv-CD3ζ	Phase I
Seminal vesicle cancer	MUC1	T cell	MUC1 scFv-Fc-IgD Hinge-CD28 TM-4-1BB-CD3ζ	NCT02587689
Ovarian cancer	FRα	T cell	FRα scFv-CD16γ TM&Signal domain	Phase I
DLBCL	CD19	T cell	CD19 scFv-CD8a Hinge&TM-4-1BB-CD3ζ	NCT02445248
Non-Hodgikin lymphoma, CLL	CD19	NK cell	iCasp9-2A-CD19 scFv-CD28-CD3ζ-2A-IL15	NCT03056339

[52]. Soon after, human iPSCs have been successfully established from fully differentiated somatic cells, even from cells in the urine [53–55]. Human immune cells can also be differentiated and generated from human iPSCs for immune cell therapy, especially to treat tumors that are incurable by conventional approaches. CRISPR/Cas9 system as the gene-editing technology can be used to modify genes associated with immune responses during the production of human pluripotent stem cell-derived immune cells. Notably, the primary immune cells are refractory to gene editing and difficult to expand afterward. Compared with primary immune cells, human pluripotent stem cells can easily be edited by transfection, and could be an ideal source for CAR-immune cell generation. Moreover, deleting MHCs will offer a universal source for "off-the-shelf" immunotherapeutic cell differentiation [6]. More studies established hypoimmunogenic universal donor iPSCs to avoid immune rejection after adoptive transfer [56, 57]. Employing the advantage of amenable and expandable features, universal iPSCs were designed by deleting immunogenic MHCs, offering the possibility to generate universal CAR-immune cells for all patients. MHC I plays a core role in mediating immune rejection after allogeneic transplantation. The deletion of the *B2M* gene leads to the loss of

MHC I and avoid attacks from CD8$^+$ T cells [57]. The resultant cells could be still attacked by both macrophages and NK cells via innate immune mechanisms that recognize and attack MHC I-null cells. Thus overexpression of immune-tolerant genes avoids attacks from NK cells (via HLA-E single-chain dimers fused to *B2M*) [58] and macrophages (via CD47). Both T cells and NK cells do not express MHC II while macrophages do express MHC II. The expression of MHC II will provoke attacks from CD4$^+$ T cells and potentially challenge the development of the CAR-macrophage approach [59]. The deletion of the *CIITA* gene results in the loss of MHC II and is expected to free CAR-macrophages from the CD4$^+$ T cells [57]. Knocking out the genes encoding TCR α and β subunits prevented the occurrence of GVHD [60] (Fig. 9.3).

T cells play pivotal roles in the adaptive immune system and form the keystone of cellular immunity. They can recognize foreign molecules expressed on the surface of antigen-presenting cells via the interaction between TCR and MHC. CD4$^+$ T helper cells can secret a series of cytokines to regulate other immune cell activity, such as CD8$^+$ T cytotoxic cells, macrophages, and B cells. CD8$^+$ T cytotoxic cells can recognize antigens presented by MHC I or tumor common antigens with the help of their TCRs. TCR α, β subunits together with CD3 γ, δ, ε and ζ subunits constitute the core part of T cell signal transduction [61]. Upon binding to foreign antigens, CD8$^+$ T cytotoxic cells secret perforin, granzymes, and granulysin to trigger the target cell's apoptosis. In addition, activated CD8$^+$ T cytotoxic cells can also induce apoptosis of FAS-expressing cells by FAS ligand expression [62]. The differentiation protocols from human pluripotent stem cells to functional T cells have been invented by several groups. The stromal cell line, such as the mouse bone marrow-derived OP9 cell line, is employed for the differentiation from human pluripotent stem cells to CD34$^+$ hematopoietic cells. Notch signaling

Fig. 9.3 The strategy to generate hypoimmunegenic, universal donor PSCs

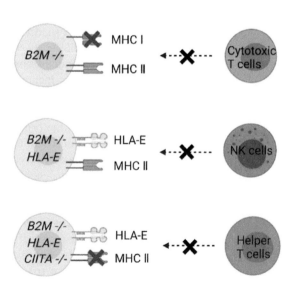

determines the further differentiation from CD34$^+$ hematopoietic cells to mature functional T cells. Therefore, the OP9-DLL1 cell line was established by transducing Notch ligand Delta-like ligand 1 into the OP9 cell line. With the help of the OP9-DLL1 cell line, human pluripotent stem cells-derived CD34^{hi}CD43^{lo} cells have the potential to differentiate into CD4$^+$ and CD8$^+$ double-positive TCRαβ T cells. Using OP9-DLL4 in place of OP9-DLL1 cell line for T cells differentiation was reported to be further efficient [63, 64]. However, since TCR rearrangements are random during in vitro differentiation, it is difficult to know their antigen specificity and HLA restriction of these T cells. The advent of CAR technology circumvents this limitation because CARs could redirect T cell specificity in an HLA-independent fashion [65]. The Sadelain group successfully produced CAR-T targeted to CD19 from iPSCs and demonstrated that these iPSCs-derived CAR-T cells potently inhibited tumor progression. The pairwise correlation analysis based on gene expression microarray results suggested that these iPSCs-derived CAR-T cells were more similar to fresh or activated γδ T cells [66].The Crooks group established PSC/ATO (pluripotent stem cells/artificial thymic organoid) system to generate mature functional T cells from human PSCs in vitro system. This 3D organoid system facilitates the differentiation from PSCs to embryonic mesoderm through hematopoietic specification, and then induces T cell lineage commitment to become naïve CD3$^+$CD8αβ$^+$ and CD3$^+$CD4$^+$ conventional T cells. This system can also be used to produce antitumor antigen-specific CD3$^+$CD8αβ$^+$ T cells by the introduction of MHC I-restricted in PSCs [67]. The Nakauchi group reported that antigen-specific CD8$^+$ T cells from HIV-1-infected patients showed exhausting phenotypes. However, after reprogramming to pluripotency and redifferentiating into CD8$^+$ T cells, these rejuvenated cells recovered antigen-specific killing capacity and possessed a high proliferative activity [68]. This discovery monumentally provides new insight and ideas for cancer immunotherapy. FT819 as a dual-targeted CAR-T candidate (CD19/CD16) made from a master iPSC cell line is being evaluated in a clinical study [69].

NK cells belong to the innate immune system because of their lack of receptors for antigen specificity and form the first line of defense against tumor cells and virus-infected cells, and they show promising potential in cancer immunotherapy. The activation of NK cells is decided by a balance between activating and inhibitory signals, which does not have a somatically rearranged and antigen-specific TCR [70]. The activating receptors of NK cells include CD94/NKG2C, NKG2D, NKp30, NKp44, and NKp46, which recognize the different ligands expressed on various target cells. The inhibitory receptors of NK cells include polymorphic inhibitory killer cell immunoglobulin-like receptors (KIRs) that bind with MHC class I [71]. The antitumor efficacy of NK cells is limited because NK cells are highly susceptible to the immunosuppressive microenvironment. Upon activation, NK cells localize the site of infection and perform functions by cytokine secretion, the release of cytolytic granules, and death receptor-mediated cytolysis [72]. The cytokines secreted from NK cells include IFNγ, TNFα, GM-CSF, RANTES, and some chemokines, which can regulate the functions of the innate and adaptive immune system [73, 74]. In addition, NK cells can lyse target cells by secreting perforins and granzymes [75, 76]. They can also express specific ligands to activate death receptors on their target cells [77]. Compared with T cells, NK cells do not depend on HLA matching to perform their function. They can be easily transferred across HLA barriers without causing GVHD. The protocols to differentiate NK cells from hPSCs have been invented. In the early protocols, mouse stromal cells (S17 or M210) were used for hematopoietic differentiation. The differentiated cells were selected and seeded onto EL08-1D2 stromal cells in presence of IL-3, 7, 15, and FLT3L, then $CD45^+CD56^+$ NK cells were generated [78, 79]. The generated NK cells were able to eradicate human tumor cells by direct cell-mediated killing and secreting antibodies. Considering the use of hPSCs-derived NK cells in clinic for disease treatment, a xeno-free and serum-free protocol needs to be developed. Spin embryoid body method was used for $CD34^+CD43^+$ hematopoietic progenitor cells generation and the resultant cells were further differentiated using membrane-bound IL-21-expressing artificial antigen-presenting cells [80, 81].The Kaufman group generated CAR-NK from human iPSCs. Human iPSCs were transfected with a plasmid encoding scFv targeted to human mesothelin, 2B4 costimulatory domain and CD3ζ chain. These genetically modified human iPSCs were differentiated to functional CAR-NK cells. Compared with CAR-T cells, CAR-NK cells displayed similar antitumor efficacy, but with less overall toxicity [82]. Nowadays, the design strategy of fourth-generation CAR-T has also been tested in CAR-NK generation [6].

Macrophages belong to the innate immune system with a high infiltration rate and play indispensable roles in inflammation and the protection of our body from outside invaders and tumor cells. The yolk sac, fetal liver, and bone marrow are all the sites for macrophage origination. Yolk sac-derived macrophages not only form microglia in the brain but also populate the fetal liver which produces most of the self-renewing tissue-resident macrophages (TRMs) [83, 84]. After postnatal, macrophages originate from bone marrow myeloid progenitor cells, occurring through differentiation of circulating monocytes in an MCSF- or GMCSF-dependent manner [85]. In general, the life span of bone marrow-derived macrophages is shorter than TRMs [86]. Macrophages are highly plastic cells that perform diverse functions in different organs, including clearance of cell debris, elimination of pathogens, modulation of inflammatory responses, and tissue homeostasis maintenance [87]. Macrophages may undergo M1 or M2 polarization in different tissues encountering different microenvironment stimuli and signals. M1 phenotype which is highly expressed in inflammatory cytokines has strong anti-microbial and tumor activity, while M2 phenotype can promote tumor growth and tissue remodeling [88, 89]. Macrophages can directly recognize outside invaders via pattern recognition receptors (PRRs). PRRs include Toll-like receptors, NOD-like receptors, C-type lectin receptors, and cytoplasmic proteins [90]. After receptors are activated, macrophages provoke intracellular signals to induce actin polymerization and phagocytic cup formation [91]. Then macrophages phagocytose outside invaders or tumor cells and move to lymph nodes to present antigens to T cells, subsequently triggering a series of T cells downstream responses. Compared with other immune cells, macrophages can penetrate solid tumors easily and interact with almost all cellular components in the tumor microenvironment, which endows them with profound advantages to be developed into CAR-macrophage [88]. The feeder- and xeno-free protocol about the differentiation from hPSCs to functional macrophages has been reported. First, iPSCs were exposed to morphogens and cytokines such as BMP4 and VEGF step-by-step, after specifying the lateral plate mesoderm organoids, the organoids were then exposed to hematopoietic cytokines such as SCF, IL-6, and FLT3 to specify immune cells. The resultant mesoderm organoids will generate CD34$^+$ FLK1$^+$ endothelial cells (so-called hemogenic endothelium) that will derive the innate immune cells including macrophages [92]. The hPSCs-derived macrophages have the capacity of phagocytosis and polarization, and they can also secret cytokines in response to LPS, indicating the same characteristic and function as macrophages that develop naturally in the body. It has been reported that CAR-macrophages could destroy the extracellular matrix (ECM) of the tumor and facilitate the penetration of T cells into the tumor, thus playing an antitumor role [93]. The Zhang group successfully established CAR-macrophages from human iPSCs. CAR expression endowed iPSCs-derived antigen-dependent macrophages with enhanced phagocytosis of tumor cells and in vivo antitumor activity [36]. The Gill group evaluated the antitumor potential of CAR-macrophages in different animal models and found that they could effectively reduce tumor burden. Moreover, in humanized mouse models, CAR-macrophages were demonstrated to strengthen T cell's antitumor activity and facilitate the formation of a pro-inflammatory environment. For the intracellular domain of CAR-macrophages, the Gill group used CD3ζ chain similar with CAR-T cells [59], while the Tonald Vale group applied the cytosolic domains from Megf10 and FcRγ as the intracellular domain of CAR-macrophages, which showed robust phagocytosis capacity [94] (Table 9.2).

9.4   Potential and Perspectives of CAR-Immune Cells in Cancer Treatment

CAR-T therapy as the earliest CAR-immune cells therapy has achieved great success and become a powerful immunotherapeutic source in hematologic cancer treatment. FDA has already approved four CAR-T-related drugs Kymriah, Yescarta, Tecarta, and Breyanzi from 2017 to 2021 [97, 98]. Lately, CAR-NK therapy has emerged as an alternative therapy option to CAR-T therapy. Compared with CAR-T therapy, allogeneic CAR-NK therapy has reduced risk for GVHD, CRS, and neurotoxicity [99, 100]. That is because activated T cells predominantly produce more cytokines associated with CRS and severe neurotoxicity than activated NK cells [101]. CAR-NK cells may be able to eliminate tumor cells via both CAR-dependent and NK cell receptor-dependent mechanisms. Therefore, CAR-NK cells can form a second line of defense in case tumor cells escape T cells recognition by MHC downregulation. The use of NK cell lines such as NK92 and allogeneic NK cells with CAR engineered functions have been

Table 9.2 The study of CAR-immune cells derived from pluripotent stem cells

Type of CAR	CAR target	CAR structure	Cell line used (in vitro)	Cancer type (in vivo)
CAR-T [66]	CD19	1928z-T-iPSC	EL4 cells (CD19^{high})	Burkitt lymphoma
CAR-NK [82]	mesothelin	SS1-NKG2D-2B4-CD3ζ	K562 cells (meso^{high}), A1847 cells (meso^{high})	Ovarian cancer
CAR-macrophage [36]	CD19/meso	scFv-CD8α-CD86-FcγRI	K562 cells (CD19^{high}), OVCAR3/ASPC1 cells (meso^{high})	Ovarian cancer
CAR-T (FT819) [69]	CD19 and CD16	not shown	N/A	Clinical study
CAR-NK [95]	GPC3	G2-CD8α-CD28-4/1BB-CD3ζ	SK-Hep-GPC3	Ovarian cancer
CAR-NK [96]	CD19	1928z-NK-autologous HSC	N/A	Clinical study (NCT03579927)
CAR-NK [96]	CD19	1928z-NK-allogeneic UCB	N/A	Clinical study (NCT03056339)
CAR-NK [96]	CD19	not shown	N/A	Clinical study (NCT04245722)

studied only recently [96]. Currently, there are more than 500 CAR-T-related and 19 CAR-NK-related clinical trials being conducted in the world [98]. The majority of CAR-T therapy under clinical evaluation still employs patient-derived autologous T cells, whereas almost all CAR-NK therapy applies to cells from allogeneic donors. The first large-scale clinical trial (NCT03056339) of CAR-NK cells has shown promising and safe results in patients with $CD19^+$ CLL and B cell lymphoma [102]. Although CAR-NK therapy possesses multiple advantages in comparison with CAR-T therapy, CAR-NK therapy still needs to be optimized to improve efficacy. Nowadays, researchers have paid great interest in developing CAR-macrophage for cancer treatment. FDA has already approved one CAR-macrophage clinical trial, which is CT-0508 from CARISMA Therapeutics with anti-HER2 CAR-macrophage in subjects with HER2 overexpressing solid tumors (NCT04660929).

Allogeneic CAR-T therapy has monumental advantages compared with autologous approaches, such as a reduced expense and timesaving production cycle as a result of the implementation of standardized and scaled-up manufacturing processes, in which a host of CAR-T cells can be generated from healthy donors, even the therapeutic CAR-T cells that have already been produced and stored in advance before patients arrive. The applicable targets for allogeneic CAR-T therapy include CD19 and CD22 in ALL and B cell lymphomas, respectively, CD30 in Hodgkin lymphoma and anaplastic large cell lymphoma, BCMA, CS1 and CD38 in multiple myeloma, and CD123, CD33, and CLL1 in AML [103]. Owing to the shorter persistence of allogeneic CAR-T cells, the approaches, such as a systematic strategy of redosing [34], the combination of CAR-T cells targeted different antigens [104] and the combination of CAR-T therapy with immune checkpoint modulators or cancer vaccine [105] can be employed to enhance CAR-T therapy efficacy. To date, the efficacy of CAR-T in solid tumors is much less satisfactory than in hematologic malignancies owing to the sturdy physical barriers, immune-suppressive tumor milieu, and the heterogeneity of inner tumor cells. CAR-T cells coexpressing catalase are able to promote their antioxidative capacity by metabolizing H_2O_2, subsequently more resilient toward the harsh tumor microenvironment caused by abundant reactive oxygen species (ROS), and perform superior over conventional CAR-T cells [106]. Moreover, gene-editing approaches reduce the sensitivity of T cells to negative immune checkpoints. The Moon group generated a new switch receptor construct which introduced truncated extracellular domain of PD-1 and costimulatory domain CD28 into CAR-T cells. They demonstrated that the application of PD-1/CD28 can enhance the antitumor activity of CAR-T cells against solid tumors [107]. The Brentjens group reported CAR-T cells which can secrete PD-1 blocking scFv increased antitumor activity [108]. Targeting chemokine receptors, such as CXCR2 [109] and CCR2B [110], allows CAR-T cells migration to the tumor site. The Dotti group revealed that CAR-T cells expressing heparanase, a heparan sulfate-degrading enzyme, could enhance tumor penetration of T cells, subsequently improving antitumor activity [28]. Constructing CAR-T cells which can secrete cytokines further promote their survival or greater activity. CAR-T cells secreting IL-12 [111], IL-18 [112], and IL-15 [113] have been reported to optimize their antitumor activity by different mechanisms. There is a multitude

Fig. 9.4 Hallmark of
modified CAR-T to target
solid tumor

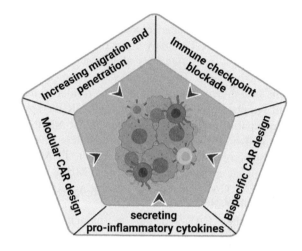

of potential modifications for CAR-T therapy, and the proper modification needs
to be selected to implement based on the individual tumor characteristics, which
can provide effective ways to eradicate tumors independent of tumor-expressing
MHC. More advanced modification techniques, such as modular CAR and dual-
targeting approach, are being used in CAR-immune cells design to circumvent
therapy resistance and avoid GvHD (Fig. 9.4).

9.5   Future Prospects

CAR-immune cell therapy holds an unprecedented potential to treat cancers that
are incurable by conventional treatments. The number of clinical trials involving CAR-immune cell therapy is increasing exponentially, indicating more and
more researchers show great enthusiasm for this area [114]. Developing more
potent, more cost-effective, and safer CAR-immune cell therapy is the critical
goal in the future. Compared with primary immune cells, human pluripotent
stem cells-derived immune cells can be easily engineered and have the capacity to proliferate indefinitely, enabling clonal selection and generation of enough
clonally-selected therapeutic cells for cancer treatment [115]. The application of
gene-editing approaches and fourth-generation CARs can generate CAR-immune
cells that are less prone to causing severe CRS [116] and subsequently optimize
therapy in terms of safety, cost and potency. However, there is no denying that the
generation and application of human pluripotent stem cells-derived CAR-T cells,
CAR-NK cells, and CAR-macrophages are still at the early stage. The manufacturing processes from human pluripotent stem cells to functional CAR-immune cells
need to be standardized. Moreover, how to improve the efficacy of CAR-immune
cells in solid tumors is an inevitable hurdle. Another great challenge in this area is
the paucity of preclinical models to carry on the safety and efficacy evaluation of
CAR-immune cells before human studies or in response to safety issues that have

been observed in early-phase clinical trials. More basic and translational research need to be dedicated to this area to improve CAR-immune cell therapy and foster new applications beyond oncology in autoimmunity, infectious diseases, and organ transplantation.

References

1. Dobosz P, Dzieciatkowski T (2019) The intriguing history of cancer immunotherapy. Front Immunol 10:2965
2. Pham T, Roth S, Kong J, Guerra G, Narasimhan V, Pereira L et al (2018) An update on immunotherapy for solid tumors: a review. Ann Surg Oncol 25(11):3404–3412
3. June CH, O'Connor RS, Kawalekar OU, Ghassemi S, Milone MC (2018) CAR T cell immunotherapy for human cancer. Science 359(6382):1361–1365
4. Gross G, Waks T, Eshhar Z (1989) Expression of immunoglobulin-T-cell receptor chimeric molecules as functional receptors with antibody-type specificity. Proc Natl Acad Sci U S A 86(24):10024–10028
5. Eshhar Z, Waks T, Gross G, Schindler DG (1993) Specific activation and targeting of cytotoxic lymphocytes through chimeric single chains consisting of antibody-binding domains and the gamma or zeta subunits of the immunoglobulin and T-cell receptors. Proc Natl Acad Sci U S A 90(2):720–724
6. Lee JM (2019) When CAR meets stem cells. Int J Mol Sci 20(8)
7. Kuwana Y, Asakura Y, Utsunomiya N, Nakanishi M, Arata Y, Itoh S et al (1987) Expression of chimeric receptor composed of immunoglobulin-derived V regions and T-cell receptor-derived C regions. Biochem Biophys Res Commun 149(3):960–968
8. Kowolik CM, Topp MS, Gonzalez S, Pfeiffer T, Olivares S, Gonzalez N et al (2006) CD28 costimulation provided through a CD19-specific chimeric antigen receptor enhances in vivo persistence and antitumor efficacy of adoptively transferred T cells. Cancer Res 66(22):10995–11004
9. Finney HM, Lawson AD, Bebbington CR, Weir AN (1998) Chimeric receptors providing both primary and costimulatory signaling in T cells from a single gene product. J Immunol 161(6):2791–2797
10. Porter DL, Levine BL, Kalos M, Bagg A, June CH (2011) Chimeric antigen receptor-modified T cells in chronic lymphoid leukemia. N Engl J Med 365(8):725–733
11. Brentjens RJ, Davila ML, Riviere I, Park J, Wang X, Cowell LG et al (2013) CD19-targeted T cells rapidly induce molecular remissions in adults with chemotherapy-refractory acute lymphoblastic leukemia. Sci Transl Med 5(177):177ra38
12. June CH, Sadelain M (2018) Chimeric antigen receptor therapy. N Engl J Med 379(1):64–73
13. Maude SL, Laetsch TW, Buechner J, Rives S, Boyer M, Bittencourt H et al (2018) Tisagenlecleucel in children and young adults with B-cell lymphoblastic leukemia. N Engl J Med 378(5):439–448
14. Neelapu SS, Locke FL, Bartlett NL, Lekakis LJ, Miklos DB, Jacobson CA et al (2017) Axicabtagene ciloleucel CAR T-cell therapy in refractory large B-cell lymphoma. N Engl J Med 377(26):2531–2544
15. Tang XY, Sun Y, Zhang A, Hu GL, Cao W, Wang DH et al (2016) Third-generation CD28/4-1BB chimeric antigen receptor T cells for chemotherapy relapsed or refractory acute lymphoblastic leukaemia: a non-randomised, open-label phase I trial protocol. BMJ Open 6(12):e013904
16. Carpenito C, Milone MC, Hassan R, Simonet JC, Lakhal M, Suhoski MM et al (2009) Control of large, established tumor xenografts with genetically retargeted human T cells containing CD28 and CD137 domains. Proc Natl Acad Sci U S A 106(9):3360–3365
17. Chmielewski M, Kopecky C, Hombach AA, Abken H (2011) IL-12 release by engineered T cells expressing chimeric antigen receptors can effectively Muster an antigen-independent

macrophage response on tumor cells that have shut down tumor antigen expression. Cancer Res 71(17):5697–5706

18. Diaconu I, Ballard B, Zhang M, Chen Y, West J, Dotti G et al (2017) Inducible caspase-9 selectively modulates the toxicities of CD19-specific chimeric antigen receptor-modified T cells. Mol Ther 25(3):580–592

19. Zhou X, Tu S, Wang C, Huang R, Deng L, Song C et al (2020) Phase I trial of fourth-generation anti-CD19 chimeric antigen receptor T cells against relapsed or refractory B cell non-Hodgkin Lymphomas. Front Immunol 11:564099

20. Chmielewski M, Abken H (2015) TRUCKs: the fourth generation of CARs. Expert Opin Biol Ther 15(8):1145–1154

21. Roybal KT, Rupp LJ, Morsut L, Walker WJ, McNally KA, Park JS et al (2016) Precision tumor recognition by T cells with combinatorial antigen-sensing circuits. Cell 164(4):770–779

22. Cho JH, Collins JJ, Wong WW (2018) Universal chimeric antigen receptors for multiplexed and logical control of T cell responses. Cell 173(6):1426–38 e11

23. Schuster SJ, Bishop MR, Tam CS, Waller EK, Borchmann P, McGuirk JP et al (2019) Tisagenlecleucel in adult relapsed or refractory diffuse large B-cell lymphoma. N Engl J Med 380(1):45–56

24. Locke FL, Ghobadi A, Jacobson CA, Miklos DB, Lekakis LJ, Oluwole OO et al (2019) Long-term safety and activity of axicabtagene ciloleucel in refractory large B-cell lymphoma (ZUMA-1): a single-arm, multicentre, phase 1–2 trial. Lancet Oncol 20(1):31–42

25. Sermer D, Brentjens R (2019) CAR T-cell therapy: full speed ahead. Hematol Oncol 37(Suppl 1):95–100

26. Lim F, Ang SO (2020) Emerging CAR landscape for cancer immunotherapy. Biochem Pharmacol 178:114051

27. Knochelmann HM, Smith AS, Dwyer CJ, Wyatt MM, Mehrotra S, Paulos CM (2018) CAR T cells in solid tumors: blueprints for building effective therapies. Front Immunol 9:1740

28. Caruana I, Savoldo B, Hoyos V, Weber G, Liu H, Kim ES et al (2015) Heparanase promotes tumor infiltration and antitumor activity of CAR-redirected T lymphocytes. Nat Med 21(5):524–529

29. Zhang E, Gu J, Xu H (2018) Prospects for chimeric antigen receptor-modified T cell therapy for solid tumors. Mol Cancer 17(1):7

30. Jamal-Hanjani M, Quezada SA, Larkin J, Swanton C (2015) Translational implications of tumor heterogeneity. Clin Cancer Res 21(6):1258–1266

31. Thommen DS, Schumacher TN (2018) T cell dysfunction in cancer. Cancer Cell 33(4):547–562

32. Zakrzewski JL, Suh D, Markley JC, Smith OM, King C, Goldberg GL et al (2008) Tumor immunotherapy across MHC barriers using allogeneic T-cell precursors. Nat Biotechnol 26(4):453–461

33. Kwoczek J, Riese SB, Tischer S, Bak S, Lahrberg J, Oelke M et al (2018) Cord blood-derived T cells allow the generation of a more naive tumor-reactive cytotoxic T-cell phenotype. Transfusion 58(1):88–99

34. Depil S, Duchateau P, Grupp SA, Mufti G, Poirot L (2020) 'Off-the-shelf' allogeneic CAR T cells: development and challenges. Nat Rev Drug Discov 19(3):185–199

35. Patel SJ, Yamauchi T, Ito F (2019) Induced pluripotent stem cell-derived t cells for cancer immunotherapy. Surg Oncol Clin N Am 28(3):489–504

36. Zhang L, Tian L, Dai X, Yu H, Wang J, Lei A et al (2020) Pluripotent stem cell-derived CAR-macrophage cells with antigen-dependent anti-cancer cell functions. J Hematol Oncol 13(1):153

37. Saetersmoen ML, Hammer Q, Valamehr B, Kaufman DS, Malmberg KJ (2019) Off-the-shelf cell therapy with induced pluripotent stem cell-derived natural killer cells. Semin Immunopathol 41(1):59–68

38. Butler CL, Valenzuela NM, Thomas KA, Reed EF (2017) Not all antibodies are created equal: factors that influence antibody mediated rejection. J Immunol Res 2017:7903471

39. Ciurea SO, Cao K, Fernandez-Vina M, Kongtim P, Malki MA, Fuchs E et al (2018) The European Society for blood and marrow transplantation (EBMT) consensus guidelines for the detection and treatment of donor-specific anti-HLA antibodies (DSA) in haploidentical hematopoietic cell transplantation. Bone Marrow Transplant 53(5):521–534

40. Frame JN, Collins NH, Cartagena T, Waldmann H, O'Reilly RJ, Dupont B et al (1989) T cell depletion of human bone marrow. Comparison of Campath-1 plus complement, anti-T cell ricin a chain immunotoxin, and soybean agglutinin alone or in combination with sheep erythrocytes or immunomagnetic beads. Transplantation 47(6):984–8

41. Champlin RE, Passweg JR, Zhang MJ, Rowlings PA, Pelz CJ, Atkinson KA et al (2000) T-cell depletion of bone marrow transplants for leukemia from donors other than HLA-identical siblings: advantage of T-cell antibodies with narrow specificities. Blood 95(12):3996–4003

42. Radestad E, Wikell H, Engstrom M, Watz E, Sundberg B, Thunberg S et al (2014) Alpha/beta T-cell depleted grafts as an immunological booster to treat graft failure after hematopoietic stem cell transplantation with HLA-matched related and unrelated donors. J Immunol Res 2014:578741

43. Abdelhakim H, Abdel-Azim H, Saad A (2017) Role of alphabeta T cell depletion in prevention of graft versus host disease. Biomedicines 5(3)

44. Zeiser R, Blazar BR (2017) Acute graft-versus-host disease—biologic process, prevention, and therapy. N Engl J Med 377(22):2167–2179

45. Baker MB, Altman NH, Podack ER, Levy RB (1996) The role of cell-mediated cytotoxicity in acute GVHD after MHC-matched allogeneic bone marrow transplantation in mice. J Exp Med 183(6):2645–2656

46. Qasim W, Zhan H, Samarasinghe S, Adams S, Amrolia P, Stafford S et al (207) Molecular remission of infant B-ALL after infusion of universal TALEN gene-edited CAR T cells. Sci Transl Med 9(374)

47. Bishop DC, Clancy LE, Simms R, Burgess J, Mathew G, Moezzi L et al (2021) Development of CAR T-cell lymphoma in two of ten patients effectively treated with piggyBac modified CD19 CAR T-cells. Blood

48. Xu X, Sun Q, Liang X, Chen Z, Zhang X, Zhou X et al (2019) Mechanisms of relapse after CD19 CAR T-cell therapy for acute lymphoblastic leukemia and its prevention and treatment strategies. Front Immunol 10:2664

49. Kailayangiri S, Altvater B, Wiebel M, Jamitzky S, Rossig C (2020) Overcoming heterogeneity of antigen expression for effective CAR T cell targeting of cancers. Cancers (Basel) 12(5)

50. Sutherland AR, Owens MN, Geyer CR (2020) Modular chimeric antigen receptor systems for universal CAR T cell retargeting. Int J Mol Sci 21(19)

51. Thomson JA, Itskovitz-Eldor J, Shapiro SS, Waknitz MA, Swiergiel JJ, Marshall VS et al (1998) Embryonic stem cell lines derived from human blastocysts. Science 282(5391):1145–1147

52. Takahashi K, Yamanaka S (2006) Induction of pluripotent stem cells from mouse embryonic and adult fibroblast cultures by defined factors. Cell 126(4):663–676

53. Takahashi K, Tanabe K, Ohnuki M, Narita M, Ichisaka T, Tomoda K et al (2007) Induction of pluripotent stem cells from adult human fibroblasts by defined factors. Cell 131(5):861–872

54. Yu J, Vodyanik MA, Smuga-Otto K, Antosiewicz-Bourget J, Frane JL, Tian S et al (2007) Induced pluripotent stem cell lines derived from human somatic cells. Science 318(5858):1917–1920

55. Zhou T, Benda C, Duzinger S, Huang Y, Li X, Li Y et al (2011) Generation of induced pluripotent stem cells from urine. J Am Soc Nephrol 22(7):1221–1228

56. Gornalusse GG, Hirata RK, Funk SE, Riolobos L, Lopes VS, Manske G et al (2017) HLA-E-expressing pluripotent stem cells escape allogeneic responses and lysis by NK cells. Nat Biotechnol 35(8):765–772

57. Deuse T, Hu X, Gravina A, Wang D, Tediashvili G, De C et al (2019) Hypoimmunogenic derivatives of induced pluripotent stem cells evade immune rejection in fully immunocompetent allogeneic recipients. Nat Biotechnol 37(3):252–258

58. Han X, Wang M, Duan S, Franco PJ, Kenty JH, Hedrick P et al (2019) Generation of hypoimmunogenic human pluripotent stem cells. Proc Natl Acad Sci U S A 116(21):10441–10446
59. Klichinsky M, Ruella M, Shestova O, Lu XM, Best A, Zeeman M et al (2020) Human chimeric antigen receptor macrophages for cancer immunotherapy. Nat Biotechnol 38(8):947–953
60. Graham C, Jozwik A, Pepper A, Benjamin R (2018) Allogeneic CAR-T cells: more than ease of access? Cells 7(10)
61. Alcover A, Alarcon B, Di Bartolo V (2018) Cell biology of T cell receptor expression and regulation. Annu Rev Immunol 36:103–125
62. Zhang N, Bevan MJ (2011) CD8(+) T cells: foot soldiers of the immune system. Immunity 35(2):161–168
63. Kennedy M, Awong G, Sturgeon CM, Ditadi A, LaMotte-Mohs R, Zuniga-Pflucker JC et al (2012) T lymphocyte potential marks the emergence of definitive hematopoietic progenitors in human pluripotent stem cell differentiation cultures. Cell Rep 2(6):1722–1735
64. Timmermans F, Velghe I, Vanwalleghem L, De Smedt M, Van Coppernolle S, Taghon T et al (2009) Generation of T cells from human embryonic stem cell-derived hematopoietic zones. J Immunol 182(11):6879–6888
65. Sadelain M, Brentjens R, Riviere I (2013) The basic principles of chimeric antigen receptor design. Cancer Discov 3(4):388–398
66. Themeli M, Kloss CC, Ciriello G, Fedorov VD, Perna F, Gonen M et al (2013) Generation of tumor-targeted human T lymphocytes from induced pluripotent stem cells for cancer therapy. Nat Biotechnol 31(10):928–933
67. Montel-Hagen A, Seet CS, Li S, Chick B, Zhu Y, Chang P et al (2019) Organoid-induced differentiation of conventional T cells from human pluripotent stem cells. Cell Stem Cell 24(3):376–89 e8
68. Nishimura T, Kaneko S, Kawana-Tachikawa A, Tajima Y, Goto H, Zhu D et al (2013) Generation of rejuvenated antigen-specific T cells by reprogramming to pluripotency and redifferentiation. Cell Stem Cell 12(1):114–126
69. From Pluripotent Stem to CAR T Cells (2018) Cancer Discov 8(6):OF5
70. Terren I, Orrantia A, Vitalle J, Zenarruzabeitia O, Borrego F (2019) NK cell metabolism and tumor microenvironment. Front Immunol 10:2278
71. Becker PS, Suck G, Nowakowska P, Ullrich E, Seifried E, Bader P et al (2016) Selection and expansion of natural killer cells for NK cell-based immunotherapy. Cancer Immunol Immunother 65(4):477–484
72. Lee SH, Miyagi T, Biron CA (2007) Keeping NK cells in highly regulated antiviral warfare. Trends Immunol 28(6):252–259
73. Roda JM, Parihar R, Magro C, Nuovo GJ, Tridandapani S, Carson WE 3rd (2006) Natural killer cells produce T cell-recruiting chemokines in response to antibody-coated tumor cells. Cancer Res 66(1):517–526
74. Fauriat C, Long EO, Ljunggren HG, Bryceson YT (2010) Regulation of human NK-cell cytokine and chemokine production by target cell recognition. Blood 115(11):2167–2176
75. Pipkin ME, Lieberman J (2007) Delivering the kiss of death: progress on understanding how perforin works. Curr Opin Immunol 19(3):301–308
76. Trapani JA, Bird PI (2008) A renaissance in understanding the multiple and diverse functions of granzymes? Immunity 29(5):665–667
77. Colucci F, Caligiuri MA, Di Santo JP (2003) What does it take to make a natural killer? Nat Rev Immunol 3(5):413–425
78. Woll PS, Martin CH, Miller JS, Kaufman DS (2005) Human embryonic stem cell-derived NK cells acquire functional receptors and cytolytic activity. J Immunol 175(8):5095–5103
79. Woll PS, Grzywacz B, Tian X, Marcus RK, Knorr DA, Verneris MR et al (2009) Human embryonic stem cells differentiate into a homogeneous population of natural killer cells with potent in vivo antitumor activity. Blood 113(24):6094–6101

80. Knorr DA, Ni Z, Hermanson D, Hexum MK, Bendzick L, Cooper LJ et al (2013) Clinical-scale derivation of natural killer cells from human pluripotent stem cells for cancer therapy. Stem Cells Transl Med 2(4):274–283
81. Bock AM, Knorr D, Kaufman DS (2013) Development, expansion, and in vivo monitoring of human NK cells from human embryonic stem cells (hESCs) and induced pluripotent stem cells (iPSCs). J Vis Exp (74):e50337
82. Li Y, Hermanson DL, Moriarity BS, Kaufman DS (2018) Human iPSC-derived natural killer cells engineered with chimeric antigen receptors enhance antitumor activity. Cell Stem Cell 23(2):181–92 e5
83. Ginhoux F, Greter M, Leboeuf M, Nandi S, See P, Gokhan S et al (2010) Fate mapping analysis reveals that adult microglia derive from primitive macrophages. Science 330(6005):841–845
84. Schulz C, Gomez Perdiguero E, Chorro L, Szabo-Rogers H, Cagnard N, Kierdorf K et al (2012) A lineage of myeloid cells independent of Myb and hematopoietic stem cells. Science 336(6077):86–90
85. Lavin Y, Merad M (2013) Macrophages: gatekeepers of tissue integrity. Cancer Immunol Res 1(4):201–209
86. Serbina NV, Jia T, Hohl TM, Pamer EG (2008) Monocyte-mediated defense against microbial pathogens. Annu Rev Immunol 26:421–452
87. Anderson NR, Minutolo NG, Gill S, Klichinsky M (2021) Macrophage-based approaches for cancer immunotherapy. Cancer Res 81(5):1201–1208
88. Chen Y, Yu Z, Tan X, Jiang H, Xu Z, Fang Y et al (2021) CAR-macrophage: a new immunotherapy candidate against solid tumors. Biomed Pharmacother 139:111605
89. Shapouri-Moghaddam A, Mohammadian S, Vazini H, Taghadosi M, Esmaeili SA, Mardani F et al (2018) Macrophage plasticity, polarization, and function in health and disease. J Cell Physiol 233(9):6425–6440
90. Takeuchi O, Akira S (2010) Pattern recognition receptors and inflammation. Cell 140(6):805–820
91. Weiss G, Schaible UE (2015) Macrophage defense mechanisms against intracellular bacteria. Immunol Rev 264(1):182–203
92. Ohta R, Sugimura R, Niwa A, Saito MK (2019) Hemogenic endothelium differentiation from human pluripotent stem cells in a feeder- and xeno-free defined condition. J Vis Exp (148)
93. Zhang W, Liu L, Su H, Liu Q, Shen J, Dai H et al (2019) Chimeric antigen receptor macrophage therapy for breast tumours mediated by targeting the tumour extracellular matrix. Br J Cancer 121(10):837–845
94. Morrissey MA, Williamson AP, Steinbach AM, Roberts EW, Kern N, Headley MB et al (2018) Chimeric antigen receptors that trigger phagocytosis. Elife vol 7
95. Ueda T, Kumagai A, Iriguchi S, Yasui Y, Miyasaka T, Nakagoshi K et al (2020) Non-clinical efficacy, safety and stable clinical cell processing of induced pluripotent stem cell-derived anti-glypican-3 chimeric antigen receptor-expressing natural killer/innate lymphoid cells. Cancer Sci 111(5):1478–1490
96. Arias J, Yu J, Varshney M, Inzunza J, Nalvarte I (2021) HSC and iPS cell-derived CAR-NK cells as reliable cell-based therapy solutions. Stem Cells Transl Med
97. Detela G, Lodge A (2019) EU regulatory pathways for atmps: standard, accelerated and adaptive pathways to marketing authorisation. Mol Ther Methods Clin Dev 13:205–232
98. Albinger N, Hartmann J, Ullrich E (2021) Current status and perspective of CAR-T and CAR-NK cell therapy trials in Germany. Gene Ther
99. Lupo KB, Matosevic S (2019) Natural killer cells as allogeneic effectors in adoptive cancer immunotherapy. Cancers (Basel) 11(6)
100. Chou CK, Turtle CJ (2019) Insight into mechanisms associated with cytokine release syndrome and neurotoxicity after CD19 CAR-T cell immunotherapy. Bone Marrow Transplant 54(Suppl 2):780–784
101. Xie G, Dong H, Liang Y, Ham JD, Rizwan R, Chen J (2020) CAR-NK cells: a promising cellular immunotherapy for cancer. EBioMedicine 59:102975

102. Liu E, Marin D, Banerjee P, Macapinlac HA, Thompson P, Basar R et al (2020) Use of CAR-transduced natural killer cells in CD19-positive lymphoid tumors. N Engl J Med 382(6):545–553

103. Salter AI, Pont MJ, Riddell SR (2018) Chimeric antigen receptor-modified T cells: CD19 and the road beyond. Blood 131(24):2621–2629

104. Fry TJ, Shah NN, Orentas RJ, Stetler-Stevenson M, Yuan CM, Ramakrishna S et al (2018) CD22-targeted CAR T cells induce remission in B-ALL that is naive or resistant to CD19-targeted CAR immunotherapy. Nat Med 24(1):20–28

105. Anguille S, Van de Velde AL, Smits EL, Van Tendeloo VF, Juliusson G, Cools N et al (2017) Dendritic cell vaccination as postremission treatment to prevent or delay relapse in acute myeloid leukemia. Blood 130(15):1713–1721

106. Ligtenberg MA, Mougiakakos D, Mukhopadhyay M, Witt K, Lladser A, Chmielewski M et al (2016) Coexpressed catalase protects chimeric antigen receptor-redirected T Cells as well as Bystander Cells from oxidative stress-induced loss of antitumor activity. J Immunol 196(2):759–766

107. Liu X, Ranganathan R, Jiang S, Fang C, Sun J, Kim S et al (2016) A chimeric switch-receptor targeting PD1 augments the efficacy of second-generation CAR T cells in advanced solid tumors. Cancer Res 76(6):1578–1590

108. Rafiq S, Yeku OO, Jackson HJ, Purdon TJ, van Leeuwen DG, Drakes DJ et al (2018) Targeted delivery of a PD-1-blocking scFv by CAR-T cells enhances antitumor efficacy in vivo. Nat Biotechnol 36(9):847–856

109. Peng W, Ye Y, Rabinovich BA, Liu C, Lou Y, Zhang M et al (2010) Transduction of tumor-specific T cells with CXCR2 chemokine receptor improves migration to tumor and antitumor immune responses. Clin Cancer Res 16(22):5458–5468

110. Moon EK, Carpenito C, Sun J, Wang LC, Kapoor V, Predina J et al (2011) Expression of a functional CCR2 receptor enhances tumor localization and tumor eradication by retargeted human T cells expressing a mesothelin-specific chimeric antibody receptor. Clin Cancer Res 17(14):4719–4730

111. Zhang L, Kerkar SP, Yu Z, Zheng Z, Yang S, Restifo NP et al (2011) Improving adoptive T cell therapy by targeting and controlling IL-12 expression to the tumor environment. Mol Ther 19(4):751–759

112. Avanzi MP, Yeku O, Li X, Wijewarnasuriya DP, van Leeuwen DG, Cheung K et al (2018) Engineered tumor-targeted T Cells mediate enhanced antitumor efficacy both directly and through activation of the endogenous immune system. Cell Rep 23(7):2130–2141

113. Krenciute G, Prinzing BL, Yi Z, Wu MF, Liu H, Dotti G et al (2017) Transgenic expression of IL15 Improves antiglioma activity of IL13Ralpha2-CAR T cells but results in antigen loss variants. Cancer Immunol Res 5(7):571–581

114. Hartmann J, Schussler-Lenz M, Bondanza A, Buchholz CJ (2017) Clinical development of CAR T cells-challenges and opportunities in translating innovative treatment concepts. EMBO Mol Med 9(9):1183–1197

115. Shi Y, Inoue H, Wu JC, Yamanaka S (2017) Induced pluripotent stem cell technology: a decade of progress. Nat Rev Drug Discov 16(2):115–130

116. Sachdeva M, Duchateau P, Depil S, Poirot L, Valton J (2019) Granulocyte-macrophage colony-stimulating factor inactivation in CAR T-cells prevents monocyte-dependent release of key cytokine release syndrome mediators. J Biol Chem 294(14):5430–5437

The Single-Cell Level Perspective of the Tumor Microenvironment and Its Remodeling by CAR-T Cells

10

Sanxing Gao and Ryohichi Sugimura

10.1 Introduction of the Tumor Microenvironment (TME)

Chimeric antigen receptor T (CAR-T) cell therapies show promising efficacy in leukemia and lymphoma [1]. However, CAR-T therapy does not demonstrate efficacy in solid tumors due to the complex milieu in solid cancers, i.e., the tumor microenvironment (TME), which hampers the tumoricidal activity of CAR-T cell [2, 3]. TME is a complicated niche consisting of tumor cells, myeloid-derived suppressor cells (MDSCs) [4, 5], tumor-associated macrophages (TAMs) [6, 7], exhausted T cells [8], immunosuppressive non-cellular components such as cytokines and extracellular matrix (Fig. 10.1) [9–11].

TME contributes to cancer progression and relapse [2, 12]. The presence of tumor-associated MDSCs such as TAMs, neutrophils, and dendritic cells is strongly associated with the failure of cancer immunotherapy. MDSCs play a pivotal role in the invasion and migration of cancer cells. For example, MDSCs interact with cancer stem cells to mediate the immunosuppressive repertoire to CAR-T therapy [4, 13, 14].

Preclinical experiments showed that CAR-T cells became dysfunctional after trafficking into solid tumors [15]. CAR-T cells in TME increased expression of immune-suppressive molecules such as diacylglycerol kinase and Src homology region 2 domain-containing phosphatase-1(SHP-1), programmed cell death protein 1 (PD-1), T cell immunoglobulin, and mucin-domain containing 3 (TIM-3), Lymphocyte-activation gene 3 (LAG-3), and natural killer cell receptor 2B4.

S. Gao · R. Sugimura (✉)
School of Biomedical Sciences, Li Ka Shing Faculty of Medicine, The University of Hong Kong, Pok Fu Lam, Hong Kong
e-mail: Rios@hku.hk

© The Author(s), under exclusive license to Springer Nature Switzerland AG 2022
P. Hays (ed.), *Cancer Immunotherapies*, Cancer Treatment and Research 183,
https://doi.org/10.1007/978-3-030-96376-7_10

275

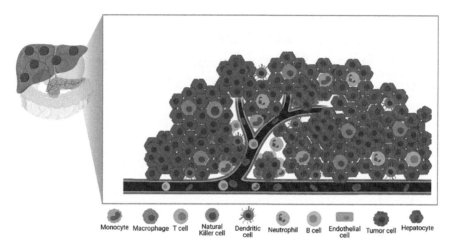

Monocyte Macrophage T cell Natural Killer cell Dendritic cell Neutrophil B cell Endothelial cell Tumor cell Hepatocyte

Fig. 10.1 Liver tumor microenvironment. In TME, tumor cells release cytokines that recruit myeloid-derived suppressive cells including monocytes, macrophages, dendritic cells, and neutrophils. T cells are exhausted and lose their antitumor function in TME

Moreover, the dysfunctional T cells could be restored when they were isolated from TME [16], which indicates that TME plays a crucial role in CAR-T immunotherapy.

Here we describe factors and cytokines in the immune-suppressive TME. TGF-β signaling represses Type 2 helper T (Th2) cells and fosters tumor growth by angiogenesis [17]. TGF-β dominant cancers enrich anti-inflammatory macrophage signatures, consistent with an immunosuppressive TME [18]. TGF-β exhausts cytotoxic T (Tc) cells by inducing the expression of PD-1 and TIM-3, differentiates $CD4^+$ T cells to regulatory T cells (Tregs), and inhibits the expression of granzyme and perforin in NK cells [19]. IL-4 fosters tumor progression through upregulating anti-apoptotic genes such as Bcl-xl and cFLIP in tumor cells [20]. IL-4 activates PI3K/Akt pathway for tumor survival and metastasis [21]. A recent study reported that the increased expression level of Notch ligand (DLL4) and receptor (NOTCH2) were responsible for immune suppression of human fetal liver and hepatocellular carcinoma [22]. In line with these results, Notch pathway activation induces IL-4 secretion and polarizes macrophages to immunosuppressive TAMs [23].

10.2 Tumor-Associated Macrophages in TME

TAMs play a key role in TME via tumor growth, immunosuppression, invasion, and metastasis (Fig. 10.2) [6, 24]. In the following, we are going to introduce how TAMs regulate TME.

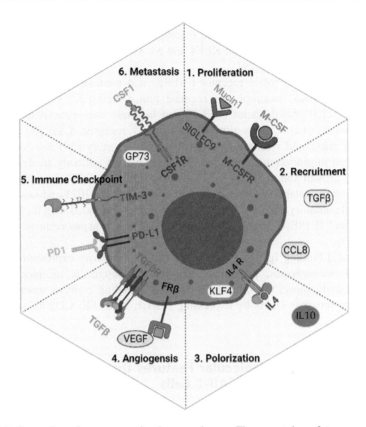

Fig. 10.2 Properties of tumor-associated macrophages. The repertoire of tumor-associated macrophages facilitates tumor progression in the TME. (1) Mucin1 induces the proliferation of TAMs and expression of anti-inflammatory markers such as M-CSFR, CD206 leading to tumor progression. (2) TAMs secrete the TGFβ and CCL8 to facilitate the recruitment of monocytes leading to the accumulation of TAMs in TME. (3) TAMs polarize surrounding macrophages into anti-inflammatory phenotype by IL4, IL10, or IL13. KLF4 is involved in the induction of the anti-inflammatory phenotype. (4) FRβ⁺ macrophages release VEGF to promote angiogenesis of tumor. And TGFβ reprogrammes macrophages into TAMs leading to angiogenesis progression. (5) TAMs express the immune checkpoints such as PD-L1 and TIM3 to exhaust cytotoxic T cells. (6) CSF1R⁺ TAMs enhance the invasion of myeloid cells, leading to the metastasis of tumor cells

Macrophages can be polarized to pro-inflammatory macrophages (M1 phenotype) induced by lipopolysaccharide of microbes or interferon γ [25]. On the other hand, macrophages become alternatively anti-inflammatory macrophages (M2 phenotype) induced by IL-4, IL-13, or TGF-β [26]. The pro-inflammatory macrophages have antitumor activity, whereas the anti-inflammatory macrophages have tumor-promoting properties.

TAMs secrete TGFβ and IL-10 to promote tumor cell growth and angiogenesis through the PI3K pathway [27]. TAMs produce CCL8 to promote the recruitment of monocytes, resulting in more macrophages becoming immunosuppressive

TAMs [28]. Hedgehog signaling facilitates the communication of TAMs and tumor cells leading to polarizing the macrophage toward anti-inflammatory phenotype. The study suggested that KLF4 and NF-kB mediate the anti-inflammatory macrophages polarization [29].

TAMs express the folate receptor β (FRβ) and mediate immune suppression in TME [30]. FRβ⁺ macrophages regulate tumor metastasis via secreting vascular endothelial growth factor (VEGF) and facilitate angiogenesis in pancreatic cancer patients [12]. Colony-stimulating factor 1 receptor (CSF-1R)-expressing TAMs are associated with tumor progression and motility [28] due to increased myeloid cell migration and invasion. The anti-CSF-1R antibody treatment inhibited tumor growth and metastasis [31]. Golgi protein 73 (GP73) is a biomarker of invasion and metastasis of hepatocellular carcinoma [32]. GP73 endows the TAMs an anti-inflammatory phenotype. GP73 expression is correlated with the expression of TIM3 and IL18Bpa, immunosuppressive markers in hepatocellular carcinoma (HCC) [33].

Sialic acid-binding Ig-like lectin 9 (SIGLEC9), primarily expressed on monocytes and macrophages, promotes cell growth through its receptor mucin 1 [34]. The study shows that SIGLEC9-mucin 1 signaling converts macrophage to immune-suppressive TAMs by expressing PD-L1, M-CSFR, CD206, and CD163 [35, 36].

10.3 Cellular and Molecular Features that Determine the Response to CAR-T Cells

Herein we describe immune checkpoint molecules that curb CAR-T cells (Fig. 10.3). PD-1 expresses on the surface of the immune cell such as T cells, B

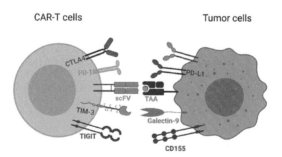

Fig. 10.3 Key immune checkpoints of CAR-T cells engagement with tumor cells. CAR-T cells recognize the tumor cells by tumor antigen-specific scFV. The main four immune checkpoints, CTLA-4, PD-1, TIM-3, and TIGIT, impair the CAR-T cells' antitumor function. CTLA4 binds to costimulation ligand B7(CD80 or CD86) leading to inhibition of T cells. PD-L1 suppresses CAR-T by engaging with PD-1, which results in the apoptosis of CAR-T cells. TIM-3 and TIGIT suppress CAR-T by interaction with galectin-9 and CD155, respectively. CAR-T cells will lose their tumor-killing function through engagement with these molecules

cells, and macrophages. Granulocyte–macrophage colony-stimulating factor (GM-CSF) induces PD-L1 on MDSCs curbing the immune activity of CAR-T cells in liver metastases. The combination of anti-GM-CSF and anti-PD-L1 antibodies restored the efficacy of CAR-T cells [5], which indicates the crucial role of GM-CSF and PD-L1 in CAR-T therapy.

A recent study shows that anti-inflammatory TAMs upregulated immunosuppressive genes such as T cell immunoreceptor with Ig and ITIM domains (TIGIT), CD305, and TIM-3 in HCC. These signals limit the $CD8^+$ T cell infiltration directed to the tumor and are associated with poor clinical prognosis 37]. The low-level expression of PD-1 and CTLA-4 signal in the primary HCC patients correlate with the low efficacy of anti-PD-1 and anti-CTLA-4 immunotherapy in clinical settings [38]. Targeting TIGIT and TIM-3 combined with PD-1 or CTLA-4 may enhance the prognosis of HCC.

TIM-3 is another crucial immune checkpoint molecular [39]. A recent study suggested that TIM-3 induces the exhaustion of $CD8^+$ tumor-infiltrating lymphocytes exhausted in advanced non-small cell lung cancer (NSCLC) patients. The high expression of TIM-3 correlated with the poor efficacy of anti-PD-1 therapy [8]. Clinical study shows that TIM-3 is upregulated on patients' peripheral $CD4^+$ and $CD8^+$ T cells [40]. Combination of anti-TIM-3 and anti-PD-1 therapy increase IFNγ-secreting $CD8^+$ cells and IFNγ^+ TNFα effector T cells in TME leading to improve survival of glioblastoma [41].

Nuclear receptor subfamily 4A (NR4A) activates the nuclear factor of activated T cell (NFAT) leading to the $CD8^+$ T cell exhaustion. CAR-T cells with NR4A deletion reduced the expression of the PD-1 and TIM-3 and enhanced antitumor efficacy [42].

10.4   Single-Cell Sequencing Combined with the Different Approaches Uncovers TME

Bulk RNA sequencing informs the transcriptome of total cells on average, which could have a bias due to the heterogeneity of cells. If some cell populations play a pivotal role in TME but their proportion is low, bulk RNA sequencing could not be informative [43]. Single-cell sequencing could provide a solution to decipher the heterogeneity of cells in TME. The single-cell level perspective of TME provides knowledge about the nature of the tumor property and may lead to innovative cancer therapies [44].

Single-cell transcriptomes identified that Tregs accumulate in brain metastases and resulted in T cell dysfunction by secreting IL-10 and IL-4 to shift TAMs to an immunosuppressive phenotype in TME [45]. A small population of TAMs interacts with $CD40^+CCR7^+LAMP3^+$ dendritic cells and immune stimulation in colorectal cancer patients. The results indicate that targeting these subpopulations can enhance the therapy [46].

Mass cytometry analysis of renal cell carcinoma demonstrated the distribution of PD-1, CTLA-4, and TIM-3 in the TME. This would open up the precision medicine of cancer immunotherapy to the patients. For example, patients predominantly expressing PD-1/PD-L1 in T cells could choose anti-PD-1 therapy as a preference [47]. Pembrolizumab, a humanized anti-PD-1 drug, shows improved efficacy in PD-L1+ non–small-cell lung cancer patients [48].

Integration of flow cytometry with immunofluorescence imaging on brain tumors demonstrated that T cells with high expression of immune checkpoints such as PD-1, LAG-3, TIM-3, and TIGIT were dysfunctional. Advanced brain metastases accumulated Tregs reflecting the immune-suppressive milieu, while early-stage glioma accumulated immature NK cells reflecting potentially immunologically active state [45, 49].

10.5   Strategies of CAR-T Remodel the TME

CAR-T cell immunotherapy can be improved by applying insights from single-cell RNA sequencing of TME. Blocking highly expressed immune checkpoint molecules such as CTLA-4, PD-1, LAG-3, TIGIT, VISTA in CAR-T cells could rescue them from exhaustion in TME, or rewire surrounding immune cells by converting immunosuppressive signals to stimulant signals. Arming the CAR-T cells with Th1 triggering cytokines such as IL-7, IL-12, IL-15, IL-18, IL-21, or JAK-STAT signal switches the TME to a pro-inflammatory state [50]. This could reprogram surrounding TAMs to pro-inflammatory phenotype, and subsequently remodel the TME to an antitumor niche [51]. Moreover, conveying the T cells with two single-chain variable fragments, i.e., bispecific T cell engagers (BiTEs) could enhance the specificity to target tumor CAR-T cells and could be engineered to secrete BiTEs [52]. In the following, we summarize four approaches to remodel the TME (Fig. 10.4).

First, endowing CAR-T cells with immune checkpoint blockades allows for CAR-T cells to be engineered and secrete anti-PD-1 scFv, which could engage bystander T cells with antitumor activity [53]. They found that PD-1 scFV-secreting CAR-T cells show stronger antitumor efficacy in both Raji-PD-L1 hematologic and SKOV3-PD-L1 solid tumor-bearing mouse models compared to the single CAR-T approach due to the escort of bystander T cells from PD-1 scFV-secreting CAR-T cells.

Second, CAR-T cells can be engineered to secrete antitumor cytokines. IL-12 enhances CAR-T cell responses by sustaining T cell cytotoxicity [54]. Intratumoral delivery of IL-12 in the combination with tumor-targeted CAR-T cell therapy remodeled the TME into a pro-inflammatory state by the production of pro-inflammatory cytokines IFN-γ and TNF, decreasing regulatory T cells and polarization to inflammatory macrophages [55]. CAR-T cells expressing IL-7 and CCL19 showed superior antitumor activity [56]. CAR-T cells coexpressing IL-15 remodeled the TME by activating NK cells and reduced anti-inflammatory macrophages [57]. CAR-T cells expressing the p40 subunit of IL-23 enhanced the tumoricidal function by upregulating the granzyme B and downregulating PD-1

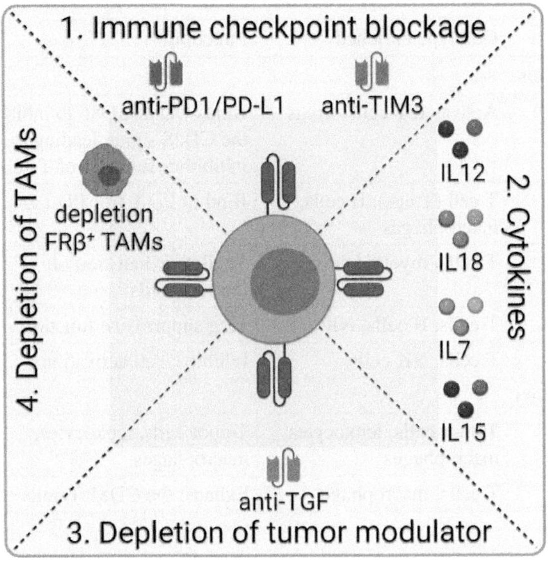

Fig. 10.4 Strategies to remodel TME. The major strategies are blockage or depletion of immuno-suppressive factors in the TME by CAR-T cells. (1) CAR-T cells secrete the anti-PD1 antibody, which blocks the PD-1 signal of immune cells, leading to both protecting the CAR-T cell and restoring the bystander T cell. (2) Secreting the immune priming cytokines such as IL12, IL18 can boost the T cell activation and convert the TAMs to a pro-inflammatory state. (3) CAR-T cells block the immune-suppressive cytokines such as TGFβ to improve the enrichment of cytotoxic T cells in the TME. (4) Targeting the immunosuppressive TAMs by CAR-T. Elimination of FRβ+ TAMs increased the infiltration of cytotoxic T cells in the TME

expression [58]. CAR-T cells releasing IL-18 showed superior efficacy of expansion and antitumor by increasing the cytotoxic T cells [59], as well as reversing the exhausted T cell to a tumoricidal Tbet ^{high} FoxO1^{low} T cells [60].

Third, engineering CAR-T cells to antagonize immune-suppressive cytokine. TGF-β, secreted by tumor cells, shapes an immunosuppressive TME, leading to resistance to immunotherapy [61]. Anti-TGF-β therapy reduced the epithelial-to-mesenchymal transition of tumor cells and improved the penetration of T cells into tumors [62]. Selective inactivation of TGF-β1 by SRK-181 antibody facilitated the antitumor activity by enriching the CD8$^+$ T cell and the memory cell in the TME [17, 63]. Co-expression of a dominant-negative TGF-β RII with anti-prostate specific membrane antigen CAR can be resistant to TGF-β dominant TME in PC3-PSMA tumor-bearing mouse model [64]. Anti-TGF-β CAR-T cells protect T cells from immunosuppressive TGF-β into an immunostimulatory phenotype. And what is more, Anti-TGF-β CAR-T cells can reverse the TGF-β from an immunosuppressive molecule toward a stimulator of T cell proliferation in vitro [65].

Table 10.1 Key molecules determine the response of CAR-T cells

Molecular	Cell type enriched	Function	References
Immune checkpoint			
CTLA-4 (CD152)	Activated T cells, Tregs	Binds CD80/CD86 to inhibit the CD28 signal leading to inhibitory function of T cell	[69, 70]
PD-1 (CD279)	T cell (Tregs), B cells, macrophages	Bind to PD-L1 or PD-L2	[70–72]
TIM-3(CD366)	T cells, myeloid cells	Mediate exhaustion of immune cells	[8, 39, 73]
LAG-3(CD223)	T cells, B cells, NK cells	Treg suppressive function	[74, 75]
TIGIT	T cells, NK cells	Inhibit T cell activation	[37, 76]
Cytokines or factors			
TGFβ	Tumor cells, leukocytes, macrophages	Tumor cells, leukocytes, macrophages	[17, 18]
NR4A	T cells, macrophage	Exhaust the CD8$^+$ T cells	[42, 77]

Fourth, targeting TAMs by CAR-T cells. Abolishing FRβ$^+$ subpopulation of TAMs improved T cell-mediated antitumor immune responses [66].

10.6   Prospective

Precision medicine of cancer immunotherapy will be a major goal of CAR-T technology. In this review, we discussed the molecules and cells which play key roles in the tumor microenvironment and CAR-T therapy. Based on the findings of single-cell sequencing in TME and CAR-T cells, we believe that the identification of novel immune checkpoint molecules and cytokines that hinge the activity of CAR-T cells will offer new targets in cancer immunotherapy. We summarized the four approaches to engineer CAR-T cells to remodel the TME. The insight from the new single-cell technologies will pave the avenue for improving CAR-T immunotherapy to benefit the patients [67]. The spatial multi-omics can define both the transcriptome and proteome of the TME [68]. By defining the TME, one could engineer CAR-T cells to precisely target immune-suppressive molecules in the TME for each patient (Table 10.1).

References

1. Kingwell K (2017) CAR T therapies drive into new terrain. Nat Rev Drug Discov 16:301–304
2. Lei X et al (2020) Immune cells within the tumor microenvironment: biological functions and roles in cancer immunotherapy. Cancer Lett 470:126–133
3. Tang TCY, Xu N, Dolnikov A (2020) Targeting the immune-suppressive tumor microenvironment to potentiate CAR T cell therapy. Cancer Rep Rev 4
4. Gabrilovich DI (2017) Myeloid-derived suppressor cells. Cancer Immunol Res 5:3–8

5. Burga RA et al (2015) Liver myeloid-derived suppressor cells expand in response to liver metastases in mice and inhibit the antitumor efficacy of anti-CEA CAR-T. Cancer Immunol Immunother 64:817–829
6. Mantovani A, Marchesi F, Malesci A, Laghi L, Allavena P (2017) Tumour-associated macrophages as treatment targets in oncology. Nat Rev Clin Oncol 14:399–416
7. Noy R, Pollard JW (2014) Tumor-associated macrophages: from mechanisms to therapy. Immunity 41:49–61
8. Sanmamed MF et al (2021) A burned-out CD8+ T-cell subset expands in the tumor microenvironment and curbs cancer immunotherapy. Cancer Discov
9. Noel A et al (2012) New and paradoxical roles of matrix metalloproteinases in the tumor microenvironment. Front Pharmacol 3:140
10. Li MO et al (2021) Innate immune cells in the tumor microenvironment. Cancer Cell 39:725–729
11. Bejarano L, Jordão MJC, Joyce JA (2021) Therapeutic targeting of the tumor microenvironment. Cancer Discov 11:933–959
12. Kurahara H et al (2012) Clinical significance of folate receptor beta-expressing tumor-associated macrophages in pancreatic cancer. Ann Surg Oncol 19:2264–2271
13. Biswas SK, Allavena P, Mantovani A (2013) Tumor-associated macrophages: functional diversity, clinical significance, and open questions. Semin Immunopathol 35:585–600
14. Kielbassa K, Vegna S, Ramirez C, Akkari L (2019) Understanding the origin and diversity of macrophages to tailor their targeting in solid cancers. Front Immunol 10:2215
15. Crespo J, Sun H, Welling TH, Tian Z, Zou W (2013) T cell anergy, exhaustion, senescence, and stemness in the tumor microenvironment. Curr Opin Immunol 25:214–221
16. Moon EK et al (2014) Multifactorial T-cell hypofunction that is reversible can limit the efficacy of chimeric antigen receptor-transduced human T cells in solid tumors. Clin Cancer Res 20:4262–4273
17. Liu M et al (2020) TGF-beta suppresses type 2 immunity to cancer. Nature 587:115–120
18. Thorsson V et al (2018) The immune landscape of cancer. Immunity 48:812–830 e814
19. de Charette M, Houot R (2018) Hide or defend, the two strategies of lymphoma immune evasion: potential implications for immunotherapy. Haematologica 103:1256–1268
20. Li Z et al (2008) Endogenous IL-4 promotes tumor development by increasing tumor cell resistance to apoptosis. Can Res 68:8687–8694
21. Venmar KT, Carter KJ, Hwang DG, Dozier EA, Fingleton B (2014) IL4 receptor ILR4alpha regulates metastatic colonization by mammary tumors through multiple signaling pathways. Can Res 74:4329–4340
22. Sharma A et al (2020) Onco-fetal reprogramming of endothelial cells drives immunosuppressive macrophages in hepatocellular carcinoma. Cell 183:377–394 e321
23. Lin X et al (2019) miR-195-5p/NOTCH2-mediated EMT modulates IL-4 secretion in colorectal cancer to affect M2-like TAM polarization. J Hematol Oncol 12:20
24. Ngambenjawong C, Gustafson HH, Pun SH (2017) Progress in tumor-associated macrophage (TAM)-targeted therapeutics. Adv Drug Deliv Rev 114:206–221
25. Hume DA (2015) The many alternative faces of macrophage activation. Front Immunol 6:370
26. Jayasingam SD et al (2019) Evaluating the polarization of tumor-associated macrophages Into M1 and M2 phenotypes in human cancer tissue: technicalities and challenges in routine clinical practice. Front Oncol 9:1512
27. Cui X et al (2018) Hacking macrophage-associated immunosuppression for regulating glioblastoma angiogenesis. Biomaterials 161:164–178
28. Cassetta L et al (2019) Human tumor-associated macrophage and monocyte transcriptional landscapes reveal cancer-specific reprogramming, biomarkers, and therapeutic targets. Cancer Cell 35:588–602 e510
29. Petty AJ et al (2019) Hedgehog signaling promotes tumor-associated macrophage polarization to suppress intratumoral CD8+ T cell recruitment. J Clin Invest 129:5151–5162

30. Puig-Kroger A et al (2009) Folate receptor beta is expressed by tumor-associated macrophages and constitutes a marker for M2 anti-inflammatory/regulatory macrophages. Can Res 69:9395–9403
31. Lohela M et al (2014) Intravital imaging reveals distinct responses of depleting dynamic tumor-associated macrophage and dendritic cell subpopulations. Proc Natl Acad Sci U S A 111:E5086-5095
32. Bao YX et al (2013) Expression and prognostic significance of golgiglycoprotein73 (GP73) with epithelial-mesenchymal transition (EMT) related molecules in hepatocellular carcinoma (HCC). Diagn Pathol 8:197
33. Wei C et al (2019) Tumor microenvironment regulation by the endoplasmic reticulum stress transmission mediator Golgi protein 73 in mice. Hepatology 70:851–870
34. Tanida S et al (2013) Binding of the sialic acid-binding lectin, Siglec-9, to the membrane mucin, MUC1, induces recruitment of beta-catenin and subsequent cell growth. J Biol Chem 288:31842–31852
35. Beatson R et al (2016) The mucin MUC1 modulates the tumor immunological microenvironment through engagement of the lectin Siglec-9. Nat Immunol 17:1273–1281
36. Allavena P, Mantovani A (2012) Immunology in the clinic review series; focus on cancer: tumour-associated macrophages: undisputed stars of the inflammatory tumour microenvironment. Clin Exp Immunol 167:195–205
37. Ho DW et al (2021) Single-cell RNA sequencing shows the immunosuppressive landscape and tumor heterogeneity of HBV-associated hepatocellular carcinoma. Nat Commun 12:3684
38. Huppert LA, Gordan JD, Kelley RK (2020) Checkpoint inhibitors for the treatment of advanced hepatocellular carcinoma. Clin Liver Dis (Hoboken) 15:53–58
39. Dixon KO et al (2021) TIM-3 restrains anti-tumour immunity by regulating inflammasome activation. Nature
40. Han S et al (2014) Tim-3 on peripheral CD4(+) and CD8(+) T cells is involved in the development of glioma. DNA Cell Biol 33:245–250
41. Kim JE et al (2017) Combination therapy with anti-PD-1, anti-TIM-3, and focal radiation results in regression of murine gliomas. Clin Cancer Res 23:124–136
42. Chen J et al (2019) NR4A transcription factors limit CAR T cell function in solid tumours. Nature 567:530–534
43. Saviano A, Henderson NC, Baumert TF (2020) Single-cell genomics and spatial transcriptomics: discovery of novel cell states and cellular interactions in liver physiology and disease biology. J Hepatol 73:1219–1230
44. Qian J et al (2020) A pan-cancer blueprint of the heterogeneous tumor microenvironment revealed by single-cell profiling. Cell Res 30:745–762
45. Friebel E et al (2020) Single-cell mapping of human brain cancer reveals tumor-specific instruction of tissue-invading leukocytes. Cell 181:1626–1642 e1620
46. Zhang L et al (2020) Single-cell analyses inform mechanisms of myeloid-targeted therapies in colon cancer. Cell 181:442–459 e429
47. Chevrier S et al (2017) An immune atlas of clear cell renal cell carcinoma. Cell 169:736-749.e718
48. Garon EB et al (2015) Pembrolizumab for the treatment of non-small-cell lung cancer. N Engl J Med 372:2018–2028
49. Woroniecka K, Fecci PE (2018) T-cell exhaustion in glioblastoma. Oncotarget 9:35287–35288
50. Hong M, Clubb JD, Chen YY (2020) Engineering CAR-T cells for next-generation cancer therapy. Cancer Cell 38:473–488
51. Rochman Y, Spolski R, Leonard WJ (2009) New insights into the regulation of T cells by gamma(c) family cytokines. Nat Rev Immunol 9:480–490
52. Choi BD et al (2019) CAR-T cells secreting BiTEs circumvent antigen escape without detectable toxicity. Nat Biotechnol 37:1049–1058
53. Rafiq S et al (2018) Targeted delivery of a PD-1-blocking scFv by CAR-T cells enhances antitumor efficacy in vivo. Nat Biotechnol 36:847–856

54. Boulch M et al (2021) A cross-talk between CAR T cell subsets and the tumor microenvironment is essential for sustained cytotoxic activity. Sci Immunol 6

55. Agliardi G et al (2021) Intratumoral IL-12 delivery empowers CAR-T cell immunotherapy in a pre-clinical model of glioblastoma. Nat Commun 12:444

56. Adachi K et al (2018) IL-7 and CCL19 expression in CAR-T cells improves immune cell infiltration and CAR-T cell survival in the tumor. Nat Biotechnol 36:346–351

57. Lanitis E et al (2021) Optimized gene engineering of murine CAR-T cells reveals the beneficial effects of IL-15 coexpression. J Exp Med 218

58. Ma X et al (2020) IL-23 engineering improves CAR T cell function in solid tumors. Nat Biotechnol 38:448–459

59. Hu B et al (2017) Augmentation of antitumor immunity by human and mouse CAR T cells secreting IL-18. Cell Rep 20:3025–3033

60. Chmielewski M, Abken H (2017) CAR T cells releasing IL-18 convert to T-bet(high) FoxO1(low) effectors that exhibit augmented activity against advanced solid tumors. Cell Rep 21:3205–3219

61. Derynck R, Turley SJ, Akhurst RJ (2020) TGFbeta biology in cancer progression and immunotherapy. Nat Rev Clin Oncol

62. Mariathasan S et al (2018) TGFbeta attenuates tumour response to PD-L1 blockade by contributing to exclusion of T cells. Nature 554:544–548

63. Martin CJ et al (2020) Selective inhibition of TGFbeta1 activation overcomes primary resistance to checkpoint blockade therapy by altering tumor immune landscape. Sci Transl Med 12

64. Kloss CC et al (2018) Dominant-negative TGF-beta receptor enhances PSMA-targeted human CAR T cell proliferation and augments prostate cancer eradication. Mol Ther 26:1855–1866

65. Chang ZL et al (2018) Rewiring T-cell responses to soluble factors with chimeric antigen receptors. Nat Chem Biol 14:317–324

66. Rodriguez-Garcia A et al (2021) CAR-T cell-mediated depletion of immunosuppressive tumor-associated macrophages promotes endogenous antitumor immunity and augments adoptive immunotherapy. Nat Commun 12:877

67. Goliwas KF, Deshane JS, Elmets CA, Athar M (2021) Moving immune therapy forward targeting TME. Physiol Rev 101:417–425

68. Liu Y et al (2020) High-spatial-resolution multi-omics sequencing via deterministic barcoding in tissue. Cell 183:1665–1681 e1618

69. Walunas TL, Bakker CY, Bluestone JA (1996) CTLA-4 ligation blocks CD28-dependent T cell activation. J Exp Med 183:2541–2550

70. Syn NL, Teng MWL, Mok TSK, Soo RA (2017) De-novo and acquired resistance to immune checkpoint targeting. Lancet Oncol 18:e731–e741

71. Agata Y et al (1996) Expression of the PD-1 antigen on the surface of stimulated mouse T and B lymphocytes. Int Immunol 8:765–772

72. Iwai Y et al (2002) Involvement of PD-L1 on tumor cells in the escape from host immune system and tumor immunotherapy by PD-L1 blockade. Proc Natl Acad Sci U S A 99:12293–12297

73. Blackburn SD et al (2009) Coregulation of CD8+ T cell exhaustion by multiple inhibitory receptors during chronic viral infection. Nat Immunol 10:29–37

74. Huang CT et al (2004) Role of LAG-3 in regulatory T cells. Immunity 21:503–513

75. Schoutrop E et al (2021) Mesothelin-specific CAR T cells target ovarian cancer. Can Res 81:3022–3035

76. Chiu DK et al (2020) Hepatocellular carcinoma cells up-regulate PVRL1, stabilizing PVR and inhibiting the cytotoxic T-cell response via TIGIT to mediate tumor resistance to PD1 inhibitors in mice. Gastroenterology 159:609–623

77. Martinez GJ et al (2015) The transcription factor NFAT promotes exhaustion of activated CD8(+) T cells. Immunity 42:265–278

Clinical Development and Therapeutic Applications of Bispecific Antibodies for Hematologic Malignancies

11

Priya Hays

Abbreviations

Ab	Antibody
AE	Adverse event
ALL	Acute lymphocytic leukemia
AML	Acute myeloid leukemia
BCMA	B-cell maturation antigen
BCP-ALL	B-cell precursor acute lymphocytic leukemia
Ph-R/R	Philadelphia relapsed/recurrence
BiTEs	Bispecific T cell engagers
BsAbs	Bispecific antibodies
CAR-T cells	Chimeric antigen receptor-T cells
CD	Cluster of differentiation
CLL	Chronic lymphocytic leukemia
CR	Complete response
CRS	Cytokine release syndrome
CTCAE	Common terminology criteria for adverse events
CTLA-4	Cytotoxic T-lymphocyte associated antigen-4
DART	Dual-affinity re-targeting antibody
DLBCL	Diffuse large B-cell lymphoma
EMA	European Medicines Agency
Fc	Fragment crystallizable
FDA	Food and Drug Administration
HR	Hazard ratio
ICANs	Immune effector cell-associated neurotoxicity syndrome

P. Hays (✉)
Hays Documentation Specialists, LLC, San Mateo, CA, USA
e-mail: priya.hays@outlook.com

© The Author(s), under exclusive license to Springer Nature Switzerland AG 2022
P. Hays (ed.), *Cancer Immunotherapies*, Cancer Treatment and Research 183,
https://doi.org/10.1007/978-3-030-96376-7_11

ICIs	Immune checkpoint inhibitors
Ig	Immunoglobulin
IL	Interleukin
IV	Intravenous
kDa	Kilodalton
LAG-3	Lymphocyte-activation gene 3
MHC	Major histocompatibility complex
MM	Multiple myeloma
MRD	Minimal residual disease
NK	Natural killer
NSCLC	Non-small cell lung cancer
ORR	Objective response rate
OS	Overall survival
PD	Pharmacodynamic
PD-1/PD-L1	Programmed death-1/programmed death ligand-1
Ph+ R/R BCP-ALL	Philadelphia chromosome positive relapsed/recurrence B-cell precursor acute lymphocytic leukemia
PI3	Phosphoinositol-3
PK	Pharmacokinetic
R/R MM	Relapsed/recurrent multiple myeloma
R/R NHL	Relapsed/recurrence non-Hodgkin lymphoma
scRvs	Single-chain variable fragments
SNVs	Single nucleotide variations
SOC	Standard of care
TandAb	Tandem diabody
TCR	T-cell receptor
TEAEs	Treatment emergent adverse events
TKIs	Tyrosine kinase inhibitors
VH	Heavy chain
VL	Light chain

11.1 Introduction

Immune checkpoint inhibitors and chimeric antigen receptor T cells have been discussed in great depth in this volume. As shown, they form a class of T cell-based cancer immunotherapies that focus on immunosuppressive factors and immunostimulatory pathways, respectively [1]. ICIs have demonstrated clinical efficacy against solid tumors such as melanoma and NSCLC by releasing the blockade of T cells from PD-1/PD-L1 and CTLA-4 immunosuppressive molecules in "hot" tumors. CAR-T cells therapies, also discussed in this volume, are genetically engineered T cells generated ex-vivo from patients for re-infusion to direct T cell activity for tumor cell destruction.

This chapter focuses on a third class of T cell-based cancer immunotherapy known as bispecific antibodies, which are composed of two monoclonal antibodies that link cell surface molecules on T cells to tumor-associated antigens to lead to cancer cell lysis. This class of therapies has shown clinical efficacy in Hematologic malignancies mainly and is discussed in this chapter.

11.2 The Construction of Bispecific Antibodies and Their Cellular Properties

Bispecific antibodies comprise a therapeutic class of agents that are designed to target tumor cells by directing T cells to the antigens on these tumor cells. They recognize and bind to two distinct antigens. The majority of bispecific antibodies fall into the category of bispecific T cell engagers or BiTEs. One of the first FDA-approved agents is blinatumomab, a bispecific T cell engager with CD19 and CD3 epitopes. CD3 is invariably utilized as a surface epitope on the T cell receptor as a target, and BsAbs are developed with CD3 targets, which are in turn categorized as BsAbs with or without Fc domains. Bispecific antibodies are constructed to bring two different antigens on different cells together and bring cytotoxic T cells in contact with tumor cells, thus destroying them. More than 100 different BsAb formats have been invented, making them much more complex than monoclonal antibodies as a result of innovative advances in protein and gene engineering [2]. Based on the structure of the five classes of antibodies, IgG, IgM, IgA, IgD, and IgE, these antibodies are composed of the antigen-binding fragments (Fab) and the fragment crystallizable region (Fc). The Fc domain confers stability, high half-life, and a relatively uncomplicated purification process, but can lead to non-specific immune response from interaction of the Fc domain receptors with other cytotoxic immune cells such as natural killer cells, monocytes, and macrophages [1]. According to Huang, the various BsAb formats can be distinguished into two categories depending on the presence of an Fc domain. They can also be divided into the Fc architecture and the Fc less architecture. The latter include BiTE, DART, and TandAbm, which hold benefits of high yield and are more able to penetrate tissues. However, they have short in vivo half-lives and decreased stability.

Tumor-associated antigens are presented by MHC molecules expressed on tumor cells, allowing for T cells to be activated and destroy these cells. ICIs restriction is due to MHC restriction impairment, inhibiting T cell presentation to tumor cells. BiTE, on the other hand, can lead to interaction between cytotoxic T cells and tumor cells independently of MHC restriction, which leads to immunological synapse and the secretion of perforins and granzymes. BiTE can overcome the limitations of CAR T-cells, being produced in a simple and fast process. The coupling of tumor-associated antigen with the CD3 complex of T cells leads to T cell engagement with malignant cells and a T cell response. Tumor lysis results since the BiTE design bypasses the MHC barrier, thus bypassing this "common evasion mechanism" of tumor cells [3].

11.3   Clinical Outcomes of BiTE in Hematologic Malignancies

Hematologic malignancies benefited from the development of BsAbs since many of these cancers are amenable to treatment of BsAbs. In 2005, a clinical trial established blinatumomab as an effective therapeutic agent for non-Hodgkin's lymphoma [4]. BiTEs therapy depends on the identification of an antigen that is tumor cell-specific; CD19 has abnormal expression on malignant cells of B cell lineage and is selected as a target, being a glycoprotein that has stable expression on B cell precursor cells, particularly malignant cells of B cell origin [3].

As of 2020, 123 BsAbs are being clinically evaluated, of which bispecific T cell engagers or BiTEs, remain the largest category, which targets the two different surfaces of the immune cell and tumor cell and thereby engaging them toward tumor cell destruction. Among B cell malignancies, acute lymphoblastic leukemia, multiple myeloma, chronic lymphocytic leukemia, non-Hodgkin lymphoma, BiTEs are being evaluated for their treatment strategies for these malignancies. Standard of care approaches includes the anti-CD20 monoclonal Ab rituximab and the Bruton tyrosine kinase ibrutinib, and autologous stem-cell transplantation, with considerations for minimal residual disease for determining efficacy [4] (Fig. 11.1).

"The agent blinatumomab is considered the "CD19-CD3" canonical BiTE construct" with a clinical efficacy for ALL [1].

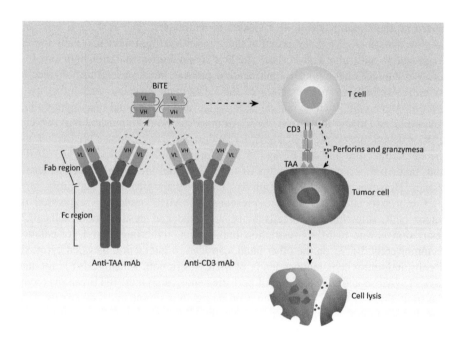

Fig. 11.1 The schematic representation of structure and mechanism of action of canonical bispecific T-cell engager (BiTE). mAb: Monoclonal antibody; CH: heavy chain variable region; VL: light chain variable region; TAA: tumor-associated antigen. Adapted from Zhou (2021)

11.4 Development of Blinatumomab

Blinatumomab was first evaluated in R/R NHL and CLL patients and intravenously administered in phase I studies. Dosage was 0.5–90.0181 ug/m^2/day to accommodate its short half-life (2 h). CRS and neurological sequelae were observed. BCP-ALL patients were also among the first to be evaluated as well and achieved complete remission with minimal residual disease. Orphan drug designation in 2008 and Breakthrough Therapy (2014) and Priority Review (2014) designations soon followed in the US. Marketing authorization was initially given for Ph-R/R BCP-ALL as a result of two phase II studies that were open-label, single-arm, and multicenter. In turn, the drug received accelerated approval pathway (2014) followed by regular FDA approval granted for patients as a result of the clinical benefit demonstrated as a result of the TOWER study, as discussed below. The ALCANTRA trial, also discussed below, provided the clinical data for full FDA approval for both Ph- and Ph+ R/R BCP-ALL. According to Stein et al., as of 2017, the agent was approved in 53 countries in R/R patients, and as of 2018, BCP-ALL pediatric patients were included for blinatumomab treatment in the phase II BLAST study who are in remission but have MRD [5, 6].

11.5 Acute Lymphoblastic Leukemia

11.5.1 Administration and Dosing Schedules

Blinatumomab dosing is dependent on the type of malignancy and evidence of tumor burden. Short-term infusion scheduled 2–4 h for 1–3 times/week demonstrates no clinical response as measured by B cell depletion as observed in ALL. Recommendations include two induction cycles followed by maintenance and consolidation cycles with hospitalization for 9 days in the first cycle and first 2 days of the second cycle for R/R ALL patients. An induction cycle of 28 days is followed by a 14-day treatment-free interval and accompanied by hematologic CR and MRD. To minimize CRS, patients are premedicated with dexamethasone or a similar corticosteroid, especially in R/R cases with a dosing regimen consisting of step-up blinatumomab in R/R disease >25% blasts in the bone marrow especially for NHL. A phase 2 study showed that a stepwise dosing schedule was tolerated to also avoid adverse neurological side effects. After a period of four treatment-free weeks, patients experiencing CR, PR, or stable disease were administered further consolidation cycle; efficacy was monitored during these weeks as progression occurs rapidly [7–9]. Some disadvantages of IV perfusion are lack of convenience for patients and costs. BsAbs such as blinatumomab are known for their short half-lives that lead to the need for more frequent infusion. The BsAbs with longer half-lives could have the potential for greater toxicity, which is being corroborated by ongoing clinical trials [10–12].

11.5.1.1   Clinical Efficacy of Blinatumomab in B-ALL

Blinatumomab, the anti-CD19-CD3 bispecific T cell engager, was approved by the FDA in 2014 and the EMA in December 2015 for the treatment of relapsed refractory B cell precursor ALL [13–15]. Katarajan et al. showed in a multi-institutional phase 3 trial that blinatumomab had better outcomes when compared with chemotherapy. Out of 405 evaluable patients, 271 received blinatumomab while 134 received chemotherapy. OS was the primary endpoint and blinatumomab OS was 7.7 months, compared to 4.0 months for chemotherapy (HR: 0.71). Grade 3 or higher adverse events were 87% versus 92% for blinatumomab and chemotherapy, respectively [16]. In a phase 2 single-group trial, Foa et al. evaluated dasatinib, a tyrosine kinase inhibitor, with glucocorticoids followed by two cycles of blinatumomab in 63 patients with (Ph)-positive ALL with no upper age limit. Sustained molecular response was the primary endpoint. At day 85, 29% had a molecular response and this percentage increased to 60% after two cycles of blinatumomab at median follow-up of 18 months; OS was 95% and disease-free survival was 88% (Table 11.1) [17].

11.5.2   BCP-ALL in CR with MRD and R/R BCP-ALL [3]

A series of trials established the clinical utility of blinatumomab for BCP-ALL. The phase 2 BLAST study evaluated adults with BCP-ALL in hematologic CR with MRD. Median OS was 38.9 months versus 12.5 months ($P = 0.002$) in patients who did or did not have complete MRD response within one cycle of treatment, respectively [18]. However, cure was eventually achieved by these patients after five years, as median OS was not reached for patients who experienced MRD in cycle 1 [19]. Since BLAST was a single-arm study, control data was provided through a historical comparator that led to an analysis of Hematologic relapse-free survival [20] and eventual approval in this patient population.

The phase III TOWER study examined a randomized Ph- R/R BCP-ALL patient population that compared blinatumomab ($n = 271$) with standard treatment chemotherapy ($n = 134$) [21], and was considered requisite since the drug received accelerated approval by the FDA. These patients were heavily pretreated with intensive combination therapy for initial or subsequent salvage treatment [22]. Deep, durable outcomes were achieved by the blinatumomab cohort compared to chemotherapy to the point that the trial was halted due to robust OS benefits [21]. OS served as the primary endpoint with CR with complete Hematologic recovery as secondary endpoints. The blinatumomab cohort was found to have superior OS relative to SOC, with median OS at 8 and 4 months, respectively. [HR 0.71 (95% CI 0.55–0.93); $p = 0.01$]. According to Stein et al., "[r]emission within 12 weeks following initiation of treatment was also significantly higher in the blinatumomab group versus the SOC group: CR with full hematologic recovery (34% vs 16%, p < 0.001) and CR with full, partial, or incomplete hematologic recovery (44% vs. 25%; $p < 0.001$)" [5]. In terms of salvage therapy, blinatumomab was effective, especially in first salvage, and led to doubling of the median survival compared to

Table 11.1 Blinatumomab for Acute Lymphoblastic Leukemia (Kantarjian 2017; NEJM) (Dasatinib study Foa NEJM 2020)

Trial	Comparator	Patient population	Primary endpoint	Clinical outcomes	Adverse events
Multi-institutional phase 3 trial (Kantarjian 2017; NEJM)	Blinatumomab versus chemotherapy; 2:1 ratio	405 patients: 271 blinatumomab; 134 chemotherapy	Overall survival	7.7 months versus 4.0 months. HR: 0.71; 95% CI (0.55–0.93; $P = 0.01$); full hematologic recovery (34% vs. 16%): 34% in each treatment group allogeneic stem-cell transplantation	87% versus 92%; Grade 3 or higher
Phase 2 single-group trial	Dasatinib plus glucocorticoids followed by two cycles of blinatumomab	63 patients with (Ph)-positive ALL with no upper age limit	Sustained molecular response in bone marrow after treatment	At the end of dasatinib induction therapy (day 85) 29% had a molecular response and this percentage increased to 60% after two cycles of blinatumomab; at median follow-up of 18 months, OS was 95% and DFS was 88%	

standard of care chemotherapy. Salvage status was the driving factor for survival in these responders independent of subsequent allogeneic stem-cell transplantation. Additionally, blinatumomab also led to significant quality of life that was health-related [23–25]. Adverse events included pyrexia, CRS, and infusion site reactions at a greater than 5% incidence, and serious neurotoxicity was associated in the blinatumomab arm. After adjusting treatment exposure, however, the overall incidence of these events in the blinatumomab was significantly lower (349 vs. 642 events per 100 patient-years of exposure) [22]. After adjusting for time on treatment, Grade ≥ exposure adjusted event rates were less for blinatumomab compared with the SOC arm (11 vs. 45 events per patient-year; $p < 0.001$). "For specific Grade ≥ 3 events of clinical interest, the exposure adjusted event rates for blinatumomab versus SOC were lower for infections (2 vs. 6 events per patient-year; $p < 0.001$), cytopenias (4 vs. 20 events per patient-year; $p < 0.001$), and neurologic events (0.4 vs. 1 event per patient-year; $p = 0.008$), and higher for CRS (0.2 vs. 0 events per patient-year; $p = 0.038$)" [5].

Ph+ positive with ALL is associated with poor prognosis. TKIs in combination with chemotherapy is considered the standard frontline treatment for Ph+ BCP-ALL in adults. The phase 2 ALCANTRA study was a single-arm trial that evaluated blinatumomab therapy in Ph+ BCP-ALL patients that were unresponsive to second-generation TKIs or imatinib [26, 27]. The results were that CR or CR with partial hematologic recovery was achieved in 16/45 of 36% of patients. Additionally, blinatumomab treatment was seen to be highly effective in leading to the elimination of detectable MRD in 12/14 or 86% of responders that had complete MRD response [27].

11.5.3 Predictive Indicators for Blinatumomab Treatment in R/R B-ALL

Blinatumomab has been the most intensively studied example of BiTE with substantial clinical outcomes demonstrated, especially for high tumor burden disease as represented by ≥50% bone marrow blasts R/R B-ALL. As of 2020, it is the only FDA and EMA-approved BiTE therapy [28–31]. Tumor burden or percentage of bone marrow blasts is predictive of CD19 BiTE therapy. In Ph-BCP-ALL patients from previous trials underwent subgroup analysis and those populations with <50% bone marrow blasts were observed to have the greatest OS and remission rates when treated with blinatumomab [21].

According to one phase 3 study that showed statistical significance, CR for these patients in terms of exhibiting full or partial hematologic recovery with the percentage of bone marrow blasts served as a predictive indicator: 65.5% for less than 50% and 34.4% for ≥50%. ($P = 0.039$) [28, 31]. As mentioned earlier, dexamethasone has a cytoreductive effect for blinatumomab therapy [32]. Extramedullary disease or EMD can serve as a "surrogate for disease burden", being indicative of progressive disease. A retrospective historical study evaluated baseline and treatment measures of EMD and demonstrated lower CR rates associated with EMD ($P = 0.005$ and $P = 0.05$, respectively). MRD has a similar role and one study showed at the day 15, bone marrow MRD in children receiving blinatumomab could predict "complete MRD response" with significant accuracy (up to 95% for the first two treatment cycles) [33]. At day 15, 59 patients were evaluated for complete MRD response: "among 46 MRD positive patients, 44/46 patients had no complete MRD response with an accuracy of 96%, meanwhile, 12/13 patients achieved complete MRD response with an accuracy of 92% for 13 MRD negative patients." [33].

Wei et al. examined prognostic and predictive biomarkers associated with blinatumomab or chemotherapy in adults with Ph-negative R/R ALL. Patients were randomized 2:1 and administered blinatumomab or chemotherapy. After evaluating baseline blood samples, platelets, tumor burden, and T cell percentage were found to be prognostic markers: platelets were associated with improved 6-month survival, decreased tumor burden was prognostic for remission, and CD3+ T cell percentage was prognostic for minimal residual disease. CD45+, CD3+, and CD8+ T cells were found to be associated with Hematologic remission after receiving blinatumomab [34].

11.5.4 Adverse Events Associated with Blinatumomab in ALL

Blinatumomab treatment is associated with AEs, and the most concerning are CRS and neurotoxicity, which have been observed as Grade ≥3, which range from 0 to 6% for B cell malignancies, can occur within the first several days and are dose-limiting. Higher tumor burden and disease are associated with higher incidence of CRS [28, 29, 35–39]. CRS in particular can range from mild symptoms resembling the flu to fatal multi-organ failure. The mechanism of CRS is not completely understood and is mainly thought of as a product of distinct immune signatures such as T lymphocytes, monocyte and macrophage activation, leading to the massive release of inflammatory cytokines such as IL-6 and Interferon-gamma that is initiated by T cell activation [39]. The systemic production of these toxic cytokines is massive that is facilitated by monocyte and macrophage activation. T cell interferon-gamma, IL-6, IL-10, and tumor necrosis factor-alpha facilitate this cytokine production [40]. Symptomology presents as fever, fatigue, chills, headache, and more serious events such as hypotension, tachycardia, and other cardiovascular events such as vascular leaks and circulatory collapse and during and post-administration of the medication that appears mainly in the first cycle whose severity does not impact response. CRS is generally managed by steroids and IL-6 blockade, and more complicated management strategy involves disassociating tumor cell lysis from cytokine release based on the "two distinct thresholds for T cell activation based on the number of TCR peptide MHC complexes formed:" [4, 41, 42]

> An alternative way to avoid CRS-related problems is to dissociate tumor cell destruction and cytokine release. There are two distinct thresholds for T cell activation based on the number of TCR- peptide-MHC (pMHC) complexes formed. The formation of two TCR-pMHC complexes is sufficient between a T cell and an Ag-presenting cell, to trigger T cell-mediated cell lysis. On the other hand, 10 TCR-pMHC complexes are required for the formation of a complete immune synapse and cytokine secretion. Thus, adjusting the binding characteristics for the CD3-binding arm, a BsAb could more closely mimic the natural TCR-pMHC induced T cell activation. Consequently, new CD3-binding Abs have been generated that bind to multiple epitopes on CD3 with a wide range of affinities and agonist activities. Functional studies were realized with BsAbs that integrated the different CD3-binding domains. A BsAb with a new T cell-engaging domain could be created that elicited strong in vivo tumor cell killing and low levels of cytokine release [4].

The second most common event associated with BsAbs is neurotoxicity, for which symptomology ranges from personality changes, tremors, confusion, and focal neurological episodes. More serious episodes such as ataxia, encephalopathy, convulsions, and delirium may also result. As in CRS, these neurotoxicity episodes may be precipitated by inflammatory cytokines. 10–20% of patients treated with blinatumomab experience Grade 3 or higher adverse events, which are considered reversible after stopping the perfusion and corticosteroid initiation. Additionally, these events may be avoided by implementing a progressive dosing regimen and prophylactic administration of dexamethasone, but this constitutes a double-edged sword as the application of steroids could potentially lead to mitigated immune response. However, no inhibition of the cytotoxic capabilities of T cells was observed when reduced levels of inflammatory cytokines were produced as a result of dexamethasone-treated T cells, indicating that dexamethasone does not interfere with therapeutic efficacy of BsAbs [43].

Neurotoxicity can also lead to death. Both are usually managed with corticosteroids and supportive therapy and in severe cases of CRS, the interleukin (IL)-6 receptor inhibitor tocilizumab. More milder cases are treated with dexamethasone as prophylaxis "combined with stepwise administration of blinatumomab is useful to decrease the risk of severe CRS" [1]. Other adverse events are neutropenia, elevated liver enzymes, and infection [28, 35, 44–46].

Immune-effector cell-associated neurotoxicity syndrome, or ICANS, is also associated with T cell engaging therapies. Grade ≥3 events range from 5.5 to 24% for blinatumomab [28, 29, 35, 36, 38, 47–50]. Neurotoxicity, in general, occurs in treatment cycle 1 and its risk is increased when higher dosage of blinatumomab administration. Most common symptoms manifest as dizziness, tremor, confusion, and encephalopathy [45]. Administration of blinatumomab has also led to other adverse effects, such as tumor lysis syndrome, cytopenias, pyrexia, and anemia [51]. CTCAE or Common Terminology Criteria for Adverse Events apply to blinatumomab as well and approximately 5% of R/R BCP-ALL or MRD-positive BCP-ALL experienced a serious CRS event with (CTCAE Grade ≥3) [52, 53]. During phase 2 studies, BCP-ALL patients receiving blinatumomab, close to 53% experienced neurological events of any Grade, and up to 13% had Grade ≥3/4, with no associated deaths [18, 54]. However, these events are manageable as a result of blinatumomab's pharmacokinetics (high clearance rate) and interruption of treatment is sufficient [55].

The pathogenic mechanisms behind neurotoxicity remain unclear and are characterized as "complex and incompletely understood" [55]. An analysis of five clinical trials showed that selected patients exhibited adhesion of T cells to endothelial cells, leading to neurotoxicity, which was supported by in vitro experiments and preclinical evidence [56]. According to Zhou et al., blinatumomab led to peripheral T cell recruitment to the brain through this process: T cells attached to the cerebral microvascular endothelium, endothelial cells were activated leading to an increased level of Ang-2 (a marker of endothelial cell activation), T cells transmigrated across the blood-brain barrier into the brain that in turn led to the release of cytokines and severe immunological response and neurotoxicity as a result of these T cells destroying resident B cells [56]. Perhaps then agents that inhibit this adhesion between T cells and blood vessel endothelium could potentially be developed to mitigate neurotoxicity.

Other avenues are being pursued in clinical trials to reduce risk of the systemic toxicity of CRS such as developing novel routes of administration for B-ALL, such as subcutaneous administration which could improve convenience and compliance and reduce overall costs versus intravenous infusion [57, 58]. Management strategies include pretreatment with steroids and dose adjustments [53, 58].

Other adverse events, such as medication errors, elevated liver enzymes, and infections have been reported, especially in clinical trial settings. Medication errors usually result from incorrect setting of the infusion rate and malfunction of the flow rate in the pump leading to accidental increase of dose, and usually occurred at Grade 1 or 2 in severity. Additionally in this immunocompromised patient population, treatment-related infections occurred, such as sepsis and pneumonia and opportunistic infections. As a result of B cell depletion and associated decrease in serum immunoglobulins, risk for infection is higher. In the MT103-211 phase 2 study 32% or 60/189 had serious infections including sepsis, pneumonia, and catheter site infections with 9% or 17/189 leading to death [58]. In the TOWER study, transient elevation of liver enzymes during cycle 1 was observed in both cohorts, with 22% in the blinatumomab, arm and 25% in the chemotherapy arm; Grade ≥ 3 TEAEs were reported for 13% and 15%, respectively. Three serious elevated liver enzyme events and one treatment discontinuation were reported in the blinatumomab arm. No fatal events due to elevation of liver enzymes were reported during this trial [5].

11.5.5 Resistance to Blinatumomab and BITEs

Non-responders form a significant portion of patients receiving BiTEs that implicate loss of CD19 antigen and immunosuppressive factors. The PD-1/PD-L1 axis plays a role in the suppression of anti-tumor activity, as their blockade through antibodies led to significant clinical outcomes. One case study demonstrated positivity to this action in a patient receiving blinatumomab, leading to less tumor cell destruction accompanied by lower levels of interferon-gamma [59]. As a result, ICI administration has been proposed as a way of overcoming this resistance [60]. The immune environment with Tregs also contributes to non-response, since increased levels of Tregs have been observed in R/R ALL [61, 62]. As CR was observed in blinatumomab this also accompanied by relapse, approximately 8–50% experience CD19-negative relapse as a result of antigen loss, which can be "interpreted as the loss of antigen expression and the loss of antigen-binding to targeted antibodies or cells, the presence of either situation or both can lead to the CD19-negative relapse. A study analyzed data from four B-ALL patients who had been treated with blinatumomab and experienced CD19-negative relapse and found that CD19 trafficking from the intracellular space to the membrane of B cells was prevented with the lack of CD81 that provided docking sites for CD19 signal transduction, resulting in absent CD19 expression" [31, 63, 64]. This is as a result of CD19 mutations such as in-frame deletions, SNVs, and nonsense mutation, which were observed in R/R ALL patients with CD19-negative relapse. Other mechanisms are responsive for loss of CD19 expressions such as CD19 mutant allele-specific expression and low CD19 expression of mRNA [32]. Alternative splicing led to CD19 release "which caused antigen escape by changing CD19 epitopes and ultimately disrupting the binding of blinatumomab to CD19 molecule rather than reducing CD19 expression" [1].

These mutations and alternative splicing were observed to occur in parallel leading to antigen loss. Lineage transformation has also been observed where B lymphocytes turn into cells of myeloid lineage, as myeloid marker levels "upregulate", including CD33 [65, 66]. This lineage switch was thought to be associated with the existence of subclones that had significant selective advantage that did not express CD19 and were shown to have KMT2A/AFF1 and ZNF384 gene rearrangements, generating the need for multitargeted therapies to overcome antigen loss such as one drug that can concomitantly target multiple tumor antigens or in combination with other immunotherapies [67, 68].

11.6 Resistance Mechanisms

11.6.1 T Cell Exhaustion/Dysfunction

Other causes of resistance may occur such as T cell exhaustion or dysfunction as a result of persistent antigen exposure. Their proliferation and cytotoxicity are impacted, and inhibitor receptors such as PD-1, CTLA-4, and LAG-3 (discussed in this volume) become overexpressed in tumor cells, the most central on being the PD-1/PD-L1 axis. This inhibitory pathway is targeted for blocking immunosuppressive signals and leads to more enduring T cell activation. T cells do not become completely inactive, but are not as effective in promoting cell lysis [69–71].

The mechanism of T cell activation and proliferation requires antigen recognition by the T cell receptor, co-stimulation, and consequent release of cytokines by the T cells, and then followed by T cell expansion. BsAb meets the first requirement, but the development of BiTEs may be enhanced to trigger "effective immunological synapse" obviating the need for co-stimulation. [72]. Additionally, CD28 or 4-1BB could lead to co-activation and further affect T cell activation as a result of the bispecific T cell engager [73, 74]. Other BsAbs have been constructed to include the IL-15 cytokine [75]. Further, blocking the PD-1/PD-L1 axis can reactivate T cells, but sustainable responses have not been observed in patients for this therapy since other inhibitory pathways are present. A balance must be achieved between enabling sufficient BsAb targeting activities and mitigating lethal autoimmune adverse events, as resistance and evasion mechanisms that sustain dysfunction among T cells are a major concern, but must still be considered in the development and clinical utility of BiTEs [76]. Since by design BsAbs lead to T cell activation, ancillary T cells may also be activated, such as regulatory T lymphocytes or Tregs, which is predictive of treatment resistance and preventing of tumor cell lysis. T cell depletion pre-therapy is suggested [77].

11.6.1.1 CD20 BsAbs

A number of CD20-based BsAbs are in development including REGN1979, Mosunetuzumab, and RG6026. REGN1979 is a fully-humanized IgG4 Ab that since it is has a similarity with natural human Abs is conferred with stability and stable pharmacokinetics and low immunogenicity. In a phase I trial on R/R NHL patients, a 100% overall response was observed in follicular lymphoma and demonstrated CR in CAR-T nonresponders [78]. Mosunetuzumab also has similarities with native B structure being a full-length humanized IgG molecule. In aggressive NHL, an ORR of 37.1% with a CR rate of 19.4% was observed, which was even higher in indolent NHL, with an ORR of 62.7% and CR of 43.3% [79]. RG6026 is unique in that it was constructed in a 2:1 format, providing better TAA binding affinity. A short flexible linker ties the CD3 binding arm with the CD20 binding arm. An extended half-life is conferred through its modified heterodimeric Fc region that prevents binding to FcgRs, which leads to an extended circulatory half-life. Substantive clinical activity was shown in in vitro and in vivo models, even on cells that have low CD20 expression. It has the further advantage of bypassing rituximab resistance since it remains active in the presence of competing for anti-CD20 monoclonal antibodies. Its safety profile is also significant with low cytotoxicity activity. Each of these compounds is undergoing initial clinical investigation to evaluate efficacy [80, 81].

11.7 Acute Myeloid Leukemia

The development of bispecific antibodies for the treatment of AML has been limited due to the lack of tumor-associated antigens on leukemic cells for targeting, (that would be selectively expressed on leukemic cells but spare healthy hematopoietic cells, similar to blinatumomab for B cell destruction) [82]. CD33 and CD123 have been implicated in acute myeloid leukemia, being a mediator of myeloid cell proliferation and differentiation. CD33 is also known as sialic acid-binding Ig-like lectin 3 and is a 7-kDa transmembrane cell surface glycoprotein with expression on leukemic cells [83, 84]. In the initial stages of development of BsAbs for AML, four agents were under investigation one being AMG330 human BiTE tandem single-chain antibody with the N-terminal specific for human CD33 and C-terminal directed toward CD3 [85]. This agent showed anti-leukemic activity in in vitro and in clinical models. Once daily IV infusion was conducted in a phase I study for R/R AML. GEM333 is a CD3 × CD33 BsAb in a phase I study for R/R AML. GEM333 is a humanized antibody with a single-chain bispecific antibody with variable light and heavy chains targeting both CD3 and CD33 that is linked in unique fashion through a tandem format arrangement [86]. In preclinical models, the GEM333 construct efficiently redirected cytotoxic T cells toward CD33+ AML blasts and led to the destruction of AML cell lines and AML blasts in patients. Of interest is that this agent spared normal human CD34+ hematopoietic and progenitor stem cells in vitro [87]. A CD33 × CD3 bispecific antibody was constructed and evaluated in AML patients. 55 patients were dosed at 0.5–720 μg/d continuously through infusion; among 42 evaluable patients, 3 CR and 4 CRis (incomplete Hematologic recovery) were observed the dose of ≥120 μg/d [88, 89]. Grade 3 or higher CRS at a rate of 13% was also observed [88].

CD123 BsAb targets have also been developed for AML. CD123, also known as IL-3 receptor alpha chain, is considered the low-affinity binding subunit of the IL3 receptor. Its mechanism of action is that it "triggers CD123 heterodimerization with the granulocyte-macrophage stimulating factor and IL5 receptor complex" leading to PI3 kinase activity and anti-apoptotic protein upregulation [90, 91]. CD123 expression on AML blasts is associated with lower CR rates and poor prognosis concurrent with higher blast counts. JNJ-63709178 is a CD3 × CD123 construct that contains a bispecific IgG1 antibody created through Genmab Duo-Body technology which employs a process termed Fab-arm exchange [82]. Since they retain the Fc region, their effector functions and in vivo stability are enhanced. In murine models, the compound exerted anti-tumor effects and led to tumor regression in a human peripheral blood T cell environment [92]. This compound is undergoing phase I trials for relapsed and refractory patients. XmAb14045 is another CD123 BsAb that also possesses a unique Fc region and undergoes spontaneous formation of stable heterodimers facilitating its manufacturing. In a preclinical monkey model, this agent strongly activated T cells to stimulate CD123+ cell destruction [93]. Other BsAbs for AML are undergoing evaluation. MCLA-117 is a human full-length IgG1 BsAb that targets CLEC12A, a myeloid antigen expressed on AML cells [82]. CLEC12A has selective expression, being expressed on leukemic stem cells but sparing normal hematopoietic cells [56, 57]. In an HL-60 cell line, MCLA-117 led to efficient CLEC12A antigen-dependent T cell activity and targeted tumor cell lysis. The agent also induced "T cell-mediated lysis of AML blasts in an ex vivo culture system and is currently being investigated in a phase I clinical study" to assess safety, tolerability, and efficacy in AML adult patients [82]. These compounds have associated adverse events as a result of their T cell redirecting therapy, which is similar to blinatumomab, including CRS, with high levels of inflammatory cytokines such as IL-6 and IL-2, and associated with flu-like symptoms and quite possibly elevated fever, end-organ dysfunction, or even more threatening complications such as renal failure, hepatic failure and cardiac dysfunction [94]. Neurotoxicity may also result including mild confusion, headaches, to even severe encephalopathy, aphasia, seizures, and delirium [95–97].

11.8 Multiple Myeloma

B cell maturation antigen or BCMA is likewise expressed on multiple myeloma tumor cells, with very little expression on normal cells, leading to development of anti-BCMA bispecific antibodies [1]. BCMA is a membrane antigen that has selective expression on malignant cells but it is not expressed in naïve B cells nor other normal tissue cells. It is considered a crucial target of study for the development of a BsAb for multiple myeloma. It is prognostic of poor clinical outcomes and is highly expressed on multiple myeloma cells. According to Lejeune et al., "a rapid re-emergence of B cell immunity after the end of the anti-BCMA treatment would be possible since this [antigen] is not expressed early in B cell development" and "the lack of BCMA expression in other bone marrow populations prevents off-tumor toxicities." [4].

One such example, AMG420 showed in a clinical study favorable efficacy and safety profiles in R/R MM patients. In this trial, 42 R/R MM patients with more than two lines of prior therapies "were enrolled and received 6-week cycles of AMG 420 at the dose of 0.2–800 µg/d." [1]. Objective response rate was robust, being 70% with 5 CR (MRD-negative), 1 very good PR, and 1 PR. Infections constituted the most common adverse events, with a rate of 33%. CNS toxicity was not observed; Grade ≥ 3 CRS was 2%, leading to FDA consideration for approval [98]. AMG420 and 701 are BCMA-CD3 BiTEs that have short-life and IV infusion administration like blinatumomab, being administered for 4 weeks followed by 2 weeks treatment-free. AMG420 targets BCMA-positive MM selectively while avoiding BCMA-negative MM cells in both in vitro and in vivo models. In clinical trials evaluating 42 refractory MM patients, a 70% response rate was observed with 70% MRD-negativity; adverse events included infections and neuropathy [99]. AMG701 has single-chain variable fragments of AMG420 with a half-life extension and is undergoing evaluation for toxicity and response through once-weekly dosing.

11.8.1 Clinical Development for MM (CD38-CD3)

A number of tumor-associated antigen BsAbs have been studied for MM. BsAbs that target CD38, such as humanized anti-CD38/CD3 XmAbs with differing affinities for CD38 and CD3. AGM424 has been studied in in vitro and in vivo models and led to significant tumor cell destruction in the presence of soluble CD38. It has lower affinity for CD3 and is associated with uncontrolled CRS. It is currently in phase I studies to evaluate safety and tolerability, and PK, PD and efficacy in R/R MM [100]. GBR 1342 is another anti-CD38/CD3 BsAb that has a complete Fc domain and was shown in preclinical models to have superiority to the anti-CD38 monoclonal antibody daratumumab. T cells and CD38+ T cell depletion were induced in both blood and bone marrow. A phase 1 study began to evaluate its tolerability.

IgG2a-based BCMA-CD3 (PF-06863135) is also a humanized BsAb that has an IgG2a backbone with mutations in the Fc region that lead to heavy chain heterodimer formation which reduces FcG receptor binding [48]. This compound is undergoing a phase I study for safety but has shown anti-myeloma activity in in vivo models [101].

11.8.2 Clinical Development for MM (FcRL5-CD3 and GPRC5D-CD3)

These anti-myeloma compounds comprise two new targets developed as part of MM-related target: Fc Receptor-Like 5 (FcRL5) and G-protein coupled receptor family C group 5 member D (GPRC5D). FcRH5 contains an exclusive surface marker from B cell lineage but is detected starting from early pre-B cell stage development [102]. It remains unique among other B cell-specific surface proteins in that FcRL5 is preserved in both normal and tumor B cells, which enables further activity in other B cell tumors, such as CLL, DLBCL, and follicular lymphoma [102, 103].

GPRC5D on the other hand is "expressed on the surface of malignant cells involved in multiple myeloma without being expressed at appreciable levels by normal hematopoietic cells, such as T cells, NK cells, monocytes, granulocytes and bone marrow progenitors, including hematopoietic stem cells" [104]. Additionally, mRNA expression of the marker was only expressed in MM patients with low expression in normal tissues, which was associated with poor outcomes [105]. This profile lends itself as a suitable target for MM patients. Two BsAbs are in development against these targets and are currently in phase I clinical studies: RG6160 which targets FcRL5 and the DuoBody JNJ-64407564. Both target GPRC5D, and led to encouraging results from in vitro and in vivo models that demonstrated B cell depletion and tumor growth suppression in myeloma models [104, 106].

Teclistamab, a B cell maturation antigen × CD3 bispecific antibody showed clinical efficacy in R/R MM patients in an open-label phase 1 multicenter trial. Teclistamab was initially considered an investigational bispecific T cell engager that has structural differences from AMG 420 with "promising efficacy" (Table 11.2).

Table 11.2 Teclistamab and multiple myeloma

Trial	Patient population	Primary endpoint	Clinical results	Adverse events
Open-label, single-arm phase 1	R/R MM (n = 157) patients intolerant to established therapies	Dosing in part 1/safety and tolerability in part 2	58% achieved a very good partial response or better 22 (n = 85%) of 26 responders were alive and continuing treatment after 7.1 months median follow-up	Cytokine release syndrome in 28 and neutropenia in 26 patients (Grade 3 or 4)

Source Usmani [118]

Table 11.3 Blinatumomab in R/R diffuse large B cell lymphoma [107]

Trial	Patient population	Comparator	Primary endpoint	Clinical results	Adverse events
Phase 2 study	21 patients with a median of three prior lines of therapy with stepwise dosing	21 patients with a median of three prior lines of therapy with stepwise dosing	Overall response rate; complete response	ORR: 43% after 1 blinatumomab cycle; CR 9%	Tremor (48%); pyrexia (44%); fatigue (26%); edema (21%); Grade 3 neurologic events encephalopathy and aphasia (each 9%); mostly resolved

11.9 Non-Hodgkin Lymphoma

Blinatumomab has demonstrated clinical efficacy for R/R DLBCL, as shown in Table 11.3. Other bispecific antibodies are being studied for determining safety and tolerability in DLBCL and NHL.

Pharmacokinetic and pharmacodynamic analysis revealed B cell depletion rates for NHL patients receiving blinatumomab. B cell depletion took place within 48 h after continuous IVD infusion doses of greater than 5 ug/m^2/day that took place in first-order kinetics. A 50% reduction in tumor size was a result of dosage of 47 ug/m^2/day for 28 days. The authors concluded that B-lymphocyte depletion was dependent on exposure while adverse cytokine elevation was transient but also increased with dose. Overall a PK/PD relationship was established in this medication dose selection [108].

11.10 Dual Affinity BsAbs and Tandem Diabodies

Dual affinity bispecific antibodies and tandem diabodies, other types of BsAb constructs have been discussed earlier, but are worth specifying further here. MGD011 (duvortuxizumab) is a CD19 × CD3 DART with a silenced, human IgG1 Fc domain that confers it with a relatively long circulating half-life (approximately 14.3–20.6 days), which is similar to conventional mAbs that allow for a every 2-weeks administration. The benefit of this DART is that its humanized Ab arms have a much greater affinity for CD19 than for CD3 which lends itself to preferential binding to targets. However, due to high neurotoxicity in phase I studies for B cell malignancies such as NHL and CLL, its clinical development was stopped [109].

"AFM11 is a tetravalent bispecific TandAb with two binding sites for CD3 and two for CD19". With its increased binding affinity, its high potential for treatment efficacy was anticipated but phase I studies for ALL and R/R NHL was suspended as it was further revealed that AFM11 potency was not correlated with the CD19 density on target cell surfaces and additionally severe neurotoxicity with one fatality occurred [110].

AMV564 is a tetravalent anti-CD33-anti-CD3 tandem diabody construct that forms a homodimer from two VH and VL chains that are composed of antigen-binding single-chain variable fragments (scFvs). Two binding sites for each epitope are created, thus increasing the avidity of the antibody to its targets. It also has a longer half-life compared to BiTEs owing to its molecular weight of 106 kDa [111]. This tandem diabody is being studied in a first-in-human phase I trial in R/R AML patients at a 14-day continuous infusion every 28 days and may hold potential over the two constructs discussed above since preliminary evidence has shown evidence of T cell activation by increased cytokine levels. 13–38% reduction in bone marrow blasts was evidence of biological activity in 10/16 patients. Safety profiles were also favorable with no Grade 3–4 toxicities present. A Grade 2 CRS at the 50 ug/day dose in a single patient was observed [112].

Table 11.4 is a summary of bispecific antibodies for Hematologic malignancies.

11.11 Combination and Sequential Therapies [3]

PD-1 expression has been associated with resistance to blinatumomab [113] since T cell exhaustion is observed when PD-1 is overexpressed. This association led to studies combining blinatumomab with immune checkpoint inhibitors, case in point being a 12-year patient with refractory ALL achieved remission was administered pembrolizumab and blinatumomab since pembrolizumab enhanced T cell function [113]. Table 11.5 from Lejeuene et al. shows the clinical studies as of 2019 combining BsAbs with immune checkpoint inhibitors. These studies have been extrapolated to design BsAbs that concomitantly target two immune checkpoints such as the dual blockade of PD-1 and LAG-3 (also discussed in this volume). MGD013 is an anti-PD-1/anti-LAG-3 DART that binds specifically to both PD-1 and LAG-3 (142) and enhances T cell pathways. However, these positive outcomes have been deterred to some extent by increases in adverse events due to over-activation of the immune system, which investigators are studying to overcome. A new approach was developed that "consists in the deletion of the PD-1 pathway via high-affinity PD-1 binding while inhibiting CTLA-4 with a low-affinity binding arm. This construct inhibits CTLA-4 in double-positive T cells while reducing the binding to peripheral T lymphocytes expressing CTLA-4, resulting in better tolerability" [114].

In R/R BCP-ALL patients, blinatumomab is combined with PD-1 and CTLA-4 inhibitors to determine efficacy and tolerability. Preliminary findings from a study evaluating nivolumab and blinatumomab have demonstrated feasibility with acceptable toxicity. Heavily pretreated patients showed a 80% MRD complete response rate [115]. Deep and durable remissions are also anticipated from a study combining pembrolizumab with blinatumomab in a phase 1/2 study in R/R BCP-ALL patients with a high percentage of bone marrow blasts as a result

Table 11.4 Bispecific antibodies for Hematologic malignancies

Agent	Tumor	Type of bispecific antibody	Tumor-associated antigen	Dosing	Status
Blinatumomab	R/R Ph-/Ph+ B-ALL; NHL	BiTE	CD19	0.06 mg/day IV infusion	FDA-approved
IL-3Ralpha	Cd34+/CD38-AML, HL	BiTE	CD123		In clinical trials
MGD006	AML	DART	CD123	Continuous infusion	Phase I/II clinical trials
XmAb14045 (Vibecotamab)	AML	BsAb	CD123	Intermittently administered	Phase I clinical study
AMG330	CD33+ AML	BiTE	CD33	IV infusion	Ongoing phase I clinical trial
AMV564	AML, MDS	BiTE	CD33	Intravenous injection	Phase I clinical trial
AMG 420	BCMA+ MM	BiTE	BCMA	Continuous intravenous infusion	Dose-escalation trial
AMG 701	R/R MM	BiTE	BCMA	Persistent intravenous infusion	Phase 1 clinical trial
AMG 424	MM	BsAb	CD38	N/A	Clinical trials
REGN1979	R/R B-NHL	BsAb	CD20	N/A	Phase I clinical trial
RG6026	R/R NHL	BsAb	CD20	N/A	Under investigation
Mosunetuzumab	NHL	BsAb	CD20	N/A	Under investigation
CD20-CD47SL	NHL	BsAb	CD47/CD20	N/A	Preclinical models
AFM13	CD30+ HL	Bispecific NK cell engager	CD30	N/A	Phase I trial
KHK2823	MDS	BsAb	CD123	N/A	Clinical trials

(continued)

Table 11.4 (continued)

Agent	Tumor	Type of bispecific antibody	Tumor-associated antigen	Dosing	Status
MGD006	MDS	DART	CD123	N/A	Phase I clinical trial
AMV564	MDS	Tetravalent BsAb	CD33	N/A	

Source Tian [119]

Table 11.5 Clinical trials for combination therapies for blinatumomab

Name (sponsors)	Targets	Diseases indication	Phase (NCT#)
Combinations with immune modulators			
Combination of binatumomab and nivolumab (anti-PD-1 mAb) ± Ipilimumab (anti-CTLA4 mAb) [National Cancer Institute (NCI)]	CD3 × CD19 × PD-1 (× CTLA4)	B-ALL	Phase I (NCT02879695)
Combination of binatumomab and pembrolizumab (anti-PD-1 mAb) (Merck Sharp & Dohme Corp., Amgen)	CD3 × CD19 × PD-1	B-ALL	Phase I/II (NCT03160079)
Combination of binatumomab and pembrolizumab (anti-PD-1 mAb) (Amgen)	CD3 × CD19 × PD-1	NHL	Phase I (NCT03340766)
Combination of binatumomab and pembrolizumab (anti-PD-1 mAb) (Merck Sharp & Dohme Corp., Amgen)	CD3 × CD19 × PD-1	ALL	Phase I/II (NCT03512405)
Combination of Binatumomab and Pembrolizumab (anti-PD-1 mAb) (Children's Hospital Medical Center, Cincinnati)	CD3 × CD19 × PD-1	B-cell lymphoma and leukemia	Phase 1 (NCT03605589)
Combination of BTCT4465A and atezolimumab (anti-PD-L1 mAb) (Genentech)	CD3 × CD20 × PD-L1	CLL, NHL	Phase 1 (NCT02500407)
Combination of REGN1979 and REGN2810 (anti-PD-L1 mAb) (Genentech)	CD3 × CD20 × PD-1	Lymphoma	Phase 1 (NCT02651662)
Combination of REGN1979 and REGN2810 (anti-PD-L1 mAb) (Hoffmann-La Roche)	CD3 × CD20 × PD-L1	NHL	Phase 1 (NCT03533283)
Combination with mAb			
Combination of JNJ-64407564/JNJ-64007957 and daratumumab (Janssen)	CD3 × BCMA or GPRC5D × CD38	MM	Phase 1 (NCT04108195)

(continued)

of resistance to blinatumomab [116]. Limited data are also arriving from studies combining TKIs such as ponatinib, dasatinib, and bosatinib, achieving 50% hematologic response rates [117].

Bispecific antibodies have made enormous progress in the treatment of Hematologic malignancies, the most prominent being the bispecific T cell engager blinatumomab for acute lymphoblastic leukemia. Acute myeloid leukemia and multiple myeloma are also benefiting from BsAb treatment. They have proved to be a formidable alternative to immune checkpoint inhibitors and adoptive cellular

Table 11.5 (continued)

Name (sponsors)	Targets	Diseases indication	Phase (NCT#)
Combination with ADC			
Combination of BTCT4465A and polatuzumab vedotin (anti-CD79b × MMAE) (Hoffmann-La Roche)	CD3 × CD20 × ADC	B-cell NHL	Phase 1 (NCT03671018)

Adapted from Lejeune et al. [4]

Data available as of November 05, 2019. Molecules are classified based on target antigens

ADC: Antibody-drug conjugate; BCMA: B-cell maturation antigen; CTLA4: cytotoxic T-lymphocyte-associated protein 4; GPRC5D: G protein-coupled receptor family C group 5 member D; mAb: monoclonal antibody; MMAE: monomethyl auristatin E; PD-1: programmed cell death 1; PD-L1: programmed cell death 1 ligand

therapies, and are vying with each for clinical and commercial viability. They are most often compared with chimeric antigen receptor therapies in terms of clinical efficacy and response, toxicity and manufacture, and several studies are underway for combining both of them for more deep and durable responses. Other types of bispecific antibodies other than the canonical CD3 T cell construct are in development, such as those for directing natural killer cells to leukemic and myeloid targets through their cell surface antigens and mitigating tumor escape as tumor cells lose their antigens. Others are being developed with longer half-lives so that the intravenous administration of these compounds is more facile and there is less possibility of medication errors happening. Also, one of the major advances of bispecific antibody-drug development is that their clinical effectivity runs the gamut for all patient populations: pediatric, adult, and elderly. As regulatory bodies are continuously receiving data for the hundreds of bispecific antibodies, particularly bispecific T cell engagers, under study and ongoing development, the landscape for the treatment of hematologic malignancies by these agents will expand most certainly.

References

1. Zhou S, Liu M, Ren F, Meng X, Yu J (2021) The landscape of bispecific T cell engager in cancer treatment. Biomarker Res 9:38
2. Huang S, van Duijnhoven SMJ, Sijtx AJA, van Elsas A (2020) Bispecific antibodies targeting dual tumor-associated antigens in cancer therapy. J Cancer Res Clin Oncol 146:3111–3122
3. Viardot A, Locatelli F, Stieglmaier J, Zaman F, Jabbour E (2020) Concepts in immune-oncology: tackling B cell malignancies with CD19-directed bispecific T cell engager therapies. Ann Hematol 999:2215–2229
4. Lejeune M, Cem Kose M, Duray E, Einsele H, Beguin Y, Caers J (2020) Bispecific, T-cell recruiting antibodies in B-cell malignancies. Front Immunol 11(762)
5. Stein A, Franklin JL, Chia VM, Arrindell D, Kormany W et al (2019) Benefit-risk assessment of blinatumomab in the treatment of relapsed/refractory B-cell precursor acute lymphoblastic leukemia. Drug Saf 42:587–601
6. Poussin M, Sereno A, Wu X, Huang F, Manro J et al (2021) Dichotomous impact of affinity on the function of T cell engaging bispecific antibodies. J Immunother Cancer 9:e002444

7. Nagorsen D, Kufer P, Baeuerle PA, Bargou R (2012) Blinatumomab: a historical perspective. Pharmacol Ther 136(3):334–342
8. BLINCYTO® (blinatumomab) [prescribing information] (2019) Amgen, Thousand Oaks, CA
9. Viardot A, Goebeler ME, Hess G, Neumann S, Pfreundschuh M, Adrian N, Zettl F, Libicher M, Sayehli C, Stieglmaier J, Zhang A, Nagorsen D, Bargou RC (2016) Phase 2 study of bispecific T-cell engager (BiTE) antibody blinatumomab in relapsed/refractory diffuse large B cell lymphoma. Blood 127(11):1410–1416
10. Klinger M, Brandl C, Zugmaier G, Hijazi Y, Bargou RC, Topp MS et al (2012) Immunopharmacologic response of patients with B-lineage acute lymphoblastic leukemia to continuous infusion of T cell-engaging CD19/CD3-bispecific BiTE antibody blinatumomab. Blood 119:6226–6233
11. De Gast GC, Van Houten AA, Haagen IA, Klein S, De Weger RA, Van Dijk A et al (1995) Clinical experience with CD13 x CD19 bispecific antibodies in patients with B cell malignancies. J Hematother 4:433–437
12. Kontermann RE, Brinkmann U (2015) Bispecific antibodies. Drug Discov Today 7
13. Topp MS, Gokbuget N, Stein AS, Zugmaier G, O'Brien S, Bargou RC et al (2015) Safety and activity of blinatumomab for adult patients with relapsed or refractory B-precursor acute lymphoblastic leukaemia: a multicentre, single-arm, phase 2 study. Lancet Oncol 16:57–66
14. Topp MS, Gokbuget N, Zugmaier G, Klappers P, Stelljes M, Neumann S et al (2014) Phase II trial of the anti-CD19 bispecific T cell-engager blinatumomab shows hematologic and molecular remissions in patients with relapsed or refractory B-precursor acute lymphoblastic leukemia. J Clin Oncol 32:4134–4140
15. Topp MS, Kufer P, Gokbuget N, Goebeler M, Klinger M, Neumann S et al (2011) Targeted therapy with the T-cell-engaging antibody blinatumomab of chemotherapy-refractory minimal residual disease in B-lineage acute lymphoblastic leukemia patients results in high response rate and prolonged leukemia-free survival. J Clin Oncol 29:2493–2498
16. Kantarjian H, Stein A, Gokbuget N, Fielding AK, Schuh AC et al (2017) Blinatumomab versus chemotherapy for advanced acute lymphoblastic leukemia. N Engl J Med 376(9):836–847
17. Foa R, Bassan R, Vitale A, Elia L, Piciocchi A (2020) Dasatinib-blinatumomab for Ph-positive acute lymphoblastic leukemia in adults. N Engl J Med 383(17):1613–1623
18. Gökbuget N, Dombret H, Bonifacio M, Reichle A, Graux C, Faul C, Diedrich H, Topp MS, Brüggemann M, Horst HA, Havelange V, Stieglmaier J, Wessels H, Haddad V, Benjamin JE, Zugmaier G, Nagorsen D, Bargou RC (2018) Blinatumomab for minimal residual disease in adults with B-cell precursor acute lymphoblastic leukemia. Blood 131(14):1522–1531
19. Gökbuget N, KelshM, Chia V, Advani A, Bassan R, Dombret H, Doubek M, Fielding AK, Giebel S, Haddad V, Hoelzer D, Holland C, Ifrah N, Katz A, Maniar T, Martinelli G, Morgades, M, O'Brien S, Ribera JM, Rowe JM, Stein A, Topp M, Wadleigh, M, Kantarjian H (2016) Blinatumomab vs historical standard therapy of adult relapsed/refractory acute lymphoblastic leukemia. Blood Cancer J 6(9):e473
20. Jen EY, Xu Q, Schetter A, Przepiorka D, Shen YL, Roscoe D, Sridhara R, Deisseroth A, Philip R, Farrell AT, Pazdur R (2019) FDA approval: blinatumomab for patients with B-cell precursor acute lymphoblastic leukemia in morphologic remission with minimal residual disease. Clin Cancer Res 25(2):473–477
21. Dombret H, Topp MS, Schuh AC, Wei AH, Durrant S, Bacon CL, Tran Q, Zimmerman Z, Kantarjian H (2019) Blinatumomab versus chemotherapy in first salvage or in later salvage for B-cell precursor acute lymphoblastic leukemia. Leuk Lymphoma 60(9):2214–2222
22. Kantarjian HM, Thomas D, Ravandi F, Faderl S, Jabbour E, Garcia-Manero G et al (2010) Defining the course and prognosis of adults with acute lymphocytic leukemia in first salvage after induction failure or short first remission duration. Cancer 116(24):5568–5574
23. Minson KA, Prasad P, Vear S, Borinstein S, Ho R, Domm J, Frangoul H (2013) t(17;19) in children with acute lymphocytic leukemia: a report of 3 cases and a review of the literature. Case Rep Hematol 563291:1–4

24. Jabbour EJ, Gokbuget N, Kantarjian HM, Thomas X, Larson RA, Yoon SS, Ghobadi A, Topp MS, Tran Q, Franklin JL, Forman SJ, Stein AS (2019) Transplantation in adults with relapsed/refractory acute lymphoblastic leukemia who are treated with blinatumomab from a phase 3 study. Cancer 125(23):4181–4192

25. Topp MS, Zimmerman Z, Cannell P, Dombret H, Maertens J, Stein A, Franklin J, Tran Q, Cong Z, Schuh AC (2018) Health related quality of life in adults with relapsed/refractory acute lymphoblastic leukemia treated with blinatumomab. Blood 131(26):2906–2914

26. Pulte ED, Vallejo J, Przepiorka D, Nie L, Farrell AT, Goldberg KB, McKee AE, Pazdur R (2018) FDA supplemental approval: blinatumomab for treatment of relapsed and refractory precursor B-cell acute lymphoblastic leukemia. Oncologist 23(11):1366–1371

27. Martinelli G, Boissel N, Chevallier P, Ottmann O, Gökbuget N, ToppMS, FieldingAK,Rambaldi A, Ritchie EK, Papayannidis C, Sterling LR, Benjamin J, Stein A (2017) Complete hematologic and molecular response in adult patients with relapsed/refractory Philadelphia chromosome–positive B-precursor acute lymphoblastic leukemia following treatment with blinatumomab: results from a phase II, single-arm, multicenter study. J Clin Oncol 35(16):1795–1802

28. Martinelli G, Boissel N, Chevallier P, Ottmann O, Gökbuget N, Topp MS et al (2017) Complete hematologic and molecular response in adult patients with relapsed/refractory Philadelphia chromosome–positive B-precursor acute lymphoblastic leukemia following treatment with blinatumomab: results from a phase II, single-arm, multicenter study. J Clin Oncol 35:1795–1802

29. Gökbuget N, Zugmaier G, Dombret H, Stein A, Bonifacio M, Graux C et al (2020) Curative outcomes following blinatumomab in adults with minimal residual disease B cell precursor acute lymphoblastic leukemia. Leuk Lymphoma 61:2665–2673

30. von Stackelberg A, Locatelli F, Zugmaier G, Handgretinger R, Trippett TM, Rizzari C et al (2016) Phase I/phase II study of blinatumomab in pediatric patients with relapsed/refractory acute lymphoblastic leukemia. J Clin Oncol 34:4381–4389

31. Aldoss I, Song J, Stiller T, Nguyen T, Palmer J, O'Donnell M et al (2017) Correlates of resistance and relapse during blinatumomab therapy for relapsed/refractory acute lymphoblastic leukemia. Am J Hematol 92:858–865

32. Zhao Y, Aldoss I, Qu C, Crawford JC, Gu Z, Allen EK et al (2021) Tumor-intrinsic and - extrinsic determinants of response to blinatumomab in adults with BALL. Blood 137:471–484

33. King AC, Bolanos R, Velasco K, Tu H, Zaman F, Geyer MB et al (2019) Real world chart review of blinatumomab to treat patients with high disease burden of relapsed or refractory B cell precursor acute lymphoblastic leukemia. Blood 134:5079

34. Brown P, Zugmaier G, Gore L, Tuglus CA, Stackelberg A (2019) Day 15 bone marrow minimal residual disease predicts response to blinatumomab in relapsed/refractory pediatric B-ALL. Brit J Haematol. 188:e36–e39

35. Gökbuget N, Zugmaier G, Klinger M, Kufer P, Stelljes M, Viardot A et al (2017) Long-term relapse-free survival in a phase 2 study of blinatumomab for the treatment of patients with minimal residual disease in B-lineage acute lymphoblastic leukemia. Haematologica 102:e132–e135

36. Topp MS, Gökbuget N, Zugmaier G, Klappers P, Stelljes M, Neumann S et al (2014) Phase II trial of the anti-CD19 bispecific T cell–engager blinatumomab shows hematologic and molecular remissions in patients with relapsed or refractory B-precursor acute lymphoblastic leukemia. J Clin Oncol 32:4134–4140

37. Kujawski M, Li L, Bhattacharya S, Wong P, Lee W-H et al (2019) Generation of dual specific bivalent BiTEs (dbBIspecific T cell engaging antibodies) for cellular immunotherapy. BMC Cancer 19:882

38. Coyle L, Morley NJ, Rambaldi A, Mason KD, Verhoef G, Furness CL et al (2020) Open-label, phase 2 study of blinatumomab as second salvage therapy in adults with relapsed/refractory aggressive B cell non-Hodgkin lymphoma. Leuk Lymphoma 61:2103–2112

39. Viardot A, Goebeler M, Hess G, Neumann S, Pfreundschuh M, Adrian N et al (2016) Phase 2 study of the bispecific T cell engager (BiTE) antibody blinatumomab in relapsed/refractory diffuse large B cell lymphoma. Blood 127:1410–1416
40. Lee DW, Gardner R, Porter DL, Louis CU, Ahmed N, Jensen M et al (2014) Current concepts in the diagnosis and management of cytokine release syndrome. Blood 124:188–195. https://doi.org/10.1182/blood-2014-05-552729
41. Li J, Piskol R, Ybarra R, Chen Y-JJ, Li J, Slaga D et al (2019) CD3 bispecific antibody induced cytokine release is dispensable for cytotoxic T cell activity. Sci Transl Med 11:eaax8861
42. Barrett DM, Teachey DT, Grupp SA (2014) Toxicity management for patients receiving novel T cell engaging therapies. Curr Opin Pediatr 26:43–49
43. Shimabukuro-Vornhagen A, Godel P, Subklewe M, Stemmler HJ, Schlosser HA, Schlaak M et al (2018) Cytokine release syndrome. J Immunother Cancer 6:56
44. Brandl C, Haas C, d'Argouges S, Fisch T, Kufer P, Brischwein K et al (2007) The effect of dexamethasone on polyclonal T cell activation and redirected target cell lysis as induced by a CD19/CD3-bispecific single-chain antibody construct. Cancer Immunol Immunother 56:1551–1563
45. Seckinger A, Delgado JA, Moser S, Moreno L, Neuber B, Grab A et al (2017) Target expression, generation, preclinical activity, and pharmacokinetics of the BCMA-T cell bispecific antibody EM801 for multiple myeloma treatment. Cancer Cell 31:396–410
46. Costa LJ, Wong SW, Bermúdez A, de la Rubia J, Mateos M-V, Ocio EM et al (2019) First clinical study of the B cell maturation antigen (BCMA) 2+1 T cell engager (TCE) CC-93269 in patients (Pts) with relapsed/refractory multiple myeloma (RRMM): interim results of a phase 1 multicenter trial. Blood 134(Suppl. 1):143
47. Wei AH, Ribera J-M, Larson RA, Ritchie D, Ghobadi A (2021) Biomarkers associated with blinatumomab outcomes in acute lymphoblastic leukemia. Leukemia 35:2220–2231
48. Panowski SH, Kuo TC, Zhang Y, Chen A, Geng T, Aschenbrenner L et al (2019) Preclinical efficacy and safety comparison of CD3 bispecific and ADC modalities targeting BCMA for the treatment of multiple myeloma. Mol Cancer Ther 18:2008–2020
49. Liu D, Zhao J, Song Y, Luo X, Yang T (2019) Clinical trial update on bispecific antibodies, antibody-drug conjugates, and antibody-containing regimens for acute lymphoblastic leukemia. J Hematol Oncol 12:15
50. Locatelli F, Zugmaier G, Mergen N, Bader P, Jeha S, Schlegel P et al (2020) Blinatumomab in pediatric patients with relapsed/refractory acute lymphoblastic leukemia: results of the RIALTO trial, an expanded access study. Blood Cancer J 10:1–5
51. Goebeler M, Knop S, Viardot A, Kufer P, Topp MS, Einsele H et al (2016) Bispecific T cell engager (BiTE) antibody construct blinatumomab for the treatment of patients with relapsed/refractory non-Hodgkin lymphoma: final results from a phase I study. J Clin Oncol 34:1104–1111
52. Stein AS, Larson RA, Schuh AC, Stevenson W L-M, Tran Q, Zimmerman Z, Kormany W TMS (2018) Exposure adjusted adverse events comparing blinatumomab with chemotherapy in advanced acute lymphoblastic leukemia. Blood Adv 2(13):1522–1531
53. Frey N, Porter D (2016) Cytokine release syndrome with novel therapeutics for acute lymphoblastic leukemia. Hematol Am Soc Hematol Educ Program 2016(1):567–572
54. von Stackelberg A, Locatelli F, Zugmaier G, Handgretinger R, Trippett TM, Rizzari C, Bader P, O'Brien MM, Brethon B, Bhojwani D, Schlegel PG, Borkhardt A, Rheingold SR, Cooper TM, Zwaan CM, Barnette P, Messina C, Michel G, DuBois SG, Hu K, Zhu M WJA, Gore L (2016) Phase I/Phase II study of blinatumomab in pediatric patients with relapsed/refractory acute lymphoblastic leukemia. J Clin Oncol 34(36):4381–4389
55. Jain T, Litzow MR (2018) No free rides: management of toxicities of novel immunotherapies in ALL, including financial. Blood Adv 2(22):3393–3403
56. Klinger M, Zugmaier G, Nägele V, Goebeler M, Brandl C, Stelljes M et al (2020) Adhesion of T cells to endothelial cells facilitates blinatumomab-associated neurologic adverse events. Cancer Res 80:91–101

57. Matasar MJ, Cheah CY, Yoon DH, Assouline SE, Bartlett NL, Ku M et al (2020) Subcutaneous mosunetuzumab in relapsed or refractory B cell lymphoma: promising safety and encouraging efficacy in dose escalation cohorts. Blood 136:45–46
58. Lesokhin AM, Levy MY, Dalovisio AP, Bahlis NJ, Solh M, Sebag M et al (2020) Preliminary safety, efficacy, pharmacokinetics, and pharmacodynamics of subcutaneously (SC) administered PF-06863135, a B cell maturation antigen (BCMA)-CD3 bispecific antibody, in patients with relapsed/refractory multiple myeloma (RRMM). Blood 136:8–9
59. Ribas A, Wolchok JD (2018) Cancer immunotherapy using checkpoint blockade. Science 359:1350–1355
60. Feucht J, Kayser S, Gorodezki D, Hamieh M, Döring M, Blaeschke F et al (2016) Tcell responses against CD19$^+$ pediatric acute lymphoblastic leukemia mediated by bispecific T cell engager (BiTE) are regulated contrarily by PDL1 and CD80/CD86 on leukemic blasts. Oncotarget 7:76902–76919
61. Duell J, Dittrich M, Bedke T, Mueller T, Eisele F, Rosenwald A et al (2017) Frequency of regulatory T cells determines the outcome of the T-cell engaging antibody blinatumomab in patients with B-precursor ALL. Leukemia 31:2181–2190
62. Ghiringhelli F, Larmonier N, Schmitt E, Parcellier A, Cathelin D, Garrido C et al (2004) CD4+CD25+ regulatory T cells suppress tumor immunity but are sensitive to cyclophosphamide which allows immunotherapy of established tumors to be curative. Eur J Immunol 34:336–344
63. Jabbour E, Düll J, Yilmaz M, Khoury JD, Ravandi F, Jain N et al (2018) Outcome of patients with relapsed/refractory acute lymphoblastic leukemia after blinatumomab failure: no change in the level of CD19 expression. Am J Hematol 93:371–374
64. Braig F, Brandt A, Goebeler M, Tony H, Kurze A, Nollau P et al (2017) Resistance to anti-CD19/CD3 BiTE in acute lymphoblastic leukemia may be mediated by disrupted CD19 membrane trafficking. Blood 129:100–104
65. Gardner R, Wu D, Cherian S, Fang M, Hanafi L, Finney O et al (2016) Acquisition of a CD19-negative myeloid phenotype allows immune escape of MLL rearranged B-ALL from CD19 CAR-T-cell therapy. Blood 127:2406–2410
66. Rayes A, McMasters RL, O'Brien MM (2016) Lineage switch in MLL-rearranged infant leukemia following CD19-directed therapy. Pediatr Blood Cancer 63:1113–1115
67. Ruella M, Barrett DM, Kenderian SS, Shestova O, Hofmann TJ, Perazzelli J et al (2016) Dual CD19 and CD123 targeting prevents antigen-loss relapses after CD19-directed immunotherapies. J Clin Invest 126:3814–3826
68. Dai H, Wu Z, Jia H, Tong C, Guo Y, Ti D et al (2020) Bispecific CAR-T cells targeting both CD19 and CD22 for therapy of adults with relapsed or refractory B cell acute lymphoblastic leukemia. J Hematol Oncol 13:30
69. Woo S-R, Turnis ME, Goldberg MV, Bankoti J, Selby M, Nirschl CJ et al (2012) Immune inhibitory molecules LAG-3 and PD-1 synergistically regulate T cell function to promote tumoral immune escape. Cancer Res 72:917–927
70. Johnston RJ, Comps-Agrar L, Hackney J, Yu X, Huseni M, Yang Y et al (2014) The immunoreceptor TIGIT regulates anti-tumor and antiviral CD8(+) T cell effector function. Cancer Cell 26:923–937
71. Day CL, Kaufmann DE, Kiepiela P, Brown JA, Moodley ES, Reddy S et al (2006) PD-1 expression on HIV-specific T cells is associated with T cell exhaustion and disease progression. Nature 443:350–354
72. Ruella M, Maus MV (2016) Catch me if you can: leukemia escape after CD19-directed T cell immunotherapies. Comput Struct Biotechnol J 14:357–362
73. Arndt C, Feldmann A, von Bonin M, Cartellieri M, Ewen E-M, Koristka S et al (2014) Costimulation improves the killing capability of T cells redirected to tumor cells expressing low levels of CD33: description of a novel modular targeting system. Leukemia 28:59–69
74. Liu R, Jiang W, Yang M, Guo H, Zhang Y, Wang J et al (2010) Efficient inhibition of human B cell lymphoma in SCID mice by synergistic anti-tumor effect of human 4–1BB ligand/anti-CD20 fusion proteins and anti-CD3/anti-CD20 diabodies. J Immunother 33:500–509

75. Schmohl JU, Felices M, Oh F, Lenvik AJ, Lebeau AM, Panyam J et al (2017) Engineering of Anti-CD133 trispecific molecule capable of inducing NK expansion and driving antibody-dependent cell-mediated cytotoxicity. Cancer Res Treat 49:1140–1152

76. Khan S, Gerber DE (2019) Autoimmunity, checkpoint inhibitor therapy and immune-related adverse events: a review. Semin Cancer Biol S1044-579X(19)30019-7

77. Duell J, Dittrich M, Bedke T, Mueller T, Eisele F, Rosenwald A et al (2017) Frequency of regulatory T cells determines the outcome of the T cell engaging antibody blinatumomab in patients with B-precursor ALL. Leukemia 31:2181–2190

78. Bannerji R, Brown JR, Advani RH, Arnason J, O'Brien SM, Allan JN et al (2016) Phase 1 study of REGN1979, an anti-CD20 x anti-CD3 bispecific monoclonal antibody, in patients with CD20+ B cell malignancies previously treated with CD20-directed antibody therapy. Blood 128:621

79. Schuster SJ, Bartlett NL, Assouline S, Yoon S-S, Bosch F, Sehn LH et al (2019) Mosune-tuzumab induces complete remissions in poor prognosis Non-Hodgkin lymphoma patients, including those who are resistant to or relapsing after chimeric antigen receptor T cell (CAR-T) therapies, and is active in treatment through multiple lines. Blood 134(Suppl. 1):6

80. Reusch U, Burkhardt C, Fucek I, Le Gall F, Le Gall M, Hoffmann K et al (2014) A novel tetravalent bispecific TandAb (CD30/CD16A) efficiently recruits NK cells for the lysis of CD30+ tumor cells. MAbs 6:728–739

81. Bacac M, Colombetti S, Herter S, Sam J, Perro M, Chen S et al (2018) CD20-TCB with obinutuzumab pretreatment as next-generation treatment of hematologic malignancies. Clin Cancer Res 24:4785–4797

82. Guy DG, Uy GL (2018) Bispecific antibodies for the treatment of acute myeloid leukemia. Curr Hematol Malig Rep 13:417–425

83. Dinndorf PA, Andrews RG, Benjamin D, Ridgway D, Wolff L, Bernstein ID (1986) Expression of normal myeloid-associated antigens by acute leukemia cells. Blood 67(4):1048–1053

84. Hauswirth AW, Florian S, Printz D, Sotlar K, KrauthMT, Fritsch G et al (2007) Expression of the target receptor CD33 in CD34+/CD38-/CD123+ AML stem cells. Eur J Clin Inv 37(1):73–82

85. Krupka C, Kufer P, Kischel R, Zugmaier G, Bogeholz J, Kohnke T et al (2014) CD33 target validation and sustained depletion of AML blasts in long-term cultures by the bispecific T cell-engaging antibody AMG 330. Blood 123(3):356–365

86. Stamova S, Cartellieri M, Feldmann A, Arndt C, Koristka S, Bartsch H et al (2011) Unexpected recombinations in single-chain bispecific anti-CD3-anti-CD33 antibodies can be avoided by a novel linker module. Mol Immunol 49(3):474–482

87. Arndt C, von Bonin M, Cartellieri M, Feldmann A, Koristka S, Michalk I et al (2013) Redirection of T cells with a first fully humanized bispecific CD33-CD3 antibody efficiently eliminates AML blasts without harming hematopoietic stem cells. Leukemia 27(4):964–967

88. Crocker PR, Paulson JC, Varki A (2007) Siglecs and their roles in the immune system. Nat Rev Immunol 7:255–266

89. Ravandi F, Walter RB, Subklewe M, Bueecklein V, Jongen-Lavrencic M, Paschka P et al (2020) Updated results from phase I dose-escalation study of AMG 330, a bispecific T cell engager molecule, in patients with relapsed/refractory acute myeloid leukemia (R/R AML). J Clin Oncol 38:7508

90. Reddy EP, Korapati A, Chaturvedi P, Rane S (2000) IL-3 signaling and the role of Src kinases, JAKs and STATs: a covert liaison unveiled. Oncogene 19(21):2532–2547

91. Blalock WL, Weinstein-Oppenheimer C, Chang F, Hoyle PE, Wang XY, Algate PA et al (1999) Signal transduction, cell cycle regulatory, and anti-apoptotic pathways regulated by IL-3 in hematopoietic cells: possible sites for intervention with anti-neoplastic drugs. Leukemia 13(8):1109–1166

92. Gaudet FNJ et al (2016) Development of a CD123xCD3 bispecific antibody (JNJ-63709178) for the treatment of acute myeloid leukemia (AML). Blood 128(22):2824

93. Chu SY, Pong E, Chen H, Phung S, Chan EW, Endo NA et al (2014) Immunotherapy with long-lived anti-CD123 × anti-CD3 bispecific antibodies stimulates potent Tcell-mediated

killing of human AML cell lines and of CD123+ cells in monkeys: a potential therapy for acute myelogenous leukemia. Blood 124(21):2316

94. Billiau AD, Roskams T, Van Damme-Lombaerts R, Matthys P, Wouters C (2005) Macrophage activation syndrome: characteristic findings on liver biopsy illustrating the key role of activated, IFN-g producing lymphocytes and IL-6- and TNF-alpha-producing macrophages. Blood 105(4):1648–1651

95. Davila ML, Riviere I, Wang X, Bartido S, Park J, Curran K et al (2014) Efficacy and toxicity management of 19-28z CAR T cell therapy in B cell acute lymphoblastic leukemia. Sci Transl Med 6(224):224ra25

96. Lee DW, Kochenderfer JN, Stetler-Stevenson M, Cui YK, Delbrook C, Feldman SA et al (2015) T cells expressing CD19 chimeric antigen receptors for acute lymphoblastic leukemia in children and young adults: a phase 1 dose-escalation trial. Lancet (London, England). 385(9967):517–528

97. Maude SL, Frey N, Shaw PA, Aplenc R, Barrett DM, Bunin NJ et al (2014) Chimeric antigen receptor T cells for sustained remissions in leukemia. N Engl J Med 371(16):1507–1517

98. Topp MS, Duell J, Zugmaier G, Attal M, Moreau P, Langer C et al (2020) Anti-B cell maturation antigen BiTE molecule AMG 420 induces responses in multiple myeloma. J Clin Oncol 38:775–783

99. Hipp S, Tai Y-T, Blanset D, Deegen P, Wahl J, Thomas O et al (2017) A novel BCMA/CD3 bispecific T cell engager for the treatment of multiple myeloma induces selective lysis in vitro and in vivo. Leukemia 31:1743–1751

100. Zuch de Zafra CL, Fajardo F, Zhong W, Bernett MJ, Muchhal US, Moore GL et al (2019) Targeting multiple myeloma with AMG 424, a novel Anti-CD38/CD3 bispecific t cell-recruiting antibody optimized for cytotoxicity and cytokine release. Clin Cancer Res 25:3921–3933

101. Lesokhin AM, Raje N, Gasparetto CJ, Walker J, Krupka HI, Joh T et al (2018) A phase I, open-label study to evaluate the safety, pharmacokinetic, pharmacodynamic, and clinical activity of PF-06863135, a B cell maturation antigen/CD3 bispecific antibody, in patients with relapsed/refractory advanced multiple myeloma. Blood 132(Suppl. 1):3229

102. Polson AG, Zheng B, Elkins K, Chang W, Du C, Dowd P et al (2006) Expression pattern of the human FcRH/IRTA receptors in normal tissue and in B-chronic lymphocytic leukemia. Int Immunol 18:1363–1373

103. Ise T, Nagata S, Kreitman RJ, Wilson WH, Wayne AS, Stetler-Stevenson M et al (2007) Elevation of soluble CD307 (IRTA2/FcRH5) protein in the blood and expression on malignant cells of patients with multiple myeloma, chronic lymphocytic leukemia, and mantle cell lymphoma. Leukemia 21:169–174

104. Kodama T, Kochi Y, Nakai W, Mizuno H, Baba T, Habu K et al (2019) Anti-GPRC5D/CD3 bispecific T cell redirecting antibody for the treatment of multiple myeloma. Mol Cancer Ther 18:1555–1564

105. Atamaniuk J, Gleiss A, Porpaczy E, Kainz B, Grunt TW, Raderer M et al (2012) Overexpression of G protein-coupled receptor 5D in the bone marrow is associated with poor prognosis in patients with multiple myeloma. Eur J Clin Invest 42:953–960

106. Li J, Stagg NJ, Johnston J, Harris MJ, Menzies SA, DiCara D et al (2017) Membrane proximal epitope facilitates efficient T cell synapse formation by anti-FcRH5/CD3 and is a requirement for myeloma cell killing. Cancer Cell 31:383–395

107. Viardot A, Goebeler ME, Hess G, Neumann S, Pfreundschuh M, Adrian N et al (2016) Phase 2 study of the bispecific Tcell engager (BiTE) antibody blinatumomab in relapsed/refractory diffuselarge B-cell lymphoma. Blood 127:1410–6

108. Hijazi Y, Klinger M, Kratzer A, Wu B, Baeuerle PA et al (2018) Pharmacokinetic and pharmacodynamic relationship of blinatumomab in patients with non-hodgkin lymphoma. Curr Clin Pharm 13:55–64

109. Liu L, Lam C-YK, Long V, Widjaja L, Yang Y, Li H et al (2017) MGD011, A CD19 x CD3 dual-affinity retargeting bi-specific molecule incorporating extended circulating half-life for the treatment of B cell malignancies. Clin Cancer Res 23:1506–1518

110. Reusch U, Duell J, Ellwanger K, Herbrecht C, Knackmuss SH, Fucek I et al (2015) A tetravalent bispecific TandAb (CD19/CD3), AFM11, efficiently recruits T cells for the potent lysis of CD19(+) tumor cells. MAbs 7:584–604

111. Reusch U, Harrington KH, Gudgeon CJ, Fucek I, Ellwanger K, Weichel M et al (2016) Characterization of CD33/CD3 tetravalent bispecific tandem diabodies (TandAbs) for the treatment of acute myeloid leukemia. Clin Cancer Res 22(23):5829–5838

112. Westervelt P, Roboz, GJ et al (2018) Phase 1 first-in-human trial of AMV564, a bivalent bispecific (2x2) CD33/CD3 T cell engager, in patients with relapsed/refractory acute myeloid leukemia (AML). Presented at the 23rd Congress of the European Hematology Association (EHA), 14–17 June 2018, Stockholm, Sweden

113. Feucht J, Kayser S, Gorodezki D, Hamieh M, Doring M, Blaeschke F et al (2016) T cell responses against CD19+ pediatric acute lymphoblastic leukemia mediated by bispecific T cell engager (BiTE) are regulated contrarily by PDL1 and CD80/CD86 on leukemic blasts. Oncotarget 7:76902–76919

114. Dovedi SJ, Mazor Y, Elder M, Hasani S, Wang B, Mosely S et al (2018) Abstract 2776: MEDI5752: a novel bispecific antibody that preferentially targets CTLA-4 on PD-1 expressing T cells. Cancer Res 78(13 Suppl.):2776

115. Webster J, Luskin MR, Prince GT, DeZern AE, DeAngelo DJ, Levis MJ, Blackford A, Sharon E, Streicher H, Luznik L, Gojo I (2018) Blinatumomab in combination with immune checkpoint inhibitors of PD-1 and CTLA-4 in adult patients with relapsed/refractory (R/R) CD19 positive B cell acute lymphoblastic leukemia (ALL): preliminary results of a phase I study. Blood 132(Suppl. 1):557

116. Schwartz MS, Jeyakumar D, Damon LE, Schiller GJ, Wieduwilt MJ (2019) A phase I/II study of blinatumomab in combination with pembrolizumab for adults with relapsed refractory B-lineage acute lymphoblastic leukemia: University of California Hematologic Malignancies Consortium Study 1504. J Clin Oncol 37(15_suppl):TPS7064

117. Assi R, Kantarjian H, Short NJ, Daver N, Takahashi K, Garcia-Manero G, DiNardo C, Burger J, Cortes J, Jain N, Wierda W, Chamoun S, Konopleva M, Jabbour E (2017) Safety and efficacy of blinatumomab in combination with a tyrosine kinase inhibitor for the treatment of relapsed Philadelphia chromosome-positive leukemia. Clin Lymphoma Myeloma Leuk 17(12):897–901

118. Usmani SZ et al (2021) Teclistamab, a B cell maturation antigen x CD3 bispecific antibody, in patients with relapsed or refractory multiple myeloma (MajesTEC-1): a multicenter, open label, single-arm, phase 1 study. Lancet

119. Tian Z, Liu M, Zhang Y, Wan X (2021) Bispecific T cell engagers: an emerging therapy for management of hematologic malignancies. J Hematol Oncol 14:75